RFID *and* Auto-ID in Planning and Logistics

A Practical Guide for Military UID Applications

RFID *and* Auto-ID in Planning and Logistics

A Practical Guide for Military UID Applications

Erick C. Jones, PhD
The University of Texas at Arlington, USA

Christopher A. Chung, PhD
Seabrook, Texas, USA

CRC Press is an imprint of the
Taylor & Francis Group, an **informa** business

CRC Press
Taylor & Francis Group
6000 Broken Sound Parkway NW, Suite 300
Boca Raton, FL 33487-2742

First issued in paperback 2017

ISBN-13: 978-1-4200-9427-5 (hbk)
ISBN-13: 978-1-138-07526-9 (pbk)

Library of Congress Cataloging-in-Publication Data

Jones, Erick C.
RFID and auto-ID in planning and logistics : a practical guide for military UID applications / Erick C. Jones and Christopher A. Chung.
p. cm.
Includes bibliographical references and index.
ISBN 978-1-4200-9427-5 (alk. paper)
1. Logistics--Technological innovations. 2. Radio frequency identification systems. 3. Military supplies--Identification. 4. Armed Forces--Supplies and stores--Technological innovations. I. Chung, Christopher A. II. Title.

U168.J65 2011
623'.76--dc22 2010035465

Visit the Taylor & Francis Web site at
http://www.taylorandfrancis.com

and the CRC Press Web site at
http://www.crcpress.com

Contents

PART I History and Background

PART IV Implementation, Decision Making, and Testing Approaches

Preface

RFID AND AIT IN PLANNING AND LOGISTICS

The use of RFID technology is becoming increasingly popular due to the perceived reduction in costs that it may bring to the operational costs. In 2003, we opened the RFID and Supply Chain Logistics Lab at the University of Nebraska, one of the first RFID labs focused on RFID and logistics. We believed that RFID technology was not mature enough to enter mainstream logistics operations and that an unbiased application lab was the best way to test and prove concepts prior to operational deployment. The academic lab setting would provide the researchers an opportunity to test the applications unlike vendor-driven labs that could potentially become biased.

After the lab opened, mandates from large retailers such as Walmart and TESCO motivated manufacturers to implement RFID initiatives into their supply chain. Unfortunately, because there was no developed testing or implementation plan, most of these implementations were on a trial-and-error basis. The most common challenges to implementation were lack of knowledge of passive and active technologies, not testing of the manufacturing specifications, and not realizing the operational impact of changing technologies.

The logistics consulting background of the researchers combined with the research they have conducted over the past several years has led to innovative implementation models for applying RFID into logistics operations. This approach, for example, first provides the reader with foundational relationships in RFID such as how antennae, integrated circuitry, and substrates work together. The approach then leads the reader to reach an understanding of which technology he or she is seeking to use such as passive, semi-passive, or active tags. These foundational relationships are not explained in other books in enough detail for implementers to determine if there is a design problem with tags in their troubleshooting process. In our previous RFID symposium at University of Nebraska-Lincoln (UNL), participants have asked how to validate that manufacturers are providing accurate specifications, what are true points of failure, and how do we set up and validate our prototypes. By working in close collaboration with other researchers, manufacturers, and integrators, this book will provide some opportunities to implement these more effectively.

In 2005, universities such as the University of Nebraska-Lincoln and Oklahoma State University offered a course on RFID. The books that were available did not provide enough robust material for engineering students, engineers, and operational personnel to use for foundational concepts for integrating logistics and RFID. The foundational concept of increased inventory visibility due to the automatic data capture abilities of RFID provides an opportunity to reduce inventory levels. The proof of this concept should be validated by changing the amended economic order quantity (EOQ) model to focus on the reorder point strategy. This can affect how the logistics professional configures his warehouse management system (WMS) or enterprise resource planning (ERP) system. This book will provide the foundational understanding on how these are integrated into an overall logistics model.

DESCRIPTION OF RFID AND AIT IN PLANNING AND LOGISTICS

With *RFID and Auto-ID in Planning and Logistics*, we hope to establish the concepts and principles by which students, military personnel and contractors, and corporate managers can learn about RFID and other automatic information capture technologies, and the ways in which it can be integrated into planning and logistics functions. A comparative evaluation of RFID along with other technologies such as bar codes, satellite tags, and global positioning systems provides a complete understanding of which technologies to use in specific planning and distribution operations. Also, integration of unique identification standards that are required for military contractors will allow this book to provide a useful knowledge base for military personnel and contractors. Moreover, the description of these technologies' histories, current use, and future application will serve to educate students, academia, and military personnel and contractors on supply chain planning and logistics uses for RFID and AIT technologies.

Also, we hope to provide self-help for military personnel and contractors, individuals, and corporate managers who wish to regain productivity.

UTA RfAID

Dr. Erick C. Jones currently works at the University of Texas at Arlington. In an effort to support the marriage of industry's supply chain needs like automatic identification technology with academia's theoretical applications, he has created an industry/academia collaboration in the RFID and Supply Chain Lab (RFSCL). The purpose of this facility is to support project initiatives like radio frequency identification (RFID), logistics (supply chain engineering), manufacturing (six sigma and Lean initiatives), and information technology (ERP, WMS). The facility's goal is to enhance the industrial engineering field by utilizing some of the research methodologies to provide solutions in the areas of RFID, supply chain logistics, and engineering management. The mission of the RFSCL is "providing integrated solutions in logistics and other data driven environments through automatic data capture, real world prototypes, and analysis." Equipment used in the lab include active and passive tags/readers and software (Matrics, Alien, Samsys, IMPINJ, SERIT), hytrol conveyor, and GCS WMS, HP5555 Mobile Active Reader and Software, RF Code active tags, and SAVI Active Tags and Reader (WMRM/WORM). The methodology utilized for research in the lab is known as DFSS (design for six sigma), which is similar to the six sigma DMAIC methodology. The seven steps in this methodology are define, measure, analyze, identify, design, optimize, and verify. The RFSCL team consists of approximately 15 graduate students, with 7 being PhD students and the rest masters' students. The RFSCL has received over 40 research awards that have amounted to over $3,067,756. The lab has also received over $1,726,357 in research funding. Research projects focus on four areas: logistics systems analysis and design, supply chain modeling, material flow design and improvement, and intelligent systems. In the RFSCL, projects are either applied research or theoretical research models. RFID applied research concentrates on areas such as RFID and bar code integration into WMS and ERP systems or RFID in industrial applications such as conveyors. Theoretical research models for RFID include RFID integration into GPS/GIS or alternate active tag standard development. Supply chain applied research focuses on facility/transportation network modeling and RFID and bar code systems integration for inventory. Theoretical research models include mathematical modeling inventory polices and stochastic modeling of supply chain networks. Current projects dealing with RFID include embedded RFID license plates (DOT), ROW underground RFID tags (TxDOT), and RFID RTLS (NASA). Current projects in logistics include corporate supply chain analysis and grain terminal network analysis.

BOOK PURPOSES AND USES

RFID is an emerging technology that has come of age. Other AIT technologies have been useful in logistics, and there are numerous challenges between choosing RFID in lieu of, in conjunction with, or integrated with other AIT technologies. Many practitioners, military personnel, and researchers do not have a good reference on the workings of RFID, AIT technologies such as satellite tags and bar codes, their integration into logistics initiatives (UID standards) and mandates (i.e., Walmart and Department of Defense), and the future uses for RFID. In this book, we will present the background on RFID and AIT technologies, previous history of AIT technologies (mostly in the military), current applications of RFID and AIT, a decision model for choosing RFID and/or different AIT technologies, an integration model for integrating and developing integrated RFID and AIT applications, and traditional planning and logistics theoretical models and applications, integration of AIT in planning and logistics operations, military UID requirements, how to use RFID and AIT to meet UID requirements, and, finally, future research for RFID and AIT.

SPECIAL FEATURES OF THE BOOK

This book has the following additional features. It presents

- A decision model for choosing the best RFID and AIT technologies
- A economic comparison model for evaluating RFID and AIT technologies
- A useful way to meet UID standards with RFID and AIT
- How to integrate RFID and AIT into planning and distribution operations
- A history of multiple RFID and AIT technologies
- AIT and RFID integration future research areas
- Case studies of AIT and RFID research from Dr. Jones' RFID Supply Chain Lab
- Case studies from military contractors
- Case studies from military personnel
- Planning and scheduling classroom theory and problems
- Transportation and supply chain classroom theory and problems
- RFID training software developed by Dr. Jones

KEY BENEFITS

The key benefits of this book are that it allows students to learn more on RFID and AIT technologies, allows military personnel and contractors to learn and use RFID with respect to UID standards, and allows practitioners to learn and integrate RFID into scheduling and planning operations.

BOOK ONLINE SUMMARY

This book can serve as an academic text, practitioner's handbook, and a military contractor's UID guide for using AIT and RFID technologies and can be used to educate individuals and organizations on how to cost-justify, understand, and implement RFID and AIT initiatives.

INTENDED AUDIENCE

RFID in logistics is intended for three principal academic audiences:

1. BS and MS engineering students in a logistics course
2. MBA students in a second operations management course or supply chain management course
3. Practicing logistics managers, who will find this book useful as a training reference, source of practical ideas, and valid implementation strategy

While any professional faced with conducting an RFID implementation project is encouraged to receive formal simulation training, there are limited avenues for this to happen. To date, we know of only two formalized training programs. Practitioners with limited or dated training may need resources other than theory-based academically oriented texts or software-specific manuals. This handbook is intended to provide the practitioner with an easy-to-follow reference source for each step in conducting an RFID cost/benefit analysis and valid implementation approach to an RFID logistics project.

HOW THIS BOOK DIFFERS FROM OTHER SIMULATION TEXTS

This book differs from other RFID texts in several major ways. First, the book was written to support deeper understanding of the technology. Many currently available simulation publications give standard answers to limited applications and provide no real strategy for using RFID in a logistics setting. While these types of simulation books have value to some, they are not robust enough to support practitioner nor learning about the technology and how to effectively use it.

BOOK ORGANIZATION

After a brief introductory chapter, the book is organized into seven parts, as follows: Part I, History and Background; Part II, Overview of RFID and AIT Technologies; Part III, Overview of Equipment Interoperability Protocols and Standards; Part IV, Implementation, Decision Making, and Testing Approaches; Part V, Overview of Logistics Planning and Inventory Control; Part VI, Military RFID Initiatives and Applications; and Part VII, Other Potential Military RFID Applications and Radio Theory. In our own teaching, we generally cover all parts in order, but vary the selections of topics depending on the course. Other key elements include the RFID Lean planning framework which is an extremely important section of the book.

The instructor is also faced with the choice of how much mathematical depth to use. To assist readers who want general concepts with minimal mathematics, we have set off certain sections as mathematical details. These sections, which are labeled and indented in the text, present justifications, examples, and methodologies that rely on mathematics (nothing higher than simple calculus). They can be skipped completely without loss of continuity.

For MATLAB® and Simulink® product information, please contact:
The MathWorks, Inc.
3 Apple Hill Drive
Natick, MA, 01760-2098 USA
Tel: 508-647-7000
Fax: 508-647-7001
E-mail: info@mathworks.com
Web: www.mathworks.com

Acknowledgments

This book was the product of many long hours, hard work, and persistence not only by us but also by our support team.

My thanks go to God for giving me the strength and desire, my family for giving me the motivation, my graduate students for providing the inspiration through their many questions, and my undergraduate and graduate students for their enthusiasm toward this project and their comments and their ideas on possible uses in the future.

I especially thank my mother for her view on how older people, though apprehensive of new technology, see how RF technologies can improve their lives. Our discussions on RF technologies, which were utilized during her heart surgery, provided me some last-minute inspiration, especially the final chapter on the future.

A special thank you and acknowledgment goes to Angela Garza, my PhD student, who worked almost as hard as I did on this book. Her persistence to make sure the book chapters that she worked on were perfect was commendable. I would also like to thank Jonathon Carlson, another graduate student who is now working, for his efforts.

I would like to acknowledge the support of the military personnel and contractors who read through the material and provided extensive feedback. They include Dan Kimball, Bob Kenney, Jeffrey Fee, and Patrick Burns.

My acknowledgments also to the following for their informal feedback: Reuben Vasquez, Dr. Judy Perkins, Toby Rush, and Dwight Mosby. Finally, I would like to thank the following organizations for their feedback: Savi-Lockheed Martin, SME, SRA, USSTRANSCOM, DASH7 Alliance, AIM Global, Rush Tracking, VerdaSee Solutions, NASA, Prairie View A&M University, and the University of Nebraska-Lincoln.

Erick C. Jones

Authors

Dr. Erick C. Jones is an associate professor at the University of Texas at Arlington, Arlington, Texas and the director of the RFID and Auto-ID Deployment Lab, The University of Texas at Arlington, Industrial and Manufacturing Systems Engineering (IMSE), Arlington, Texas. He graduated from Texas A&M University with a degree in industrial engineering. He later received his master's and PhD degrees in industrial engineering from the University of Houston. Dr. Jones received most of his secondary education while working in industry. He has held positions in industry, which include engineering specialist at UPS, engineering director at Academy Sports and Outdoors, engineering consultant and project manager at Tompkins Associates, and executive manager at Arthur Anderson, LLP. He has managed teams as small as 3 people and as large as 500 people. The projects he has executed include implementing ERP system modules, designing and constructing new facilities, and reengineering fortune 1000 organizations. The operations he has managed include a large-scale distribution operation and human resource department at an executive level. Dr. Jones expertise has led to becoming an expert in the field of supply chain optimization, distribution logistics, and inventory control. His unique background led him to one of the first and largest academic RFID labs in the country. He has published one textbook on RFID and has edited two industry manuals on the subject. Currently, he is working on a military handbook for RFID and other automatic information technologies, which include GPS and satellite tags. In addition, his lab has been part of the industry academic consortium, which focuses on logistics and is sponsored by the National Science Foundation, for several years.

Dr. Jones' current focus is on RFID research for the U.S. Department of Transportation, the Department of Defense Transportation Command, and NASA JSC. His research interests include RFID, RTLS, and satellite technology development and testing with respect to inventory control. Other research areas include supply chain logistics, six sigma quality engineering management, and knowledge worker turnover.

Dr. Christopher A. Chung is an associate professor in the Department of Industrial Engineering at the University of Houston, Houston, Texas. He received his doctoral and master's degrees in industrial engineering from the University of Pittsburgh and his bachelor's degree from Johns Hopkins University, Baltimore, Maryland. Dr. Chung performs research in the areas of engineering management, simulation, and other computer-related applications. His research has been funded by the Department of Homeland Security, the Department of Justice, and various corporations. He is the author of two other books and numerous refereed journal papers.

Dr. Chung's research interests are in the areas of management, operations, and equipment training simulators, educational simulators, computer applications, and discrete event simulation, including interactive simulators for responding to bomb threats and implementing advanced manufacturing technology projects. He has also conducted research on the operation of airport security checkpoints under increased threat conditions and simultaneous service approaches to reduce airport check-in times. Dr. Chung's research has received funding from the U.S. Department of Justice, the Department of Homeland Security, the United States Coast Guard, Continental Airlines, and other corporations.

Dr. Chung is a former U.S. Army bomb disposal officer and NEST team member. He is a graduate of a citizens' police academy course. He has also worked as a manufacturing quality engineer with the Michelin Tire Corporation.

Part I

History and Background

1 Introduction and History of Military Logistics

You will not find it difficult to prove that battles, campaigns, and even wars have been won or lost primarily because of logistics.

—General Dwight D. Eisenhower

1.1 INTRODUCTION

As radio frequency identification (RFID) technology is applied to military logistics throughout this book, it is important for readers to familiarize themselves with a short history of military logistics. As the reader will note, in many ways, the fundamental concepts of military logistics today are the same as they were with the first standing armies. However, in other ways, different types of new technology, particularly computer technology, have revolutionized military logistics. The use of RFID technology has the potential to represent another revolution in military logistics in some areas, but not others. We begin this introduction with a brief definition of military logistics.

According to the North Atlantic Treaty Organization (NATO) glossary of terms and definitions, military logistics is the art and science of planning and carrying out the movement and maintenance of military forces. Perhaps the earliest specific reference to the term military logistics was made by the Swiss-born army general Antoine-Henri Jomini in the nineteenth century. General Jomini actually served in the Swiss, French, and German armies between 1798 and 1829. Jomini published a number of military-oriented books based on his experiences as an army officer. One of his books was published in 1838, which described military operations as a combination of strategy, tactics, and logistics (Jomini, 1838).

Although the specific term military logistics is less than 200 years old, the fundamental operations involving troop transportation, service, and supply has always existed in some form or another since the beginning of standing armies. Not surprisingly, the development of military logistics has been related to the presence of military activity. Those periods of small military campaigns show static development. Conversely, periods with many conflicts and wars displayed major growth and advancement in military logistics practices. Beginning with the Assyrians in 700 BC, we will proceed through the major historical periods and events in military logistics to the present day.

1.2 ASSYRIANS

Some historians place the beginning of formal military logistics activity around 700 BC with the formation of the Assyrian army. This army imposed military force in the Middle East with the use of a variety of horse-drawn chariots and iron weapons and shields. The use of mechanical devices such as the chariot introduced the concept of logistics primarily with respect to maintenance and repair. Vast quantities of the chariots and iron-based weapons were manufactured and put to use in the field. Consequentially, it was necessary to dedicate specific resources just for ensuring that the large numbers of these early weapons systems were kept in fighting order.

Another major source of logistical concern involved the large dependence on horses for both the chariots, as well as, the cavalry. The first logistical concern involved acquisition issues. The Assyrian army required up to 3000 new horses per month. To acquire such a large number of horses, the Assyrians created a special logistics branch called the musarkisus. Once acquired, the Assyrian horses needed to be maintained. Each horse could consume as much as 20 lb of food per day. This in itself presented another formidable problem, particularly for the army on the move.

An even bigger challenge to the Assyrians during this period was the transportation of supplies to the entire standing army. The primary form of transportation was based on horses and other beasts of burden such as oxen. The horses and oxen presented logistical difficulties themselves. On the one hand, oxen could transport more supplies per animal than horses. On the other hand, oxen were very slow in comparison to horses. The longer it took to reach the destination, the more food and water was required for the oxens' own use. The use of horses may deliver the supplies faster, but this was at a cost of having to feed more horses for the same amount of delivered supplies. These types of problems greatly reduced the ability of an animal-based supply chain to transport useable supplies.

The overall logistical requirements of the Assyrians early standing army exceed that which could be supplied by this type of supply chain. As a result, the army was forced to move to a new logistical base with plentiful food rather than supply moving to the location of the standing army. As the logistical base became depleted of food, it was necessary for the army to continuously relocate. Thus, much of the effort of Assyrian army was devoted toward ensuring that it was logistically satisfied.

1.3 PHILIP II OF MACEDON AND ALEXANDER THE GREAT

The next major advances in military logistics were developed by Philip II of Macedon and his son Alexander the Great. Philip II of Macedon was born in 382 BC. He ruled from 359 BC until his assassination in 336 BC. Philip is in particular credited with the invention of the sarissa or 13–21 ft-long pike. Used in combination with phalanx tactics, Philip expanded his control from Macedon to all of Greece. Philip II's son, Alexander III or Alexander the Great, was born in 356 BC. He assumed the throne of his father Philip in 336 BC. Using the same sarissa and phalanx tactics as his father, Alexander continued the expansion across Asia. Alexander the Great ruled until his death in 323 BC at the young age of 32. During their reigns, Phillip II and Alexander the Great introduced the concepts of reduction of logistical burdens, naval logistics, and advanced supply depots.

1.3.1 Reduction of Logistical Burdens

During his Persian Empire conquests, Alexander the Great traveled 4000 mi from Egypt to India. In order to conduct this campaign, Alexander realized the importance of reducing the logistical burden on his forces. Gone were many of the supporting components of wagons, servants, and women of previous fighting forces. There was some complaining on the part of the Macedonian soldiers. However, by eliminating these conventional components, Alexander was able to greatly reduce not only the logistical needs of his fighting forces but also the logistical needs of the supporting components themselves. Some estimates of as high as a two-thirds reduction in logistical requirement was accomplished by making this decision. As a result, Alexander significantly increased what is now known as the tooth-to-tail ratio of his fighting forces.

During this period, the use of the common oxen cart was also mostly eliminated in favor of only horses and mules. The replacement of the oxen cart also enabled the army to function with far fewer maintenance requirements. Further reduction in transportation logistical requirements was achieved through the increased use of soldiers to carry their own supplies and weapons as well. This particular concept remains in use to this day. Because few roads existed at the time, this decision greatly increased the speed of the army's movement.

1.3.2 Naval Logistics

Alexander also made great use of naval logistics to support his army's activities. This logistical resource was primarily provided in the form of a wooden vessel called a galley. The galleys produced during this time were up to 400 ft long and were powered by different numbers of banks of oars on both sides of the vessel. Although primarily oar powered by slaves, some galleys did utilize sails when the wind was coming from a favorable direction. However, given the downwind level of sailing technology during this period, this was probably a very small percentage of the time. During Alexander's reign, galleys were primarily used for transportation and supply. This included men, beasts of burden, food supplies, and weapons. There are some references to the use of galleys for mounted weapons of the time; however, it was not until the eighteenth century that sailing vessels were commonly armed with cannons and used as warships.

By staying within a few weeks or months reach of navigable bodies of water, Alexander was able to greatly reduce the need for a conventional overland logistics train during his campaigns. An additional advantage of this approach was that the galley naval vessels used to supply the war material had far greater storage and transportation capacities than livestock. The galley's themselves also required some logistical support of their own. If left in the water for extended periods of time, the wooden hulls would absorb water, as well as, be attacked by worms and deteriorate. Thus, it was a common practice to beach the galleys and perform repairs and preventive maintenance when possible.

1.3.3 Advanced Supply Depots

Alexander the Great was also credited with the concept of advanced supply depots. He would send an advance party forward to secure supplies for the main body of troops. These supplies were acquired through either foraging or bartering. Small garrisons of troops were left to protect the supply depots until the main forces arrived. When the main force did finally arrive, instead of having to waste time and effort securing supplies, the supplies were waiting for them. Once reprovisioned, the use of the supply depots allowed Alexander to rapidly move troops between the supply depots without the normal logistical concerns.

All in all, Alexander demonstrated that by properly planning the logistical aspects of a campaign, the success of the campaign could be better assured. For his efforts, Alexander the Great is recognized as history's first great military logistician.

1.4 MIDDLE AGES LOGISTICS: 1000–1400 AD

The middle ages were logistically marked by the return of the extensive supporting train of wagons, horses, oxen, women, children, servants, and slaves. The logistical needs of entire force increased accordingly. During this period, logistical resupply once again depended mostly on local foraging. Even though foraging parties may have consisted of mounted troops, the results of the foraging were returned directly to the main force rather than being set up as advanced supply depots as in the case of Alexander the Great. All of these issues resulted in the greatly restricted movement of large forces.

1.4.1 Early Scorched Earth Policy

King Richard the Lionhearted led the third crusade to Jerusalem. While King Richard made use of stockpiling supplies and delivering them by ship, he was also forced to conduct some foraging activities. This introduced what could be called an early version of the scorched earth policy that was made popular by the Russians in World War II. Realizing the King Richard's forces were dependent on foraging, Salidan, the Muslim general, began stripping the countryside of potential supplies. This also included the poisoning of drinking water wells. Salidan's logistical attacks on King Richard's military logistical approach ultimately resulted in forcing a treaty.

1.4.2 Siege Logistical Considerations

The middle ages also marked the development of large fortified castles. Stocked with up to a year of provisions, the castle inhabitants could withstand prolonged sieges. If an attacking force could not defeat the castle inhabitants by starvation, the siege commander might resort to a combination of battering rams, catapults, ballistas, trebuchets, and siege towers. Fortunately for the assaulting forces, their intended military targets were stationary in nature. This allowed the assaulting forces bypass the logistical requirements of transporting the large siege weapons over the poorly developed roads of the time. Instead, the assaulting forces need only bring a limited set of tools. They then might construct these types of weapons in the vicinity of the fortified castle. Raw materials such as timber were harvested from nearby forests. Design plans from previous sieges were used to manufacture weapons.

The ammunition for the siege weapons were also gathered locally. Common catapult payloads included rocks or boulders, diseased livestock, flaming bundles of wood, and rotten food. Less common, but equally effective payloads included beehives and even excrement. These in particular had the advantage of introducing a psychological demoralizing aspect to the castle defenders. Once again, logistics played an important role as many castles sieges continued until either the inhabitants ran out of supplies or the walls were eventually breached by heavy weapons.

Lastly, during this period, military food logistics is credited with the development of the cracker. This is a twice-baked bread intended to reduce the amount of moisture in the cracker. This, in turn, reduced the possibility of mold. The cracker has been issued to military personnel ever since as a staple food item.

1.5 EARLY BRITISH NAVAL LOGISTICS: 1700–1800s

The Royal British Navy is credited with helping create the British Empire that greatly influenced the world between the 1700s and World War II. During this period between 1700 and 1815, the British Navy is also credited as being the largest standing navy in the world. As such, it successfully waged war against the nations of France, Spain, and the Netherlands.

During this period, the cannon was first introduced to sailing vessels. The use of sailing vessel for both combat and supply was a significant development in naval warfare. The primary British naval vessel was the man of war. A typical man of war displaced 2000 ton, was up to 200 ft long, and carried a crew of 800 men. With three masts of sail, the Man of War could make as much as eight or nine knots. The man of war could also carry over 100 cannons of various sizes rated by the weight of the cannon ball each fired. These ranged up to 32 lb. Each man of war was essentially a small self-sufficient city with the same logistical requirements as such.

No doubt, a contributing factor in the success of the British Navy was their system of naval logistics. As today, the naval logistics of the time required the repair and overhaul of vessels themselves in addition to the replenishment of food and ammunition. The refinement of British naval logistics was such that their military vessels could remain at sea for extended periods of time now, only rivaled by nuclear-powered submarines and larger warships. For example, during the period of 1803–1805, Lord Horatio Nelson's ships actually remained continuously at sea in a blockage of Toulun France for 18 months.

1.5.1 Lord Nelson

Lord Nelson is actually credited as being both a master commander and logistician. In the greatest British Naval victory of Trafalgar on October 21, 1805, Nelson's forces overwhelmingly defeated the combined French and Spanish fleet. In this battle, recent documentation indicates that in the week prior to the battle, Lord Nelson ensured that his sailors were properly provisioned with dried fruits such as raisins. As a result of the victory at Trafalgar, the British Navy reigned supreme until the beginning of World War I.

Lord Nelson was mortally wounded by a sniper in the battle of Trafalgar. As a naval hero, it was decided that his body should be preserved. Contrary to normal burial at sea practices, Lord Nelson was preserved in a barrel of French brandy in what had to be an early attempt at naval mortuary logistics. Unfortunately, the brandy in this casket was reputed to have been consumed en route back to Great Britain.

1.5.2 Logistical Requirements to Prevent Scurvy

Aside of battle injuries, one of the serious medical issues facing the British during these periods was the disease scurvy. This was the gradual weakening of the individual, including the loss of teeth, internal bleeding, the opening of old wounds, and eventually death. Although it was not known at the time that the lack of vitamin c was the cause, the British Navy did determine that scurvy seemed to be held at bay when sailors were provided with fresh vegetables such as onions, turnips, and carrots, and fruits such as lemons or limes. As the initial supply of these items were either consumed or spoiled, ship commanders were faced with the increased possibility of scurvy. Thus, the major issue was the prevention of the spoilage of perishable vegetables.

1.5.3 Prevention of Spoilage of Supplies

In practice, spoilage was a problem onboard ship with all of the provisions. Since freezers and refrigeration obviously did not exist at the time, several approaches were taken to ensure at least some types of fresh food were readily available. The most common of these were the keeping of live animals such as cows, goats, and chickens on-board. A variety of meat, eggs, and milk were available in this manner. However, the consumption of these types of items was primarily limited to officers. The regular sailors, of which the majority of the crew consisted of, were provided for in a more limited manner. A common approach to the logistical problem of food spoilage for the crew was to store meat in salt and issue hard tack biscuits rather than bread that could become moldy.

Another aspect to the naval logistics of the time was that each sailor was entitled to a generous quantity of alcohol in one form or another. Up to eight pints of beer or wine was issued to the men onboard the typically British man of war. This was used to supplement water onboard, which like the food was subject to going bad or becoming infested with maggots. Life at sea often forced sailors to devise creative ways of dealing with this type of problem. For biscuits contaminated with maggots, sailors would put out a more attractive host such as fish. The maggots would leave the biscuits in favor of the fish. The maggot covered fish would be discarded overboard and replaced with a fresh one.

1.5.4 Cannon Logistics Considerations

The potentially long voyages and time at sea also required special logistical procedures with respect to the ships armament. A typical man of war would carry cannons rated as 12, 24, and 32 lb. The pound rating was based on the weight of the cannon ball that each size cannon could fire. Each cannon required a crew including a gun captain, a loader, a swabber, and a powder monkey. Even a small 12 lb cannon might weigh over 1 ton. The cannons onboard the ship fired either iron cannon balls, chain shot, or canister shot. Cannon balls were used to penetrate other ships hulls, chain shot was designed to bring down ships masts, and canister shot was used against other ships crews. The cannon balls, chain shot, and canister shot all weighted a significant amount. Thus, it was necessary for the ships to carry the majority of the cannons load in a shot locked deep within the ship's hull. Cannon balls that were needed for battle, which were removed from the shot locker, presented another problem. The round shape of the heavy cannon balls and the constant motion of the ship guaranteed that loose shot could cause serious damage. To prevent the cannon balls from rolling about on deck, shot garlands located near the centerline of the ship were used to secure the cannon balls until they were loaded. Until the cannon balls were fired, the salt air required that the iron cannon balls required preventive maintenance including scraping off rust and greasing to help slow corrosion.

1.5.5 Gunpowder Logistics Considerations

Whatever the cannons fired, it was necessary to use gunpowder. Due to the explosive nature of the loose gunpowder, barrels of gunpowder were stored deep within the hull in the ship's magazine. Only when a cartridge was needed was the gunpowder assembled into a cartridge in another room called the handling chamber. The assembled cartridge was then taken to the gun by a sailor called a powder monkey. The other major issue with gunpowder or more specifically with black powder was the necessity of keeping it dry. This of course was an issue to consider given the marine environment.

Thus, the creation of the world's most powerful navy for more than 200 years required significant military logistics. Without the development of specific logistical procedures to support naval operations, the British would not have been able to have been able to defeat the French and Spanish navies.

1.6 AMERICAN REVOLUTIONARY WAR

The American Revolutionary War took place between 1775 and 1783. This war began with the British Empire and the 13 colonial states. By 1778, France entered the war on the side of the United States. Over the next few years, the war expanded as Spain and the Dutch Republic also declared war against the British.

From a logistical perspective, the British forces dealt with a daunting 3000 mi-long logistical train. This extended primarily between the Irish port of Cork and the United States. With good weather, this amounted to a 40 day journey to the United States. Thus, it was entirely possible for a request for particular supplies to take many months by the time a request for supplies was received, fulfilled, and shipped back to the requesting unit.

1.6.1 Logistical Acquisition and Supply Difficulties

British forces faced many logistical challenges in this process. Among these was the fact that the war created a one-way passage for merchant vessels. Although there was payment for shipping goods to the United States by the British government, there was no guarantee of return passage. Thus, the fastest, most seaworthy merchant vessels were not readily available for use. Weather was also a force to be reckoned with. Some vessels returned to Britain, while others fell victim to pirates while waiting for weather windows before proceeding to a port.

For those vessels that survived the difficult journey to the United States, the voyage took a toll on man, beast, and supplies. Poor living conditions resulted in disease and illness. Replacement troops were not guaranteed to be in fighting health upon arrival. Horses were not often adequately supplied themselves on voyages that took longer than expected. Many had to be jettisoned overboard. Although the British government required supplies to packaged for the transatlantic journey, many were packaged poorly. So, by the time the supplies arrived at Cork for shipment, it was questionable whether they would survive the transatlantic voyage. Flour and biscuits, in particular, experience high spoilage rates. With the spoilage, by the time these types of supplies arrived in the United States, they would last only a fraction of the originally planned time.

1.6.2 Host Nation Support

The British Army was eventually forced to consider host nation support. British troops begin foraging for food and other supplies in the United States rather than relying on the limited ability of the British logistical train. This caused operational problems as not only did forces have to be allocated toward foraging, but additional troops had to be assigned covering duties. While the intention was to originally reimburse local farmers, such foraging parties frequently ended up engaging in pillaging.

Of course, this had the effect of the local population supporting the U.S. side. Because the British could not depend on their own logistical train and were forced to consider foraging as part of their ongoing operations, it was necessary to plan campaigns around foraging operations. In other words, an effective campaign could not include foraging, so supplies had to be stockpiled in order to concentrate on military operations.

1.7 EARLY 1800s

The 1800s were marked by the Napoleonic Wars between 1799 and 1815, and the Crimean War in the 1850s. During this period, armies faced the same logistical challenges as before. The common practice of staging in an area until supplies were gone and then moving to another plentiful area continued for many centuries. The heavy dependence on horses for troop and supply transportation also hindered movement as vast amounts of feed were required to sustain a distant campaign. Thus, as before, major campaigns could only be successfully launched from bases with navigable bodies of water from which supplies could be unloaded from naval vessels. Even then, if the logistics plans were not effective, disaster was a result. In his Russian campaign, for example, Napoleon planned poorly, arriving in Moscow with only 100,000 or his original 300,000 troops.

1.8 U.S. CIVIL WAR

The U.S. Civil War extended from 1861 to 1865. Much of the logistical principles pioneered by Alexander the Great still applied. In fact, the transportation of war material was still very much dependent on the horse as it was hundreds of years earlier. One mark the U.S. Civil War made on the world of military logistics was the introduction of the increased need for large quantities of ammunition. There was the first use of mechanized devices such as the six-barreled Gatling gun and widespread use of repeating small arm rifles in the form of the Henry and the Sharps. At the same time, the use of large artillery pieces became widespread. These needed vast quantities of both cannon balls and kegs of gunpowder.

1.8.1 Railways and Trains

The U.S. Civil War also introduced the first widespread use of railways and trains for military logistical operations. The U.S. and Confederate governments used rail transportation to rapidly move troops, weapons, and supplies. This increased mobility influenced how both sides conducted military operations. This resulted in the rapid development of the railway system. Sometimes new rail lines were laid as soon as new territory was captured. By the end of the war 30,000 mi of track had been laid. The importance of the railways to logistics became so important, that rail transportation hubs became specific targets. Of course, new weapon systems were developed to damage each side's rail transportation capabilities.

1.8.2 Mortuary Logistics

Lastly, the U.S. Civil War is considered one of the bloodiest wars in history. The 620,000 combatant deaths in the U.S. Civil War introduced the first significant logistical issue of military mortuary logistics. Due to the poorly developed state of military mortuary logistics, up to 42% of the civil war casualties were unidentified. Many of these were quickly buried in mass graves in close proximity to the battlefield. As a result, national cemeteries such as those at Arlington were established. The mortuary operations continued long after the end of the civil war. It is estimated that 295,000 unidentified soldiers were recovered and reburied in the 5 years after the hostilities ceased. This, of course, was a logistical nightmare.

1.9 ZULU WARS

One of the worst defeats involving the British military occurred during the Zulu wars in 1879. On January 11 of that year, approximately 1300 British and irregular soldiers were wiped out by 4000 Zulu warriors in the Battle of Isandlwana. What makes this a significant event in military logistical history is that the British soldiers who possessed modern breech-loading rifles of the period were actually defeated by poor ammunition logistics.

1.9.1 British Ammunition Logistics

Historians have determined that sufficient ammunition was available to the British troops to decimate the Zulu forces several times over. Some theorize that the slaughter was the result of the British quartermasters having difficulty in unpacking the ammunition from storage crates. Robust storage crates were required to properly protect the ammunition from the elements. The same robustness that protected the ammunition also prevented the ammunition from being readily accessible. For example, prior to obtaining access to the actual ammunition, the supply personnel had to unscrew the lids from the crates. The difficulty in removing the ammunition resulted in the Zulus, who were armed mostly with spears and only a few outdated rifles, to overcome the British. An interesting theatrical rendition of this version of the defeat is recorded in the motion picture movie Zulu starring Michael Caine.

Some historians discount the defeat as being attributed to this theory. However, interestingly enough, in future battles, the British made certain to unscrew the ammunition crates and unpack ammunition in advance. The military logistical moral in this lesson is that it is not enough to have supplies in place, they must also be readily available for use. Today, military small arms ammunition is typically stored in metal cans, which are sealed from the environment, yet readily accessible via a pivoting latch. Curiously, certain parallels between the British Zulu massacre and those of the U.S.–Iraq War can be established. In the case of the U.S.–Iraq War, photos of massive dumps of storage containers with unknown contents were reported to common. Obviously, if the contents of the storage containers are unknown, when the contents are needed, they will be difficult to find. Without technology such as RFID, supply personnel are reduced to visually scanning a manifest and physically searching for the desired container.

1.10 SPANISH–AMERICAN WAR

The Spanish–American War took place between April and August of 1898 in support of the Cuban interest to become independent of Spain. This war is logistically significant in that it illustrates the logistical problems that can arise as a result of a protracted period of significant military inactivity. Prior to the Spanish–American War, the U.S. Army was primarily involved in the expansion of the far west. It was totally unprepared for combat with another nation. At the time, the United States had fewer than 30,000 regular army soldiers. Through a massive recruiting effort, the Army was able to rapidly expand to over 275,000 soldiers. However, the on-hand level of military supplies was inadequate to properly equip this force. Although the United States was an economic world leader, its logistical infrastructure, among other things, reduced its effectiveness in conducting this war.

1.10.1 Logistical Problems at the Departure Port

Tampa Florida was chosen as a staging point for the Cuba assault forces. The port in Tampa needed to not only act as a launching point for the invasion force, but also as a receiving point for the supplies necessary for the 70,000 man army. Only two railways served Tampa and only one railway actually served the port. As a result, the limited railway capacity prevented both men and supplies from properly being deployed. So, even though the rest of the logistical system was able to provide the necessary support, the invasion force was not able to depart on schedule.

1.10.2 Logistical Problems at Sea

The departure of the invasion force was also hampered by the limited number of transport ships. This was the first overseas war for the United States and it was accordingly unprepared. The army not only had to use Atlantic Ocean merchant vessels, but also had to purchase additional steamships in order to transport the entire force. The naval logistics between Tampa and Cuba were also limited. The transports had few cooking facilities and the normal army rations provided to the troops included uncooked food. As a result, the invasion force was required to eat raw food, including beef.

1.10.3 More Logistical Problems in the Field

In the field, the army also encountered logistical problems. The raw canned beef in normal rations continued to be a problem in the field. Although the regular army troops were issued smokeless power Krag–Jorgensen magazine-fed bolt-action rifles, there were simply not enough to issue to the entire invasion force. While production was ramped up, it was necessary to issue old-fashioned black powder single-shot rifles to the remainder of the force. Another problem involved the soldiers' uniforms. The existing normal blue army uniform was issued to the soldiers. This heavy wool uniform became unbearable hot in the tropical Cuban environment. While some uniform modifications were made, these were not in time for widespread issue. What supplies the soldiers in the field did receive were hampered by the limited number of army animal-drawn wagons to transport the supplies from the landing port to the front line. Lastly, there were more casualties and deaths in the Spanish–American War from disease than enemy action. The army medical logistical system was unprepared for this as well.

Although the Spanish–American War left the U.S. Army with many logistical lessons to learn, there was at least one bright military logistical point in the war. This was the fact that the troops appeared to have a more than adequate supply of ammunition in comparison to the British in the Zulu wars. This apparently was a result of army logisticians emphasizing the priority of ammunition to be provided in the supply trains between the Cuban ports and the front lines. However, this was probably at the expense of other needed items. The balance of ammunition to other supplies to front-line combat units remains an issue even today.

1.11 WORLD WAR I

Between 1914 and 1918, World War I marked a point in military logistics where the forward forces could now move much faster due to the introduction of mechanized vehicles over improved roadways. Mechanized vehicles included both trucks and tanks. The use of mechanized vehicles required the development of petroleum, oil and lubricant (POL) logistics. This included the manufacture, storage, and transportation of POL. In addition, there was also the new need for maintenance support for the vehicles themselves. At the same time, the need for beast of burden such as horses and mules with their own logistical needs also continued. Naturally, the use of roads for military operations required engineering logistics support for the demolition and repair of roadways and bridges.

1.11.1 Use of Railways and Trains

The use of railways and trains for military logistics, which began with the U.S. Civil War, continued in importance in World War I. One innovation in this area was the first use of large artillery pieces mounted on railcars by the Germans. Prior to this time, large artillery was permanently positioned in defensive position or was mounted on naval vessels. The use of railcars enabled huge artillery pieces, which might not otherwise be deployed, to be easily and rapidly transported to any area with a railway. The so-called German Paris gun was used to bombard Paris from 75 mi away. The Parisians were so surprised that it was first thought that the city was being bombed by silent high-altitude zeppelins. Fortunately for the French, the payload of the Paris gun was relatively small and the gun could only be fired a few times before requiring refit.

1.11.2 Fortified Defensive Positions

World War I was also different from previous conflicts in that when the forward forces stopped, additional logistical resources were required. This resulted from the tendency of forces to develop significant defensive positions. These included elaborate trench networks. The construction of these trench networks necessitated not only traditional supplies such as ammunition and food, but also construction material such as lumber and barbed wire. Prior to this time, these types of materials were not typically used during normal combat operations involving advances and retreats.

1.11.3 Heavier than Air Aircraft

World War I is also noted for the first use of heavier than air aircraft for combat operations. Aircraft was first used for reconnaissance purposes and then later for fighter and bombing missions. In support of these operations came the concept of air operations logistics. Not only did vast airfields need to be constructed, but new logistical practices needed to be developed to supply fuel and spare parts for the aircraft. Other resources needed to be dedicated toward both general aviation maintenance and the repair of combat acquired damage. As larger aircraft were employed on bombing missions, these logistical operations became more complex. The practice of each side bombing the others' aerodrome required each side's logistical system to not only support the aircraft, but also repair itself.

The use of bombers to attack not only airfields, but also civilian populations and industrial areas soon followed. In 1917, the Germans produced the Gotha bomber that was capable of transporting up to 1000 lb of bombs. These aircraft were primarily used by the Germans to bomb London. The damage caused by the Gotha raids was limited and in fact, the German aircraft losses increased to the point that the raids were terminated. The primary effect of these initial bombing efforts was psychological.

1.11.4 U-Boats

Throughout World War I, both Entente Powers of England, France, and Russia and the German Empire attempted blockages of each other's countries. As the British Royal Navy was still the supreme European naval force and the German Navy was small in comparison, the Germans turned to the use of the diesel-powered submarine U-Boat. While the initial attacks were limited in success, U-boats soon began to account for a substantial volume of tonnage lost. The often unrestricted submarine warfare also resulted in several high-profile sinkings including the RMS Lusitania off of Ireland in 1915. The Lusitania went down with many passengers, including 128 Americans. This began as series of communications by President Woodrow Wilson and was a definite factor in the entrance of the United States into the war.

Initially, the submarines of the time were relatively primitive being blind underwater and having only a fraction of the speed of either merchant or surface warfare vessels. As a result, U-boats would often have to wait on the surface at night until merchant vessels happened along. Deck-mounted cannon on the submarines were often used to shell the merchant vessels. However, with increased emphasis on the U-boat, the Germans began building larger and more sophisticated underwater vessels. Development of the torpedo also continued. These underwater missiles were found to be capable of sinking vessels with a single launch. By the end of the war, torpedoes were used widely by both sides against both surface vessels and torpedos.

The ability of the U-Boat to avoid Entente blockages was soon recognized in addition to its ability to attack enemy merchant vessels. The first of these were the merchant U-boats, the Deutschland

and the Bremen in 1916. These U-boats were actually used to conduct trade with neutral countries, including the then neutral United States. The cargo capacity of the Deutschland and the Bremen was quite large at 700 ton. These were actually among the largest of all submarines manufactured in World War I. On her maiden voyage to the United States, the Deutschland carried expensive dyes and drugs. She returned back to Germany with rubber, nickel, and tin. These war-critical materials supplied Germany's needs for nearly a year. The Deutschland went on to make another voyage to the United States. However, just prior to a third planned voyage, the Unites States entered the war and the Deutschland was converted to a warfare vessel. The history of the Bremen is less significant as shortly after one voyage, the submarine captain took to attacking merchant shipping and was himself sunk.

1.11.5 Poison Gas

Finally, World War I was unique in that poison gas was widely used for the first time. These included irritants such as mustard gas and toxins like phosgene and chlorine. These gases were deployed in grenades and artillery shells and when the wind was in the correct direction, directly from storage cylinders. The use of these poisonous gases required an accompanying increase in logistical systems in order to handle this entirely new category of inherently dangerous weapons systems.

First, protection from the gas required the logistical system to include countermeasures. The first efforts of using soaked handkerchiefs and bicarbonate-soaked pads proved less than ideal. Eventually, primitive rubber protective masks, box respirators, and the filtering media required to allow their continued use. Not only were each country's military personnel provided with this type of equipment, but also animals such as horses and mules, which were used as part of the logistical transportation system.

A second logistical effect of the use of poison gas was the necessity of the military medical logistics resources to respond to the new types of casualties that they had not previously treated. The new injuries that now needed to be treated included both temporary and permanent blindness, blister on the skin, and lung-related injuries. Some of these were difficult to treat effectively at the time, in particular, mustard gas burns and lung scarring.

Lastly, World War I also continued the trend of the greatly increased use of explosive ordnance and ammunition through bombers, artillery pieces, machine guns, and small arms.

1.12 WORLD WAR II

While World War I was primarily limited to the European theater, World War II represented a far more global conflict. Logistics planning was complicated by the fact that countries now had to deal with a number of faraway fronts. Material now had to be moved regularly not only across land, but also across oceans. Marine logistics presented new challenges to military planners. Different types of supplies were required to effectively wage war in one theater versus another. The consequence of shipping the wrong supplies to a particular area could negatively influence the outcome of the battle.

1.12.1 Use of Widespread Air Transportation

World War II also saw the first widespread use of air transport for the distribution of both military supplies and personnel. For the first time in history, a distant campaign could be supplied in a matter of hours rather than days or weeks. Provided an airfield existed, significant numbers of combat troops could also be loaded and rapidly transported to distant battle zones. In a further development, airborne units of soldiers trained in the use of parachutes could be also rapidly transported over great distances. But, for the first time, it was not necessary for the transport plane to find a runway as the soldiers deployed their parachutes over the landing zone.

1.12.2 Use of Long-Range Bombers

World War II also saw the first widespread use of long-range bombers. The use of aircraft in this manner had just begun to become developed prior to the end of World War I with the German Gotha bomber. In World War II, the long-range bomber became a mature weapon, which could be used to strike deep within enemy territory. Logistical infrastructure such as railway exchanges, airfields, dams, bridges, and factories were all fair game. By the end of World War II, this included the use of atomic bombs on cities with both military and civilian concentrations.

1.12.3 Ammunition

Of course, the trend of consuming more ammunition continued. The use of fully automatic weapons at the individual soldier level became more common place with the first military use of the submachine gun on all sides. In addition to the small arms ammunition and increasingly larger artillery pieces, the military logistical chain now had to deal with larger munitions such as aircraft bombs.

1.12.4 U-Boats

As in World War I, German U-Boats were initially successful at wrecking havoc with Allied naval logistics supply efforts. Much of this was targeted toward the resupply of the British with U.S. and Canadian war material. The submarines would operate in wolf packs, where several U-Boats would work together to sink one target. As the U-Boats concentrated on civilian shipping targets supplying the Allies, it can be said that the German Navy was engaged in a logistical or economic war. Ultimately, advances in technology enabled the Allies to detect and sink the U-Boats. In the end, a total of 743 U-Boats were sunk.

The complications of increased logistical operations originated the concept of the field of operations research (OR). This involved the use of multidisciplinary teams of mathematicians, engineers, and scientists to solve both military and logistical problems. One of the most successful applications was the use of OR to hold the German Atlantic submarine wolf packs under control. In this particular application, OR scientists developed the bracketing approach to effectively depth charge German U-Boats. This involved setting the fuzing of the depth charges alternately deeper and shallower until the U-Boat was destroyed or lost.

1.12.5 Russian Campaign

Perhaps the most notable failure of a military logistics system in World War II involved the initial German Operation Barbarossa invasion of Russia in June 1941 and the subsequent Eastern Front War until May 1945. This campaign is frequently quoted as the largest and single-most bloody campaign in history with approximately 30 million deaths.

Among other things, the Russian Campaign highlighted a difference in industrial/military design philosophy. The German military vehicles such as the Panther and Tiger tanks were generally more complex and higher performing than their Russian counterparts such as the T34 tank. The greater complexity came at a high cost per vehicle and required more maintenance. In contrast, the Russians with the relatively simple T34 tank focused on low initial costs and less expensive maintenance. Thus, the German equipment required significantly more logistical support. As the Russian Campaign wore on, this became increasingly important as tanks and other armored vehicles were damaged and destroyed and needed to be repaired or replaced. This is significant from a manufacturing standpoint as much of the skilled German factory labor was forced while the Russian designs were less dependent on skilled workers.

Another significant logistical lesson from Operation Barbarossa was the necessity to tailor military logistics to account for the changing seasons. Adolf Hitler originally expected Operation

Barbarossa to take only 6 weeks. The invasion began in the spring of 1941 and continued through the first winter. The problems began in the fall as increased precipitation and the subsequent mud made forward mechanized progress more difficult. As the temperatures fell in the winter, the summer German uniforms were totally inadequate. At times, the lack of appropriate winter uniforms required German soldiers to insulate themselves with newspaper. This was not necessarily a very effective solution to dealing with the harsh Russian winter, but it was often the only solution that was available. Many German soldiers suffered and died from hypothermia as a result. The winter cold and snow caused other difficulties as many pieces of mechanized equipment were not designed to operate in this type of environment. In contrast, the Russian troops and equipment were far more acclimatized to the winter. Subsequent Russian winters had even more devastating impact on the German Army. In 1942, German winter resupply efforts were thwarted by both the weather and the Russian Army. Many German soldiers actually starved to death as a result.

During this campaign, the Russians further developed the concept of the Scorched Earth Policy, first employed in the middle ages. This was implemented by Joseph Stalin in July of 1941. The policy stated that during the Russian Army was to either remove or destroy any potential equipment, supplies, or infrastructure when withdrawing from Russian soil. This way nothing of use could potentially fall into German hands. As a result of the Scorched Earth Policy, many Russian roads, bridges, factories, and farms were destroyed. Interesting enough, the German army also began utilizing a Scorched Earth Policy as they later withdrew from Russian soil.

1.13 KOREAN WAR

For U.S. military logistics, the Korean War represented a difficult challenge. Much of this was due to the hilly terrain and the poorly developed transportation infrastructure. Out of these difficulties came one of the most interesting applications of military logistics. When presented the problem of moving supplies from the seaports to the front lines, military logisticians came up with the idea of an analogical logistical model.

1.13.1 ANTIAC

Named Anti-Automatic-Computation (ANTIAC), this model represented the actual transportation road system with a network of strings, lines, and rope suspended between two platforms. The diameter of the material corresponded to the transportation capacity of the particular road that the string, line, or rope represented. Thus, larger capacity routes were modeled with rope while smaller roads were represented by strings. To determine the most effective route to transport material, the logisticians placed sugar ants on the platform on one side of the network and sugar on the other.

After a number of exploratory paths, the sugar ants would converge on one route in order to get to the sugar. The corresponding physical route was then used to transport material and supplies to the front lines. This legendary model was also adaptable in that a critical road that was captured or otherwise put out of commission could be easily modeled using the existing network of strings and ropes. This was performed by simply cutting the corresponding set of strings and ropes from the model. With the previously existing paths now missing, the ants would have to figure out a new solution to the front. Just like the ants figured out the original solution, they would also recalculate the new solution that would then be used. Although not the secret weapon that it once was, research continues to be performed using this concept. Sometimes the ants themselves are modeled using computer-generated ants.

1.13.2 First Widespread Use of the Helicopter for Logistical Operations

Although the first production helicopters were produced in 1942, it was not until the Korean War that helicopters were in widespread use. During this conflict, the use of helicopters was primarily for logistical purposes, including the transportation of personnel and supplies and the medical evacuation of casualties. The ability of the helicopter to land in small clearings enabled it to perform these duties in the rough Korean terrain. In the transportation and medical evacuation roles, the helicopter was far superior to the only other alternative form of transportation, the truck. More than 4700 soldiers were medically evacuated from Korean battlefields in December 1950 alone. By the end of the war, over 18,000 soldiers had been evacuated by helicopter to rear areas. Other non-logistical applications of the helicopter included search and rescue operations. Helicopters that saw widespread use in logistical roles included the Bell H-13, the Sikorsky H-3, the Hiller H-23, and the Sikorsky H-19. The Bell H-13 was popularized as the helicopter used in the television series, MASH.

Like any new military technology, the widespread use of helicopters presented new logistical difficulties. One of the peculiarities of helicopters is that limited-life components are replaced on a per-hour use basis. Thus, the logistical stocking of replacement parts for military helicopters is not only based on damage due to enemy actions, but also due to the limited life of critical components. The logistical maintenance of these parts is further complicated by the fact that these parts each have different lifetimes. Thus, significant record keeping is necessary to maintain a large military helicopter force in safe working order. Helicopters are also significantly more difficult to fly than fixed-wing aircraft. As a result, there is an inherently higher rate of accidents requiring replacement or repair. Lastly, helicopters are entirely dependent on the horizontal rotor to provide lift. As a result, helicopters generally consume larger quantities of fuel than fixed-wing aircraft. The increased consumption must be supported by the military aviation logistics system.

1.14 COLD WAR

The cold war between the Warsaw Pact and NATO forces presented a new form of logistics to military planners. Prior to the cold war, military stockpiles were kept primarily in the host country's territory and transported to the area of conflict. Movement to new areas of conflict, such as in the Korean War, required a substantial sealift in order to position the required material. This was a costly and time consuming, but necessary approach as it was not specifically known where the next major war would develop.

1.14.1 Stockpiling of Supplies

In the case of the lengthy cold war, the battle fronts could be reasonably accurately predicted. This allowed substantial stockpiles of material to be prepositioned throughout Europe with a minimum garrison force. This allowed the bulk of the military troops to more inexpensively remain on home soil. From their overseas home locations, equipment light troops could more quickly and cheaply be deployed from their normal base of operation to the theater of war. This logistically based deployment approach was started in 1969 and practiced annually in the United States. Return to Forces Germany (REFORGER) exercises. Although the cold war actually ended in 1991, REFORGER continued to be conducted for a few years afterward. Note that currently the United States and many other nations participate in a similar exercise, designated Operation Bright Star. This is an annual air, sea, and land exercise conducted in Egypt to develop coordination and support among nations in the Middle East.

1.14.2 Dependence on Civilian Aircraft

Existing military air transporters were augmented with select U.S. civilian air carriers from the Civilian Reserve Air fleet. Military personnel on these civilian air carriers were only allowed a very limited amount of equipment. Little ammunition, particularly for artillery units, was taken. Once on site, the equipment light troops would be reunited with the prepositioned equipment and ammunition. The increased time to coordinate the fully equip and stock the equipment light troops was considered a minor disadvantage in comparison to the time and money saved by deploying without these basic loads.

With the termination of the cold war, military logisticians were once again forced to rethink themselves. No longer could a major conflict in a particular area in a particular way be expected. The use of vast storage sites of prepositioned equipment and ammunition was no longer an acceptable solution to inexpensively and effectively fulfilling the demands of military logistics.

1.15 VIETNAM WAR: U.S. PERSPECTIVE

While the Cold War was still in effect, the military logistics community had the opportunity to experience the low-intensity conflict of the Vietnam War. The Vietnam War is credited for the first widespread use of rotary wing or helicopter aircraft to support military operations. Though some limited use of helicopters was present in the Korean War, it was not until the Vietnam War that the helicopter became an essential piece of the logistical train. In this war, the flexibility of the helicopter introduced the ability to bring troops and supplies more rapidly and into areas that were not previously possible.

1.15.1 Rapid Removal of Battlefield Casualties

Similarly, helicopters were utilized at a widespread level to remove casualties from the battlefield. As a result, an unprecedented number of soldier's lives were saved. The increased logistical burden of saving the lives of soldiers, which would otherwise have been lost, was not lost upon the munitions and booby trap designers of the time. More and more designs were developed and deployed not to kill, but to inflict casualties to the enemy forces. For every wounded U.S. soldier, several others were required to transport the wounded soldier back to an evacuation zone and eventually to military medical facilities.

1.15.2 Increased Firepower

As in most previous wars, the rate of firepower of the individual soldier was also increased. The U.S. forces had the M16 while the communist forces fielded the AK47. The Civil War Gatling Gun design also made a reappearance as the six-barreled 7.62 mm Minigun. Not surprisingly, one noted infantry officer, Lieutenant Colonel Lones Wigger, estimated that approximately 250,000 rounds of small arms ammunition was expended for each enemy casualty in the Vietnam War versus 7000 in World War I. The increased need for ammunition logistics was a direct result of widespread poor marksmanship training on the part of the U.S. Army. Immediate efforts began to improve the ability of the individual U.S. infantryman to increase his ability to effectively use small arms. This is an early example of how increased military training can reduce the burden on the military logistics system.

1.16 VIETNAM WAR: VIET CONG PERSPECTIVE

Viet Cong forces of the 1960s are considered by some to be among the most logistically resourceful, flexible, and resilient military organizations (Holiday, 1968). Like many other military organizations, they relied heavily, as much as three quarters, on the local population to assist with logistical efforts. This included transportation, construction, food production, evacuation of wounded, ordnance manufacture, taxes, and the purchasing of supplies. Transportation, for example, included not only the expected truck, but also bicycles, pack animals, carts, and human porters. All of these were supplied by local villages. In the case of human porters, villagers were often required to submit to forced labor for several months a year.

1.16.1 Military Supply Recycling

However, the Viet Cong in particular are noted for overcoming initial limitations in logistical support by recycling waste into weapons. For example, in order to receive loaded small arms ammunition, Viet Cong had to submit expended shell casings. The casings were then reloaded into ammunition by rear area ordnance facilities. Similarly, the Viet Cong produced hand grenades, land mines, and booby traps from unexploded U.S. bombs and artillery shells. The casings for grenades were frequently made of discarded soda or food cans from U.S. or South Vietnamese troops. Even worn-out truck tires were recycled into the sole material for sandals. The ability to locally produce war material reduced the transportation logistical requirements for the Viet Cong. Instead of having to transport material from some distant point of acquisition, the Viet Cong only needed to concentrate on using their resources to move the locally produced material to the individual units.

1.16.2 Acquisition of Supplies on the Open Market

What the Viet Cong could not locally manufacture, they purchased on the open market in major cities such as Saigon. Again, the local village populace was utilized for this. In order not to arouse suspicion, when large amounts of rice, for example, were required, the Viet Cong would send a number of women villagers. Since many individual purchases were each smaller than fewer larger purchases, suspicion was not aroused. In the odd event that questions were asked, the women villages would reply that the food was required for different family celebrations.

1.16.3 Viet Cong Medical Logistics

The Viet Cong medical support system not only had to deal with casualties as a result of war, but also disease. As with their other logistical systems, Viet Cong medical logistics also depended on the local population and the forced laborers. While the combat units were required to withdraw the casualties from the battlefield to an initial medical facility, laborers were utilized for the evacuation to the final surgery areas. However, the medical transportation laborers were under Viet Cong guard.

1.16.4 Viet Cong Mortuary Logistics

The Viet Cong also had a primitive mortuary logistical system. This too consisted of a combination of combat troops and local population. Each Viet Cong soldier was issued a length of rope for the evacuation of deceased soldiers. The Viet Cong would evacuate the dead by attaching a loop on the end of this rope to the deceased's neck with a loop on the other end of the rope to the other soldier's waist. This would enable the soldier to evacuate the deceased by crawling away from the battlefield. Evacuation in this manner helped reduce the exposure of the soldier to enemy fire. Again forced labor was ultimately utilized to evacuate the dead to the predesigned burial sites once away from the battlefield. A final interesting point involving the Viet Cong mortuary logistical system is that burial sites were often predesigned and dug prior to major engagements.

If anything, the Viet Cong logistical efforts illustrated how a resource-poor and logistically disadvantaged military force can function by creating local manufacturing facilities. Rather than depending on a lengthy exposed supply train, the use of local facilities allowed the limited Viet Cong logistical system to focus on providing a reduced set of required supplies.

1.17 SOVIET–AFGHANISTAN WAR

The Soviet–Afghanistan War began in 1979 with the Soviet invasion of Afghanistan. This invasion was in response to insurgent activities and to replace the existing government with a government with more pro-Soviet doctrine. Like the Vietnam War, the Soviet Union was ill-prepared for unconventional battle. The war degenerated to both the Soviet forces and the Mujahideen concentrating on attacking each other's logistical pipelines (Grau and Jalali, 2001).

1.17.1 Use of Local Support

The Mujahideen engaged in a guerilla-type war. They attacked the Soviets truck convoys while traversing mountain passes. They attacked pipelines and power lines. In the mid-1980s, there were 600 or more of these types of attacks a year. In contrast, the Soviets could not reciprocate since the Mujahideen's logistical pipeline consisted of rural villages that supplied the Mujahideen with food and other supplies. Without a well-defined logistical pipeline, the Soviets took to destroying the villages. This led to massive evacuation of the Afghan people. Up to 5 million people or one-third of the population are believed to have fled to Pakistan and Iran.

1.17.2 Logistical Operations without Local Support

No longer able to live off of their local neighbors, the Mujahideen were forced to develop more conventional logistical systems. This included the transport of supplies from the neighboring countries of Iran and Pakistan. As their logistical system became more conventional, it was necessary for the Mujahideen to establish various levels of supply depots. The question for the Soviets now became how to identify and attack the newly developed Mujahideen supply depots. Once these supply facilities were identified, a typical Soviet attack began with air support and was followed by artillery and infantry units. Despite the Soviets having technologically superior weapons, the Mujahideen persisted. Part of this was no doubt attributable to the U.S. supplied Stinger missiles. While not actually credited with large numbers of Soviet losses, the Stinger missiles did force the Soviets to engage in less effective air operations. Eventually in 1987, the Soviets announced that they would withdraw from Afghanistan.

1.18 IRAN–IRAQ WAR

The Iran–Iraq War took place between September 1982 and August 1988. The war began with a surprise invasion into Iran by the Iraqis. The purpose of which was to prevent the spread of Islamic Revolution to the other Gulf states. Although Iraq was able to penetrate Iranian soil to some extent, these gains were lost by the middle of 1982. The Iraqis were left in a defensive position for balance of the war. Many comparisons have been made between this war and World War I. The conflicts were similar in that both saw major use of trench type warfare, barb wire between trench lines, human wave assaults, and possibly poisonous gas. As a result, the same logistical requirements for building materials were present for the trenches in this war. Similarly, the possible use of poisonous gas required additional medical logistics to treat these specific types of casualties as in World War I.

1.18.1 Use of Modern Foreign Weapons Systems

To military logisticians, the most interesting aspect of the Iran–Iraq War may be the use of modern weapons technology on both sides, neither of which actually manufactured the weapons systems. To sustain foreign-manufactured weapon systems, both routine maintenance replacement parts and battle-damaged components must be reliably acquired. This can be a difficult process, particularly if the country no longer has favorable trade agreements with the manufacturing countries. In these types of situations, complicated backdoor or blackmarket deals must be made to sustain the weapons systems.

For example, the Iranian Air Force employed F4, F5, and F14 aircraft and AH1 helicopters. These were left over from the Pro-U.S. Iranian days. On the other hand, the Iraqis utilized soviet aircraft including MIG fighters and Tupolev bombers and French Mirage fighters. At one point, this became an interesting situation as U.S. forces were faced with shooting down U.S.-made Iranian aircraft. This came to a head with the incident where the USS Vincennes shot down the commercial Iran Air flight 655 mistaking it for an Iranian F-14. Equally strange situations between counties not directly involved in the war and the munitions also occurred; for example, the USS Stark Frigate was fired up by an Iraqis Mirage with two Exocet Missiles.

1.18.2 World Involvement in Military Logistics

Both the United States and the Soviet Union sometimes supplied both Iran and Iraq with weapons systems. The United States did this indirectly through the Iran-Contra Affair. Other countries such as Spain and Yugoslavia openly provided weapons to both Iran and Iraq. The more normal situation was individual countries supporting either Iran or Iraq. The most significant of these being the North Koreans who supplied Iran. This was the cause for some logistical confusion as many North Korean-manufactured weapons systems were based on Soviet designs. So, it was then necessary for Iran to support logistical systems for both U.S. and Soviet weapons.

1.18.3 Overreliance on Foreign Military Suppliers

The reliance on foreign countries for logistical resupply developed into a serious issue for Iran toward the end of the war. In particular, the United States would not overtly supply parts for the Iran Air Force. As the Iranian U.S. planes became damaged and in need of maintenance components, it became increasingly difficult to field an effective air force. Some military experts believe that this was one of the principal causes of Iran's unsuccessful destruction on Iraq's nuclear facilities. These were the same facilities subsequently destroyed by the Israeli Air Force. By 1987, Iran was unable to effectively compete with the Iraqi air force that enjoyed continuous resupply of their air force through the Soviet Union.

1.19 FALKLAND ISLANDS WAR

The Falkland Islands War occurred in 1982 between the United Kingdom and Argentina. The war began with the invasion of the Falkland Islands by Argentina and lasted 74 days. Due to the 8000 mi distance between the United Kingdom and the Falkland Islands, many military strategists considered any attempt by the United Kingdom to retake the Islands doomed to failure. This was aggravated by the much closer proximity of the Falkland Islands to Argentina. In order to perform the invasion, the British were forced to utilize Ascension Island that was still 3800 mi away from the Falklands.

Due to the isolation and distance to the Falkland Islands, the British forces were required to transport virtually all of the supplies necessary to fight the war (Webb, 2007). This included a total of 100,000 ton of freight and 400,000 ton of fuel. Even fresh water had to be transported. The vessel, The Fort Toronto, served as the water tanker for the invasion force. The British were so logistically strapped at the time that foraging parties were sent to old whaling stations that had abandoned over 2 decades earlier.

1.19.1 British Vulcan Bombing of Port Stanley Airfield

Perhaps one of the most striking displays of British military logistics determination involved the bombing campaigns conducted by the Vulcan bombers. In order to prevent Argentine fighter planes from providing air cover, the British military command determined that the airstrip at Port Stanley needed to be destroyed. The only problem was the difficulty in providing bomber aircraft for the mission. The 3800 mi mission would be the longest bombing mission in military aviation history.

At the time, the only British aircraft remotely available for this type of mission was the 1950s era Vulcan Bomber. Even then, the normal range of the Vulcan bombers was well less than that required. Furthermore, Vulcan crews had not practiced the difficult and dangerous process of in-air refueling in over 20 years. Last, the Vulcans had been modified to carry nuclear years earlier. The components necessary to utilize conventional bombs had long been discarded. With the original bomb mounting equipment, the Vuclans were not capable of carrying the types conventional bombs required for the Port Stanley airfield mission. The British handled these challenges in a methodical manner. The refueling inexperience was gradually overcome as the British flight crews began intensive in-flight training. The British also overcame the conventional bomb mounting issue as the Vulcans were retrofitted with the appropriate parts from both scrap yards and museums.

To accomplish the mission, the British developed an exceptionally complicated, but effective in-flight refueling plan. Interestingly, the logistical calculations were not performed on a supercomputer, but rather on a hand calculator purchased for a few pounds. Essentially, the fleet of 14 air tankers would refuel both themselves and the two Vulcan bombers until only a single tanker remained full. As each individual tanker emptied itself into the other tankers and the Vulcans, they would return back to Ascension Island. The final single remaining tanker would then refuel the bombers one last time before the bombing run. The Vulcans would then execute the bombing mission at the Port Stanley airfield. After the run, additional tankers would be standing by off of South America to refuel the Vulcans for the return trip. Because the bombers were no longer carrying their payloads, they would be able to fly greater distances back to Ascension Island.

It is said that no battle plan ever survived first contact with the enemy. The same was true for the elaborate British refueling plan. Immediately after beginning the mission, 1 of the Vulcans and 2 of the 14 tankers became inoperative. Despite these losses, the British decided to continue on with the mission. Further troubles arose during the last refueling session, but the single Vulcan bomber continued on. At the target, the Vulcan dropped 21,000 lb bombs on the Port Stanley airfield. This forced the Argentine Air Force to reposition their fighter aircraft back to the mainland. The mission was a resounding success. Even the refueling for the return trip worked as planned. The return of the Vulcan to Ascension Island marked the longest most logistically complicated bombing run in military aviation history.

1.19.2 Exocet Missiles

Another highly publicized logistics-related lesson revolved around the use of the French Exocet missile. The Exocet is a guided antiship missile developed in the mid-1970s. It can be launched from a variety of platforms including ships, submarines, and aircraft. The Exocet is effective in that it skims the water to avoid radar detection and uses its own onboard radar system to maintain guidance to its target. The Exocet gained notoriety as it was used by the Argentines on May 4, 1982 to sink the British destroyer, the HMS Sheffield. The Exocet was launched by a Super Etendard aircraft and struck the HMS Sheffield on the second deck. Interestingly, the Exocet failed to explode, but sufficiently damaged the electrical and plumbing systems on the HMS Sheffield to prevent fire fighting attempts to douse the flames caused by the unexploded missile fuel.

The Argentines continued to fire a large number of Exocets at the British Navy. The effectiveness of the Exocet on the HMS Sheffield, even though it did not explode, led to British concerns about the vulnerability of the British carriers, HMS Invincible and HMS Hermes, to attack. Clearly, this was one of the first cases of a relatively inexpensive precision-guided munition being able to destroy a much larger and more expensive asset. As a result, the British intelligence resources were tasked with the mission of disrupting the Argentineans' Exocet logistical pipeline. This is another logistical example where a country's military forces are compromised by using weapons systems produced in countries other than itself. As it turns out, the French denied continued sales to the Argentineans. They were subsequently forced to seek additional Exocets on the open market. At the same time, British intelligence forces posed as Exocet buyers to both prevent and mislead the Argentineans from acquiring additional Exocets. Thus, a non-weapons-producing nation can be prevented from acquiring additional war supplies not only by the weapons-producing nations, as we have already seen, but also by other interested parties.

1.20 GULF WAR

The 1991 Gulf War was the first war in history to make extensive use of computer technology. This included not only precision-guided munitions and cruise missiles, but also computerized information systems. The Gulf War was also interesting from the standpoint that the hostilities lasted only 100 h. The hostilities during this period were also marked by rapid forces movement on a scale that has not been previously seen.

1.20.1 Lack of Resupply

One significant military logistics observation was that the initial introduction of material focused on the deployment of combat arms units such as armor and infantry to the exclusion of combat service support units with fuel, ammunition, and other required supplies. This left the combat arms units in a situation in which they were not able to be resupplied as the combat service support units were still waiting to deploy. In the event that Iraq had chosen to attack during this period, the outcome of the initial hours of the war might have been different. Similarly, had the war lasted much longer than 100 h, U.S. forces would have been forced to pause until combat service support units could arrive for resupply. Recommendations offered by military logisticians to avoid this situation focus on utilizing a more balanced approach that would allow both combat arms and combat service support units to be deployed at the same time.

1.20.2 Incompatibility of Logistical Information Systems

Once units established their position in Iraq, another unexpected logistical issue developed. This was the incompatibility of logistical information systems. This problem became so serious that units were directed to utilize unreliable manual reports. During this period, units also began hording supplies as no one actually knew the resupply situation. The stockpiling situations promoted the disorganized storage of supplies. As many of the supplies were containerized, it also became difficult to locate required items. The search for items required lengthy manual searches. In some cases, the movement of units also resulted in the abandonment of stockpiled supplies.

1.21 WAR ON TERRORISM IN AFGHANISTAN

The United States launched the War on Terrorism primarily in response to the September 11, 2001 attacks on the World Trade Towers. Although the War on Terror actually consists of multiple efforts on different continents, it is becoming increasingly identified with the military operations being conducted in Afghanistan. The focus of these operations is to extradite the al-Qaeda'a leaders responsible for the World Trade Tower attacks, including the Islamic terrorist Osama bin Laden. As this effort was not supported by the Afghanistan Taliban government, the United States initiated military action in the form of Operation Enduring Freedom. Since Operation Enduring Freedom began in 2001, there has been a buildup of troops to over 100,000 by the middle of 2010.

The tactics utilized by the Islamic Taliban are reminiscent of those used in Afghanistan during the earlier Soviet–Afghanistan War. In particular, the Afghanistan and Pakistan Taliban have taken to ambushing U.S. and NATO truck supply convoys approaching through the restricted Khyber mountain pass areas on the border of Pakistan. These include convoys carrying fuel and food. In one well-publicized incident in September 2009, NATO forces were forced to bomb two of their own fuel tankers in order to prevent them from falling into Taliban hands. This is among other things an example of how the Taliban have forced U.S. and NATO troops to expend even more logistical effort on protecting their own logistical supply system. In response to the difficulties of transporting the fuel and food through Pakistan, U.S. and NATO have investigated the use of supply routes approaching from the west rather than from the east.

1.22 IRAQ WAR

This is the Iraq War that began in 2003 as a result of UN concerns of the presence of weapons of mass destruction (WMDs), suspected terrorist support, and human rights violations. The Iraq War eventually resulted in the capture of Saddam Hussein, but conclusive evidence of WMDs was not found. The Iraq War has continued on since 2003 in the form of an occupation. While not the same as the earlier all-out conflict, U.S. and UN troops continue to take casualties during the occupation. Many of the lessons learned in the first Gulf War have been properly addressed in the Iraq War; however, new logistical issues have also come forth. These include shortages of basic individual supplies such as personal body armor and potable water.

1.22.1 Personal Body Armor

In as early as 2004, a shortage of army Interceptor personal body armor was identified. The goal of the U.S. Army was to equip all troops in Iraq with the Interceptor body armor; however, production problems in the logistical system as well as an unexpected increased number of troops in Iraq resulted in widespread body armor shortages. Thus, initially only combat arms type troops being properly protected with the Kevlar and ceramic plate body armor. Noncombat arms support troops were given low priority for the issue of body army. However, it was soon realized that in this war, military personnel of all types were actually at risk. This is most significant as the use of improvised explosive devices, in particular against U.S. military personnel while traveling in vehicles, left any type of soldier unnecessarily exposed to battlefield injuries without proper body armor.

Difficulties producing and supplying the body armor by the original suppliers prompted outcries for unprotected soldiers to be able to privately purchase and be reimbursed for body armor. Both state and federal legislation was considered to this effect. For example, the state of Ohio introduced a bill to set aside $500,000 for the purchase of body armor by Ohio National Guardsmen. On the other hand, while the Army could not provide the body armor, they were discouraging the use of untested products. This presented an unacceptable position for many of the U.S. soldiers and their families.

The logistical failure of the Army in this situation has generated several related investigations resulting in indictments. In particular, it appears that rather than initially awarding manufacturing contracts to several capable companies, the source contract was initially awarded to only one company. This company subsequently proved to be unable to manufacture the vests in the proper quantity at the proper quality level. Perhaps the logistical lesson learned here is that it is not only the logistical system that can reduce the effectiveness of a military force, but also its acquisition system. This is the same lesson that the British learned early on in the American revolutionary war.

1.22.2 Widespread Use of Military Contractors

A significant logistical milestone in the Iraq War was the largest use of military contractors in U.S. history. In 2006, it was estimated that there are up to 100,000 private contractors in Iraq (Merle, 2006). More recently, this estimate has risen to 180,000 according to the Congressional Budget Office. This includes the use of private security firms along the lines of the former Blackwater Corporation and the use of civilian contractors to run supply convoys such as Halliburton, which has 40,000 employees in Iraq alone. This has led many to question the ability of the U.S. military to successfully perform its mission without the use of civilian contractors.

1.22.3 U.S. Soldiers Forced to Steal Water

While potable water does not fall into the same logistical category as ammunition, it is still necessary for conducting effective military operations on all levels. Without water, humans cannot survive past a few days. Combat effectiveness is lost far before that. In order to maintain a proper level of hydration, approximately 1 gal of water must be consumed each day per person. In the environmental conditions in Iraq, the requirement may be increased to as much as 4 gal per person. Additional water is also required for hygiene. This can constitute a significant logistical requirement for a large military force. Not only must water be available in the proper quantities, it must also be potable. The consumption of non-potable water will quickly result in reduced military readiness from a variety of waterborne diseases.

In the Iraq War, a number of civilian organizations have been contracted to ensure the availability of potable water. However, in what can only be characterized as a disgraceful embarrassment, U.S. troops have been forced to steal potable water from different sources while warehouses of water were held by the civilian contractors. As discussed, in earlier times, this practice was called foraging. While these situations are most likely still under investigation, suffice it to say that a properly functioning civilian–military logistical system would not have required military personnel to divert this attention from their primary mission.

1.23 FUTURE OF MILITARY LOGISTICS

Even a short look at the history of military logistics requires a brief look into the future. The use of recent technology such as RFID is having and will have a profound impact on military logistics. Lessons learned from the 1991 Gulf War and the current Iraq War have reiterated the need for the military logistics concepts such as seamless logistics, distribution-based logistics, and total asset visibility. These will be covered in depth in the following sections of this book. In addition to these applications, we will also discuss other specific applications such as those used for the distribution of uniforms, mortuary operations, and military personnel records.

1.24 SUMMARY

The objective of this chapter was to provide a brief history of military logistics. Major developments in military logistics occurred during nearly all major conquests or conflicts. As we have seen, some of the logistical issues plaguing military operations in the days of early history are still present to this day. In fact, in many cases, the lessons presented earlier in history were not properly learned. The same mistakes were unnecessarily repeated again. In the words of George Santayana, "Those who do not remember the past are condemned to repeat it."

2 U.S. Department of Defense and North Atlantic Treaty Organization Supply Classes and National Stock Numbers

2.1 INTRODUCTION

Both the U.S. Department of Defense (DOD) and the North Atlantic Treaty Organization (NATO) have established different classes for commonly used logistical items. These include:

DOD and NATO use the same classes for Classes I–V. All 10 classes include

- Class I: Food
- Class II: Clothing
- Class III: Fuel and lubricants
- Class IV: Barrier or fortification materials
- Class V: Ammunition
- Class VI: Personal demand items
- Class VII: Major end items
- Class VIII: Medical supplies
- Class IX: Repair parts
- Class X: Material for nonmilitary programs

Items are further individually identified by a national stock number (NSN) for the United States and NATO countries. NSNs are a total 13 alphanumeric character designations organized in three subgroups in the following form.

ABCD-EF-GHIJKLM

The first subgroup of four characters represented by ABCD is the federal supply class (FSC). This is also known as the national supply classification number (NSC). The first subgroup FSC is further broken down into a two-digit FSC group represented by the AB and a two-digit FSC class represented by the CD. Currently, there are FSC 78 groups and FSC 643 classes. Some examples of major FSC group designations are

- 10 Weapons
- 13 Ammunition and explosives
- 19 Ships, small craft
- 30 Mechanical power transmission equipment
- 56 Construction and building materials
- 70 Automatic data processing equipment
- 91 Fuels, lubricants, oils, waxes

In general, all like category items should have the same NSC. For example, within the FSC Group of 10, the following NSC numbers are designated:

- 1005 Guns, through 30 mm
- 1010 Guns, over 30 up to 75 mm
- 1015 Guns, 75–125 mm
- 1020 Guns, over 125–150 mm
- 1025 Guns, over 150–200 mm
- 1030 Guns, over 200–300 mm
- 1035 Guns, over 300 mm
- 10xx Several others including chemical weapons, torpedoes, rockets, etc.

The second subgroup represented by EF is a two-digit country code (CC) or national codification bureau (NCB). Common CC include

- 00 and 01 United States
- 12 Germany
- 14 France
- 44 United Nations
- 99 United Kingdom

The third subgroup represented by GHIJKLM is the item number. The third subgroup is also sometimes designated in two parts. The first part consisting of three digits and the second part consisting of the remaining four digits. The second and third subgroups can also be combined to represent an item's national item identification number (NIIN).

An example of a complete NSN for the pistol, caliber .45, automatic is

1005-00-673-7965

This is determined by the fact that the pistol is less than 30 mm, thereby falling into FSC Group and Class 1005 and is produced by the United States with a CC of 00. The remaining seven digits identify this particular item from other items in the same FSC.

2.2 SPECIFIC SUPPLY CLASS DETAILS

2.2.1 Class I: Food

Class I items are directly related to food items. This includes items such as individual meals ready to eat (MREs), unitized group rations (UGRs), and enhancement items such as fresh fruit, salad, and bread. UGRs come in two forms, the UGR H&S and the UGR-A. Both contain rations for 50 soldiers. The UGR H&S (heat and serve) requires no refrigeration while the UGR-A has perishable and semi-perishable, which require refrigeration. Also included in Class I are health and comfort packages (HCP) that include razors, toothbrushes, toothpaste, and similar personal care items. Lastly, A-Rations are also included in Class I food items; however, these have additional logistical requirements including the need for refrigeration and distribution equipment. Specific subclass codes include A—nonperishable, C—combat rations, R—refrigerated, S—other non-refrigerated, and W—water.

2.2.2 Class II: Clothing

Class II supplies actually encompass a variety of items in addition to clothing. Other Class II items include individual equipment tools, personal protective overgarments and filters and unclassified maps. Specific subclass codes include A—air, B—ground support material, E—general supplies, F—clothing, G—electronics, M—weapons, and T—industrial supplies.

2.2.3 Class III: Fuel and Lubricants

Class III includes two divisions of supplies, bulk fuel, and packaged petroleum products. In the bulk fuel category, it includes mogas, diesel, and aviation gas. The packaged petroleum products category includes oil lubricants, hydraulic fluids, coolants, deicing fluids, preservatives, and different greases. Packaged petroleum products are stored in containers in drums of 55 gal and less. Specific subclass codes include A—POL for aircraft, W—POL for surface vehicles, and P—Packaged POL.

2.2.4 Class IV: Barrier or Fortification Materials

This class includes materials used by both individual units and engineering units to construct barriers and fortifications for fighting and protective positions. Class IV items also include those materials needed to maintain, upgrade, or construct roads, bridges, airfields, landing zones, and rafts and bridges for river crossings. It may also include material for maintaining, upgrading, or constructing buildings or facilities for host nations. Specific subclass codes include A—construction and B—barrier.

2.2.5 Class V: Ammunition

Class V includes items such as small arms ammunitions, mortar shells, artillery shells, rockets, missiles, bombs, explosives, mines, pyrotechnics, and associated components such as fuzing. Specific subclass codes include A–air delivery and W—ground.

2.2.6 Class VI: Personal Demand Items

Class VI items are generally personal demand items made available to troops through the Army and Air Force Exchange System (AAFES). This specifically includes personal items such as razors, toothbrushes, toothpaste, shampoo, and soap necessary for personal hygiene.

2.2.7 Class VII: Major End Items

Class VII includes major end items such as vehicles, aircraft, and various weapons systems. As major end items, items in this class are complete and ready to use. As Class VII items represent major capital, they are generally controlled with the loss of Class VII items requiring accountability. Specific subclass codes include A—air, B—ground support material, D—admin. Vehicles, G—electronics, J—racks, adaptors, pylons, K—tactical vehicles, L—missiles, M—weapons, N—special weapons, and X—aircraft engines.

2.2.8 Class VIII: Medical Supplies

Class VII includes all types of medical equipment and supplies to repair medical equipment. These include by first four NSN designation:

- 6505—Drugs, biologicals, and reagents
- 6510—Bandages
- 6515—Medical supplies and equipment
- 6520—Dental supplies and equipment
- 6525—Radiological supplies and equipment
- 6530—Surgical supplies and equipment
- 6532—Medical clothing and linen
- 6540—Ophthalmic supplies and equipment
- 6545—Medical sets, kits, and outfits
- 6550/6600—Laboratory supplies and equipment

Specific subclasses include A—medical and dental material and B—blood/fluids.

2.2.9 Class IX: Repair Parts

Class IX includes kits, subassemblies, and assemblies of components to provide maintenance and repair of different systems in the field. Specific subclasses include A—air, B—ground support materiel, D—admin. vehicles, G—electronics, K—tactical vehicles, L—missiles, M—weapons, N—special weapons, and X—aircraft engines.

2.2.10 Class X: Materials for Nonmilitary Programs

Class X materials includes material for nonmilitary programs. This includes tools for agriculture and economic development. This includes equipment such as tractors and consumable items such as seeds and plants.

2.3 SUMMARY

In this section, we described the basic 10 different U.S. DOD and the 5 NATO supply classes and NSN system. The supply classes consist of categories for food, clothing, petroleum, oil, and lubrication, barrier or fortification materials, ammunition, personal items, major end items, medical-related supplies, repair parts, and nonmilitary-related supplies. Currently, only Classes I, II, VI, and IX are required to be tagged at the shipping container, palletized unit load, and exterior container level. However, DOD plans call for the eventual tagging of all classes at these packaging levels.

3 DOD Shipping Level Containers RFID Designations

3.1 INTRODUCTION

DOD RFID policy as well as commercial policy has standardized on a construct for identifying shipping layers within logistics systems. The controlling document is MIL STD 129. Using a combination of active and passive RFID technology, the purpose of the standards is to facilitate the management of shipping and receiving processes. The container designations are levels 0–5 in increasing degree of packaging:

- Layer 0: Product item
- Layer 1: Package
- Layer 2: Transport unit
- Layer 3: Unit load
- Layer 4: Freight container
- Layer 5: Movement vehicle

3.2 LAYER 0: PRODUCT ITEM

Layer 0 is an individual product item or assembly of items. The product item is normally put into a layer 1 package.

3.3 LAYER 1: PACKAGE

This is the first level of packaging of the product item. The package may actually contain one or more product items depending on the item itself. It is also possible that the package contain multiple dissimilar parts meant to be used in conjunction with each other. An example of a layer 1 package is an item bubble pack. The layer 1 package is normally assembled into a layer 2 transport unit.

3.4 LAYER 2: TRANSPORT UNIT OR SHIPPING CONTAINER

The layer 2 transport unit is a carton, case, or box consisting of a number of layer 1 packages. The purpose of the transport unit is to facilitate the transport or storage of multiple layer 1 packages. The layer 2 transport unit is normally assembled into a layer 3 unit load. Transport units or shipping containers are designated as utilizing passive RFID tags.

3.5 LAYER 3: UNIT LOAD OR PALLETIZED UNIT LOAD

Layer 3 is the unit load. This is an assembly of multiple layer 2 transport units. Examples of unit loads are pallets or packaging held together by some binding such as shrink wrap, netting, or banding. The unit load assembled in this manner is capable of transport. A layer 3 unit load is normally assembled in a freight container for further shipment. Unit loads are designated as utilizing passive RFID tags.

3.6 LAYER 4: FREIGHT CONTAINER

Layer 4 consists of permanent freight containers designed for repeated use. These can come in a variety of forms, including

- SEAVANS (20 and 40 ft military containers moved by sea)
- MILVANS (military-owned, demountable containers)
- Quadcons (57.5 × 96 × 96 in. container with a metal frame and pallet base that can be strapped together to form 20 ft ISO containers)
- Sixcons, 8 × 8 × 4 ft fuel/water containers
- Palcons, 40 in. L × 48 in. W × 41 in. H pallet or palletized container
- 463 L pallets, standardized pallet for military air cargo made of aluminum with a balsa cored and measuring 88 in. (224 cm) wide, 108 in. (274 cm) long, and 2–1/4 in. (5.7 cm) high 463 L pallets can hold up to 10,000 lb.

Layer 4 containers are defined as being easy to fill and empty and having more than 1 m^3 in volume. In addition, they are readily transferred between transportation modes. This includes truck to train to ship and so forth. Freight containers are normally shipped on a layer 5 movement vehicle. At layer 4, shipments utilize active RFID tags.

3.7 LAYER 5: MOVEMENT VEHICLE

Layer 5 involves the movement vehicle for the shipment. Typically, this is one of the following forms:

- Ship
- Aircraft
- Truck or truck bed
- Railway train

At layer 5, active RFID tags are utilized.

3.8 SUMMARY

This section provides a brief overview of the use of both passive and active RFID tags in relationship to the shipping layer designation. These requirements are covered in depth in MIL STD 129 and the U.S. DOD suppliers' passive RFID information guide.

Part II

Overview of RFID and AIT Technologies

4 Overview of AIT Technologies

4.1 INTRODUCTION

Generally, in order to utilize technology effectively, one should initially understand it. Radio frequency identification (RFID) technologies fall into a group of technologies described as automatic information technologies (AITs), which are complex entities that can be utilized in many ways. Military operators will have to use knowledge, insight, and creativity to make effective decisions on how and when to use these systems. Two of these pervasive technologies, RFID and bar codes, are considered in this text. Due to this fact, this chapter describes the AITs and provides a historical perspective on bar codes and RFID technologies. The perspective provided will help military personnel and contractors, and other managers envision an organized plan that supports effective decisions and can be gained by reviewing historical events.

4.2 AUTOMATIC INFORMATION TECHNOLOGIES

In this text, when we describe AITs, we describe a group of technologies that are generally associated with automatic data capture (ADC) in both military and commercial applications. Other technologies may be included as AITs, but we present a list that this text focuses upon (Figure 4.1).

The AIT/ADC described include

1. Bar codes (linear/two-dimensional [2D]/3D)
2. Radio frequency identification (active/passive/semi-passive)
3. Radio frequency data capture (RFDC)
4. Real-time location systems (RTLS)
5. Satellite tags/GPS (not in diagram)
6. MEMS
7. Contact memory buttons
8. Biometrics
9. Common access card
10. Optical character recognition (OCR)

With such a large variety of technologies to choose from, it is important for one to consider that the strategic purpose or operational requirements should drive the technology that is chosen. The technology should not be chosen based upon technical bias. Below is a brief description of teach each technology. We first generally describe the technologies. We will commonly accepted definitions. In this chapter, we utilize a leading authority on automatic identification (auto ID), AIM, for defining these common ADC technologies.

4.2.1 AIM Global

AIM is the industry association and worldwide authority on auto ID and data collection technologies. AIM members are providers and users of technologies, systems, and services that capture, manage, and integrate accurate data into larger information systems that improve processes enterprise-wide. Serving members in 43 countries for 35 years, the association has developed key technical specifications and guidelines that support the use of auto ID and mobile IT solutions.

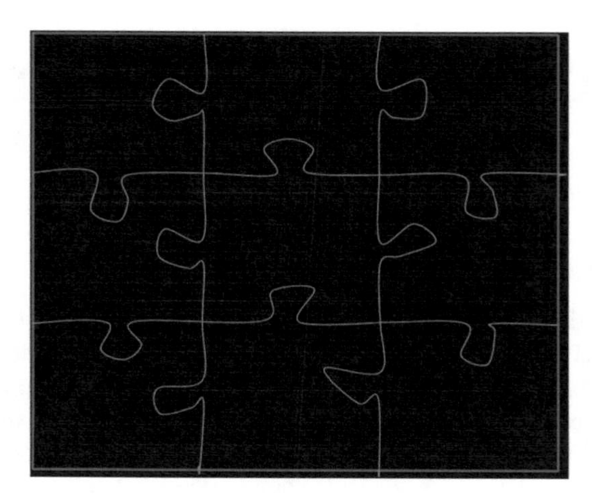

FIGURE 4.1 AIT jigsaw.

AIM actively supports the development of technology standards, guidelines, and best practices through its Technical Symbology Committee (TSC) and RFID Experts Group (REG), as well as through participation at the industry, national (ANSI), and international (ISO) levels.

AIM has an active educational and government relations focus, providing accurate and unbiased information on data collection technologies and the markets they serve. For more information about AIM and its members and services, please visit www.aimglobal.org or www.rfid.org.

4.2.1.1 Bar Codes

Generally, a bar code is an optically read representation of data in the form of lines and spacing of parallel lines. The line forms are generally categorized as one-dimensional (1D) bar codes. Other representations include geometric patterns such as dots, squares, and hexagons and categorized as 2D matrix codes. The codes are read by bar code readers that recognize the patterns. We have included direct part marking (DPM) and electronic article surveillance (EAS) samples for bar coding.

4.2.1.1.1 Direct Part Marking

DPM is a technology used to produce two different surface conditions on an item. These markings can be created by laser etching, molding, peening, etc. Traditional print quality measures are based on the assumption that there will be a measurable difference between dark and light elements of a symbol. Because DPM symbols frequently do not have sufficient contrast between elements intended to be dark and light, it is often necessary to provide specialized lighting in order to produce highlights or shadows in order to distinguish the various elements of the symbol. AIM Global shows examples of these technologies via its Web site.

4.2.1.1.2 DPM Quality Guidelines Document

Acknowledging that current ISO print quality specifications for matrix symbologies and 2D print quality are not exactly suited for DPM symbol evaluation, an ad hoc committee under the supervision of the AIM TSC developed a guideline to act as a bridge between the existing specifications and the DPM environment in order to provide a standardized image-based measurement method for DPM that is predictive of scanner performance.

The document describes modifications that are to be considered in conjunction with the symbol quality methodology defined in ISO/IEC 15415 and 2D symbology specifications. It defines alternative illumination conditions, modifications to the measurement and grading of certain parameters, and reporting the grading results.

4.2.1.1.3 EAS

EAS is a technology used to identify items as they pass through a gated area. Typically, this identification is used to alert someone of the unauthorized removal of items from a store, library, or data center.

There are several types of EAS systems. In each case, the EAS tag or label is affixed to an item. The tag is then deactivated when the item is purchased (or legally borrowed) at the checkout desk. When the item is moved through the gates (usually at a door to the premises), the gate is able to sense if the tag is active or deactivated and sound an alarm if necessary.

EAS systems are used anywhere there is a chance of theft from small items to large. By placing an EAS tag on an item, it is not necessary to hide the item behind locked doors and so makes it easier for the consumer to review the product.

Today's EAS source tagging, where the tag is built into the product at the point of manufacture or packaging, has become commonplace. This makes the labeling of goods unnecessary, saving time and money at the store.

4.2.1.2 RFID

RFID is a generic term that is used to describe a system that transmits the identity (in the form of a unique serial number) of an object or person wirelessly, using radio waves. It's grouped under the broad category of AITs. RFID is in use all around us. If you have ever chipped your pet with an ID tag, used E-ZPass through a toll booth, or paid for gas using SpeedPass, you've used RFID. In addition, RFID is increasingly used with biometric technologies for security. Unlike ubiquitous uniform product code (UPC) bar-code technology, RFID technology does not require contact or line of sight (LOS) for communication. RFID data can be read through the human body, clothing, and nonmetallic materials.

4.2.1.2.1 Components

A RFID system consists of the following main components:

- An antenna or coil
- A transceiver (with decoder) also referred to as a reader
- A transponder (RFID tag) electronically programmed with unique information (generally the EPC)
- Host computer system with a database to store the RFID reads
- Transceiver software and logic (often called Edgeware)

The antenna emits radio signals to wake up or excite the tag it may in the same signal be able to read and write data to it.

- The reader emits radio waves in ranges of anywhere from 1 in. to 100 ft or more, depending upon its power output and the tag frequency. When a tag maneuvers through the correct electromagnetic zone, which has the same frequency, it detects the reader's activation signal.
- The reader decodes the data encoded in the RFID tags integrated circuit (IC) and the data are passed to the host computer for processing.

The purpose of an RFID system is to enable data to be transmitted by a portable device, called a tag, which is read by an RFID reader and processed according to the needs of a particular application. The data transmitted by the tag may provide identification or location information, or specifics about the product tagged, such as price, color, date of purchase, etc. RFID technology has been used by thousands of companies for a decade or more. RFID quickly gained attention because of its ability to track moving objects. As the technology is refined, more pervasive and invasive uses for RFID tags are in the works.

Most manufacturers of RFID tags create the tags with microchip attached to a radio antenna mounted on a substrate. The chip generally stores 2 kB of data. To retrieve the data stored on an RFID tag, you need a reader. A typical reader is a device that has one or more antennas that emit radio waves and receive signals back from the tag. The reader then passes the information in digital form to a computer system.

4.2.1.3 RFDC

Radio frequency (RF) transmission has been with us since Guglielmo Marconi first demonstrated wireless communications a century ago. Within 30–40 years of Marconi's discovery, radios had become a fixture in nearly every U.S. household. However, it has been only within the last 6 years that wireless data transmission has come into its own in a business environment.

RFDC first appeared in warehouses and distribution centers as an enabling technology for automatic identification and data capture (AIDC) implementations, where hardwiring was unfeasible and/or real-time updating of the host database was critical. Early applications typically ran on PCs or controllers, scattered throughout a facility, which were interfaced to what was essentially a batch-oriented host. Those early systems were costly, quirky, and limited in transaction processing. However, they often made ADC a reality in environments where hard-wired systems were impossible. Further, RFDC offered certain advantages over hard-wired AIDC systems—interactivity and real-time updates of inventory, shipments, or manufacturing applications—that companies could turn to their own competitive advantage.

Technology improvements kept pace with RFDC's steady growth, so that present-day RFDC-based systems provide powerful, sophisticated, and reliable wireless solutions for a wide variety of both local area and wide area network applications.

Five frequently cited benefits to using radio frequency data communication are increased database accuracy at all times, reduced paperwork, real-time operations, higher productivity, and shorter order response times.

4.2.1.4 RTLS

Real-time visibility into exact locations of containers and cargo has never been as important as today with increased movement of cargo from offshore, the need to move it quickly to final destinations, and new security requirements. Today's wireless technology provides critical visibility into supply chain activities, delivering benefits to carriers, shippers, and customers.

RTLS are fully automated systems that continually monitor the locations of assets and personnel. An RTLS solution typically utilizes battery-operated radio tags and a cellular locating system to detect the presence and location of the tags. The locating system is usually deployed as a matrix of locating devices that are installed at a spacing of anywhere from 50 to 1000 ft. These locating devices determine the locations of the radio tags.

The systems continually update the database with current tag locations as frequently as every several seconds or as infrequently as every few hours for items that seldom move. The frequency of tag location updates may have implications for the number of tags that can be deployed and the battery life of the tag. In typical applications, systems can track thousands of tags simultaneously and the average tag battery life can be 5 or more years.

4.2.1.5 Satellite Tags with GPS

Global positioning systems (GPS) consist of a series of satellites orbiting the Earth and receivers. GPS works by calculating the distances from a receiver to a number of satellites. With each distance between a receiver and satellite, the number of possible locations is narrowed down, until there is only one possible location. A receiver must calculate its distance from at least three satellites to determine a location on the surface of the Earth. However, four satellites are usually used to increase the location accuracy (Dommety and Jain, 1996). This process of location would be controlled by the positioning module of GPS system. An average GPS positioning and navigation system would also have the following modules:

- Digital map database
- Map matching
- Route planning and guidance
- Human–machine interface
- Wireless communication

There are three positioning technologies that can be used: radio wave-based positioning, dead reckoning, and signpost. The use of GPS for navigation can have direct and indirect impacts on intelligent transportation systems (ITS). GPS navigation systems can provide information about local surroundings. Also, emergency personnel can be provided with a precise location for situations, thus reducing response times. Asset tracking is one of the most popular uses of GPS. One of the limitations of GPS is that receivers cannot communicate with satellites when indoors (Feng and Law, 2002).

4.2.1.6 MEMS

MEMS is a micro-electromechanical, micro-electro-mechanical, or microelectromechanical systems that is very small, and merges at the nanoscale or nanoelectromechanical systems (NEMS) and other nanotechnology. A good resource MEMSNET, an information resource for the MEMS and nanotechnology development community, further describes MEMS "as micromachines (in Japan), or microsystems technology—MST (in Europe). MEMS are separate and distinct from the hypothetical vision of molecular nanotechnology or molecular electronics. MEMS are made up of components between 1 and 100 μm in size (i.e., 0.001–0.1 mm) and MEMS devices generally range in size from 20 micrometers (20 millionths of a meter) to a millimeter." The promise that is available for MEMS is that together with silicon-based microelectronics and micromachining technology, the popularized idea of a computer system-on-a-chip is feasible. MEMSNET provides some futuristic ideas such as, "Microelectronic ICs can be thought of as the 'brains' of a system and MEMS augments this decision-making capability with 'eyes' and 'arms,' to allow microsystems to sense and control the environment. Sensors gather information from the environment through measuring mechanical, thermal, biological, chemical, optical, and magnetic phenomena. The electronics then process the information derived from the sensors and through some decision-making capability direct the actuators to respond by moving, positioning, regulating, pumping, and filtering, thereby controlling the environment for some desired outcome or purpose. Because MEMS devices are manufactured using batch fabrication techniques similar to those used for ICs, unprecedented levels of functionality, reliability, and sophistication can be placed on a small silicon chip at a relatively low cost" (https://www.memsnet.org/mems/what-is.html).

4.2.1.7 Contact Memory Buttons

4.2.1.7.1 Contact Memory

Contact memory technology is ideal for use in harsh industrial applications and in situations that would render bar codes unreadable or impractical. Buttons can mark hazardous and radioactive waste for long-term storage, track the maintenance of airplane brakes, and store repair diagrams. Attached to the ears of livestock, buttons track the animals from birth through processing, and carry data on feed and antibiotic use. Contact memory technology is well-suited to guard tour and access control applications in which users can access secure areas conveniently. Versatile touch/button technology can be used in healthcare to create records and match mothers and newborns, or to track items along an assembly line and to store manufacturing history.

4.2.1.8 Biometrics

Biometrics are automated methods of recognizing a person based on a physiological or behavioral characteristic. Among the features measured are face, fingerprints, hand geometry, handwriting, iris, retinal, vein, and voice.

Biometric technologies are becoming the foundation of an extensive array of highly secure identification and personal verification solutions. As the level of security breaches and transaction fraud increases, the need for highly secure identification and personal verification technologies is becoming apparent.

The need for biometrics can be found in federal, state, and local governments, in the military, and in commercial applications. Enterprise-wide network security infrastructures, government IDs, secure electronic banking, investing and other financial transactions, retail sales, law enforcement, and health and social services are already benefiting from these technologies. Biometric-based authentication applications include workstation, network, and domain access, single sign-on, application logon, data protection, remote access to resources, transaction security and Web security. Trust in these electronic transactions is essential to the healthy growth of the global economy. Utilized alone or integrated with other technologies such as smart cards, encryption keys, and digital signatures, biometrics are set to pervade nearly all aspects of the economy and our daily lives. Utilizing biometrics for personal authentication is becoming convenient and considerably more accurate than current methods (such as the utilization of passwords or PINs). This is because biometrics

- Links the event to a particular individual (a password or token may be used by someone other than the authorized user)
- Is convenient (nothing to carry or remember)
- Accurate (it provides for positive authentication)
- Can provide an audit trail
- Is becoming socially acceptable and cost effective

4.2.1.9 Common Access Cards

When we use the term “card technologies” or “smart cards,” what do we mean? The easy answer is—any technology that can be placed on a card. Typically, we think of our credit or bank card but there are other sizes and materials used for different applications. The card can be made of plastic (polyester, PVC, or some other material) or paper, or even some amalgamation of materials. The common point is that the card is used to provide “access” to something and it includes some form of AIDC technology.

There are currently three main technologies we think of when we mention card technologies:

- Magnetic stripe
- Smart cards
- Optical cards

Other technologies can be put on cards as well (such as bar codes, touch memory, etc.). Often, the card will have printing on it, which may involve technologies such as dye diffusion thermal transfer (D2T2) direct-to-card printing.

4.2.1.10 Optical Character Recognition

Optical Character Recognition (OCR) is used in high-volume financial applications such as payment processing, check reconciliation, and billing. It is also commonly used for high-volume document management in the insurance and healthcare industries. The technology is frequently found in libraries, publishing houses, and wherever printed text must be entered into a computer. OCR is also used in heavy-duty manufacturing environments for reading direct-marked, human-readable part numbers. The pharmaceuticals industry uses a variation of OCR called optical character verification (OCV) to assure that critical human-readable lot and date numbers cannot be misread.

4.2.1.10.1 Optical Mark Recognition

Optical Mark Recognition (OMR) is used for standardized testing as well as course enrollment and attendance in education. Human resource departments across industries use OMR for applications such as benefits enrollment, employee testing, change of employee status, payroll deductions, and user training. Healthcare providers use the technology for registration and surveys, and medical labs for patient evaluations and tracking supply orders and lab services. OMR is also used for time and attendance, labor tracking, inventory management, voting applications, exit surveys, polling, and all manner of questionnaires and evaluation studies. Because it is easy to use and cost-effective for opinion tracking, the technology has become a tool for on-location and direct-mail marketing.

4.2.1.10.2 Machine Vision

Traditional machine vision systems continue to be used for quality inspection, gauging, and robotic assembly in the automotive, electronics, aerospace, healthcare, and metal industries among others. These systems may also incorporate bar code reading. The next generation of 2D-dedicated vision-based scanners is being used for quality control, WIP, and high-speed sortation in industries such as electronics, automotive, and mail and package delivery. The pharmaceutical industry is also using 2D scanning systems to reconcile packaging, inserts, and labels on their packaging lines in order to satisfy the FDA's current good manufacturing practice regulations.

4.2.1.10.3 Voice

Voice recognition is commonly used in the automotive industry for various manufacturing and inspection applications. It is also used in warehousing and distribution to track material movement in real time, in the transportation industry for receiving and transporting shipments, in laboratory work, and in inspection and quality control applications across all industries.

4.2.1.10.4 Magnetic Ink Character Recognition

Magnetic ink character recognition (MICR) is most commonly used to encode and read information on checks and bank drafts to speed clearing and sorting. It is also effective for uncovering fraud, such as color copies of payroll checks or hand-altered characters on a check, both of which are easily detected by the absence of magnetic ink. Fast clearing and sorting, as well as fraud detection, benefits customers, financial institutions, and retail establishments.

Though this is a long list of AITs, it is not comprehensive due to the fact that other technologies are becoming available as this book is being published and distributed. For this book's purposes, we focus upon three main technologies: bar codes, RFID/RTLS technologies, and satellite technologies.

4.3 BAR CODES

4.3.1 Early History of the Bar Code

Retail applications were the main driver for initial bar-coding development.

Wallace Flint, whose family was in the grocery wholesale business, proposed a system using punch cards and flow racks that would automatically dispense products to customers in 1932. This automated checkout system required the need for some type of bar code system. This was the first time-documented instance of the advantages of an automated checkout and bar codes in the grocery industry. Flint emerged again, 40 years later, as the vice president of the National Association of Food Chains and an active supporter of standards for a code system. Because of his early experience,

Flint was a main supporter of bar code standardization efforts that led to the UPC. Several code formats were developed in the 1940s, 1950s, and 1960s, including a bull's-eye code, numeral codes, and various formats of bar codes.

The modern day bar code originated in 1949 when Norman Woodland and Bernard Silver, an instructor and student at Drexel Technical Institute, respectively began investigating capturing product information automatically at checkout. The story mentions that Woodland while at the beach came up with the idea to use Morse code and its dots and dashes to send information electronically and capture information on grocery products that could be communicated electronically. Woodland started to draw dots and dashes in the sand to simulate Morse code, and then extended them downward with his fingers, creating thin lines resulting from the dots and thick lines from the dashes resulting into 2D Morse code. Three years later, Silver and Woodland received a patent on what began as lines in the sand, and the linear bar code was created. Bar codes were not initially successful because the industry did not originally accept this auto ID.

4.3.2 Industry Acceptance

4.3.2.1 Railroads

Initially, bar codes were introduced as a means for tracking rail cars; they were placed on the sides of railroad freight cars. As the freight car rolled past a trackside scanner, the car, its destination, and cargo were identified. The system failed due to the fact that the freight cars were unstable and bounced as they passed the scanner. Consequently, the accuracy of the scanning was poor.

4.3.2.2 Grocery Industry

The complexity of the grocery business in the early twentieth century was due to the required stocking and inventorying of tens of thousands of items in various types and sizes of products, including perishable items from many suppliers. Perishable items required additional information needs due to inventory rotation requirements. Since many errors and inefficiencies were created from manual process, the grocery industry sought to automate data capture.

4.3.3 Universal Product Code

The best-known and most widespread use of bar codes has been on consumer products used by most grocery industries. In the late 1960s, supermarkets sought to automate point-of-sale information and testing of bar code technologies became paramount. For instance, in 1972, a Kroger store in Cincinnati operated using a bull's-eye code. The need for standardization led to the forming of a committee within the grocery industry to select a standard code to be used in the industry.

Supporters such as Wallace Flint, believing that automating the grocery checkout process could reduce labor costs, improve inventory control, speed up the process, and improve customer service, pushed for a bar code usage. Six industry associations, representing both product manufacturers and supermarkets, created an industry-wide committee of industry leaders for this initiative. Proposals were solicited from various interested parties, and on April 3, 1973, the committee selected the Universal Product Code or the UPC symbol (based on the proposal from IBM) as the industry standard. The success of the system since then has spurred on the development of other coding systems. The UPC made its first commercial appearance on a package of Wrigley's gum sold in Marsh's Supermarket in Troy, Ohio in June 1974.

4.3.3.1 Economic Impact of the UPC

Economic studies conducted for the grocery industry committee projected over $40 million in savings to the industry from scanning by the mid-1970s. Those numbers were not achieved in that time frame and there were those who predicted the demise of bar code scanning. The usefulness of the bar code required the adoption of expensive scanners by a critical mass of retailers while manufacturers simultaneously adopted bar code labels. Neither wanted to move first and results looked unpromising for the first couple of years, with Business Week eulogizing "The Supermarket Scanner That Failed."

As scanning spread, however, the $40 million projection began to look very small. A 1999 analysis by Price Waterhouse Coopers estimates the U.P.C. represents $17 billion in savings to the grocery industry annually. Even more astounding, the study concludes that the industry has not yet taken advantage of billions of dollars of potential savings that could be derived from maximizing the use of the U.P.C. The big winners—as one should have expected given the competitive nature of the markets involved—were consumers, since U.P.C. scanning generated efficiencies and productivity improvements that led to lower costs and/or greater customer service. Ironically, consumer advocates initially resisted the innovation and jeopardized its success by insisting that retailers forego substantial cost savings by continuing to mark prices on individual units. While the rise of bar-coding benefited both manufacturers and retailers, it was the retailer who benefited the most. In addition to the labor savings, retailers now had access to detailed product movement data, which they turned into a profit center by selling the data to their suppliers.

4.3.4 Technology behind the Bar Code

A linear bar code is a binary code (1s and 0s). The lines and spaces are of varying thicknesses and printed in different combinations. To be scanned, there must be accurate printing and adequate contrast between the bars and spaces. Scanners employ various technologies to "read" codes. The two most common are lasers and cameras. Scanners may be fixed position, like most supermarket checkout scanners, or hand-held devices, often used for the taking of inventories. There is a distinction between the code and the machine-readable representation of the code. The code is text that can be translated into a multiplicity of languages and/or symbols. One of the first successful bar codes, Code 39 developed by Dr. David Allais, is widely used in logistical and defense applications. Code 39 is still in use today, although it is less sophisticated than some of the newer bar codes. Code 128 and interleaved 2 of 5 are other codes that attained some success in niche markets.

Because the UPC was the first bar code symbology widely adopted, foreign interest in UPC led to the adoption of the European Article Numbering (EAN) code format, similar to UPC, in December 1976. The 2005 Sunrise and the Global Trade Item Number initiatives from the Uniform Code Council (UCC) began on January 1, 2005. This is known as the "14-digit U.P.C." For a manufacturer of a product that has an existing 8- or 12-digit UPC bar code, there was no effect; however, retailers or wholesalers with scanners had to ensure that scanners are able to decode 8-, 12-, 13-, and 14-digit bar codes and that database systems can handle the extra digits. There is a white paper by in the appendices (Sunrise GTIN) detailed the changes.

The most common UPC formats are the five versions of UPC and two versions of EAN. The Japanese Article Numbering (JAN) code has a single version identical to one of the EAN versions with the flag characters set to "49." UPC and EAN symbols are fixed in length, can only encode numbers, and are continuous symbologies using four element widths. Figure 4.2 illustrates a variety of common bar codes.

UPC version A symbols have 10 digits plus 2 overhead digits while EAN symbols have 12 digits and 1 overhead digit. The first overhead digit of a UPC version A symbol is a number related to the type of product while an EAN symbol uses the first two characters to designate the country of the EAN International organization issuing the number. UPC is, in fact, a subset of the more general EAN code. Scanners equipped to read EAN symbols can read UPC symbols as well. However, UPC scanners will not necessarily read EAN symbols.

The UPC symbology was designed to make it ideal for coding products. UPC can be printed on packages using a variety of printing processes. The format allows the symbol to be scanned with any package orientation. Omnidirectional scanning allows any package orientation provided the symbol faces the scanner. The UPC format can be scanned by hand-held wands and can be printed

FIGURE 4.2 Common bar codes.

by equipment in the store. Version A of the symbology has a first-pass read rate of 99% using a fixed laser scanner and has a substitution error rate of less than 1 error in 10,000 scanned symbols.

Nominal X dimension is 13 mils. A magnification factor of 0.8–2.0 is allowed and, as a result, makes a printable range of X dimension values of 10.4–24 mils. In other words, the nominal size of a UPC symbol is 1.469 in. wide × 1.02 in. high. The minimum recommended size is 80% of the nominal size or 1.175 in. wide × 0.816 in. high. The maximum recommended size is 200% of the nominal size or 2.938 in. wide × 2.04 in. high. Larger UPCs scan better. Smaller UPCs do not scan as well or not at all.

The UPC format can be printed using a variety of printing techniques because it allows for different ink spreading. The amount of ink spreading depends on printing press conditions, amount and viscosity of ink, and other factors that are difficult to precisely control. The UPC symbol is decoded by measuring the distance from leading edges to leading edge of bars, trailing edge to trailing edge of bars, and leading edge to leading edge of characters. Since relative distances are measured for decoding, uniform ink spread will not affect the symbol's readability. However, excessive ink spread will make the spaces very small to the point that the reader will be unable to resolve them. Since UPC is a continuous code with exacting tolerances, it is more difficult to print on any equipment except printing presses.

4.3.5 Current Level of Use

Because computing technologies have become more efficient over the last several decades, the development of applications that can utilize bar codes has increased, leading to bar codes becoming more prevalent in our society. Bar codes are currently utilized in many retail stores such as supermarkets and hardware stores, as well as many industrial and military applications such as manufacturing, warehousing, and transportation operations. With ever-increasing use, many companies have developed software to generate and manipulate bar codes. Some pundits believe as newer technologies such as RFID are developed, we may eventually see a disappearance of bar codes as we know them today.

The developers of the UPC believed that there would be fewer than 10,000 companies in the U.S. grocery industry, which would use the UPC. To the contrary, currently there are over 1 million companies in more than 100 countries in over 20 different industry sectors enjoying the benefits of scanning because of the UPC. UPC symbols are everywhere in the retail environment. They can also be found in industries as diverse as construction, utilities, and cosmetics. The UPC is extensively used in the supply chain by the suppliers of raw materials, manufacturers, and distributors. At the dawn of the twenty-first century, the Uniform Code Council, Inc., the administrator of the UPC, estimates that the U.P.C. symbol was being scanned over 5 billion times a day.

The linear bar code continues to evolve. Today, there are 2D bar codes such as PDF 417 and MaxiCode capable of incorporating the Gettysburg Address in a symbol one-quarter of an inch square. RSS and Composite symbologies will enable the bar code identification of very small items such as individual pills or a single strawberry. Rental car companies keep track of their fleet by means of bar codes on the car bumper. Airlines track passenger luggage, reducing the chance of loss. Researchers have placed tiny bar codes on individual bees to track the insects' mating habits. NASA relies on bar codes to monitor the thousands of heat tiles that need to be replaced after every space shuttle trip, and the movement of nuclear waste is tracked with a bar code inventory system. In the fashion world, designers stamp bar codes on their models to help coordinate fashion shows. The codes store information about what outfits each model should be wearing and when they are due on the runway.

4.3.6 Future Uses

The future of auto ID, however, is probably in radio frequency (RFID). Tiny transmitters embedded in items do not require a LOS to the scanner, nor are they subject to degradation by exposure. Already in use in retail stores to help prevent shoplifting and on toll roads to speed traffic, the primary deterrent to wider use of RFID has been the cost of the silicon chips required. Today, the 5-cent chip is close at hand. If the cost can be reduced to less than 1 cent a chip, in the future your breakfast cereal box will be a radio transmitter.

4.4 RFID

A sense of history in RFID is important for the following reasons. Some RFID technologies have stood the test of time and have become more pervasive in the supply chain. Other RFID technologies have been utilized in other industries such as animal tracking and present unique advantages. The convergence of RFID systems has been theorized to create innovations in current industries and lead to the creation of new industries. Given that the history of RFID is integrated with the history of other ADC devices such as bar codes, we approach chronicling RFID history in following ways.

First, we investigate the development of data acquisition device usage in the distribution and logistics. Second, we overlay the development history of RFID technologies for supply chain activities. Finally, we introduce future plans for RFID technologies in logistics operations.

The roots of RFID technology can be traced back to World War II (WWII). The radar, which had been discovered in 1935 by Scottish physicist Sir Robert Alexander Watson-Watt, was utilized by all combatants in the war (Germans, Japanese, Americans, and British) to identify aircraft, but there was no unique identification for the aircraft. The main problem with radar was there was no way to identify which planes belonged to the enemy and which were a country's own planes returning from a mission.

The Germans discovered that if pilots rolled their planes as they returned to base, it would change the radio signal reflected back. This crude method alerted the radar crew on the ground that these were German planes and not Allied aircraft. This plane "roll" created a uniquely identifiable signal that acted in essence as a unique reflected signal. This principle is what the base passive RFID systems are based upon.

Later, Watson-Watt headed a secret project by the British to develop the first active identify friend or foe (IFF) system. A transmitter was placed on each British plane. When the transmitter received signals from radar stations on the ground, it began broadcasting a signal back that identified the aircraft as friendly. RFID works on this same basic concept. A signal is sent to a transponder, which wakes up and either reflects back a signal (passive system) or broadcasts a signal (active system).

Advances in radar and RF communications systems continued through the 1950s and 1960s. Scientists and academics in the United States, Europe, and Japan did research and presented papers explaining how RF energy could be used to identify objects remotely.

Companies began commercializing antitheft systems that used radio waves to determine whether an item had been paid for or not. EAS tags, which are still used in retail packaging today at retailers such as J.C. Penney and SEARS, use a 1-bit tag. The bit is either on or off. When someone pays for the item, a cashier deactivates the tag and the bit is turned off, and a person can leave the store. In contrast, if the person doesn't pay and tries to walk out of the store, readers at the door detect the tag and sound an alarm.

4.4.1 Prior to IFF

Though many focus on the WWII as the beginning of RFID development, we will explore other events that contribute to theoretical understanding of RFID technologies. Given that some of the RFID technologies do not include both passive and active and semi-active technologies, we provide other historical events that will allow the student to investigate and create investigative thought on RFID technologies.

Many scientists believe that at the beginning of time electromagnetic energy created the universe, often referred to as the "Big Bang" Theory. Due to the fact that most RFID technologies use electromagnetic energy as the source of energy, this Big Bang may be considered the beginning of RFID technologies. Benjamin Franklin explored electromagnetism with his experiments in electricity in the 1700s. In the 1800s, Micheal Faraday and James Maxwell contributed theories on electricity; relationship of light and magnetic fields on electromagnetic energy, respectively. Michael Faraday, English scientists explored relationship of light, radio waves, and electromagnetic energy. In 1864, James Maxwell, Scottish physicist, published theory on electromagnetic fields, which concluded that electric and magnetic energy travel in transverse waves moving at the speed of light.

Later in the 1800s, in1887, Heinrich Rudolf Hertz, German physicist, confirmed Maxwell's theories and added theories about electromagnetic waves (radio waves), which showed as long transverse waves that travel at the speed of light and can be reflected, refracted, and polarized like light. Also, Hertz was the first credited for transmitting and receiving radio waves, and his demonstrations were later duplicated by Alexsander Popov of Russia. Another key breakthrough for radio transmission was when Gudliemo Marconi successfully transmitted a radiotelegraphy across the Atlantic Ocean.

Now in the twentieth Century, we have in 1906 Ernst F.W. Alexanderson discover the first continuous wave (CW) radio generation and transmission of radio signals, which signaled the beginning of modern radio communications where all aspects of radio waves are controlled.

The Manhattan project at Los Alamos Scientific Laboratory in 1922 was attributed to the birth of the radar detection. The project described how the radar sends radio waves for detecting and locating an object by the reflection of the radio waves. The reflection can determine the position and speed of an object. Given that RFID is a combination of radio broadcast technology and radar; the convergence of these disciplines allowed for future RFID development. Scottish physicist Sir Robert Alexander Watson-Watt was considered to be the inventor of the modern radar system in 1935.

In 1945, historians theorize that the first known device may have been invented by Thermin as a reporting espionage tool for the Russian government in 1945, the device was the first "bug" or covert listening device. This device was the first to use inducted energy from radio waves of one frequency to transmit an audio signal to another. This made the device difficult to detect, as it did not radiate any signal unless it was being remotely powered and listened to, and endowed it with (potentially) unlimited operational life.

This bug was embedded in a 2 ft wooden replica of the Great Seal of the United States and presented to the American ambassador in Moscow, Averell Harriman by Russian school children in 1946. This is currently on display at the National Security Agency, National Cryptologic Museum. The bug hung prominently for years, at least part of the time in the ambassador's study, before a tiny microphone was found in the eagle by a professional bug sweeper using a marta kit, which happened to catch a signal from it while it was being used.

During George F. Kennan's ambassadorship in 1952, a routine security check discovered that the seal contained a microphone and a resonant cavity, which could be stimulated from an outside radio signal. George Kennan's memoirs describe the event. In a theme now familiar, Kennan relates that Spaso House had been redecorated under Soviet supervision, without the presence of any American supervisors, giving them opportunity "to perfect their wiring of the house." "The ordinary, standard devices for the detection of electronic eavesdropping revealed nothing at all," but technicians decided to check again, in case our detection methods were out of date. The novelty of the Great Seal bug, which was hung over the desk of our Ambassador to Moscow, was its simplicity. It was simple resonate chamber, with a front wall that acted changed dimensions of the chamber when sound waves struck it. It had no power pack, wires, and no batteries. An ultrahigh frequency (UHF) signal beamed to it from a van parked near the building was reflected from the bug, after being modulated by sound waves from conversations striking the bug's diaphragm.

4.4.2 How the Great Bug Seal Worked

The Ultimate Spy Book by H. Keith Melton further details how the Great Bug Seal worked. It features a bald eagle, beneath whose beak the Soviets had drilled holes to allow sound to reach the device. Western experts were perplexed on how the device, also know as the "Thing," worked, because it had neither batteries nor electrical circuits. Peter Wright of British intelligence discovered how it operated and later produced a copy of the device for use by both British and American intelligence. The Thing was initiated when a radio beam aimed at the antenna from a source outside of the building was sent; then the sound wave struck the diaphragm causing variations in the amount of space (and the capacitance) between it and the tuning post plate. These variations altered the charge on the antenna, creating modulations in the reflected radio beam. These were picked up and interpreted by the receiver.

4.4.3 Research on RFID

One of the first works exploring RFID was the paper by Stockman (1948). This transcript not only discussed the basic problems of researching "reflected-power" communication, but also discussed the usage of the technology. It also predicted that "...considerable research and development work has to be done before the remaining basic problems in reflected-power communication are solved, and before the field of useful applications is explored." For this prediction to become valid, advances in transistor technologies, ICs, microprocessors, development of communication networks, and computing power had to happen over the next 30 years, which would spur cost economics of RFID type technologies.

Also, in the 1950s, other technical developments in radio and radar along with the IFF exploration of long-range transponder systems for identification include FL Vernon's, "Application of the microwave homdyne" and D.B. Harris, "Radio transmission systems with Modula table passive transponder." These developments also led to future patents for RFID technology.

4.4.3.1 In the Twentieth Century

In the 1960s, RF Harrington studied the electromagnetic theory related to RFID papers, "Field Measurements using active scatterers," "Theory of loaded scatterers" in 1963–1964; Robert

Richardsons "Remotely activated radio frequency powered devices" in 1963; Otto Rittenbacks's "Communication by radar beams" in 1969; H.H. Vogelmans's "Passive data transmission techniques utilizing radar beams" in 1968; and J.P. Vindings "Interrogator-responder identification system" in 1967.

Commercial activities were beginning in the 1960s. Sensormatic and Checkpoint were founded in the late 1960s. Other companies like Knogo, developed EAS equipment to counter theft using 1-bit tags. Fundamentally, the presence or absence of a tag was detected and the tags were made inexpensively and could provide effective antitheft measures. Most of these systems used microwave or inductive technology. This EAS technology was the first widespread use of RFID technology.

4.4.3.2 The First RFID Patents

The first U.S. patents for RFID tags were from Mario W. Cardullo and Charles Watson in 1973. Mario W. Cardullo received the first U.S. patent for an active RFID tag with rewritable memory on January 23, 1973. That same year, Charles Walton, a California entrepreneur, received a patent for a passive transponder used to unlock a door without a key. The electronic door lock operated with a card that communicated with an embedded transponder that communicated a signal to a reader near the door. When the reader detected a valid identity number stored within the RFID tag, the reader unlocked the door. Walton licensed the technology to lock makers and other similar companies.

The testing of these technologies was still relevant when one of the authors worked for United Parcel Services (UPS) in the early 1990s in the Strategic Systems Group. Unfortunately, the reliability and the cost effectiveness were not viable even 20 years later at a company as successful at UPS.

4.4.3.3 Toll Road and Animal Tracking

Also in the 1970s, Los Alamos National Laboratory at the request of the U.S. energy department developed a system for tracking nuclear materials. Scientists developed the idea of putting a transponder in a truck and readers at the gates of secure facilities. The gate's antenna would wake up the transponder in the truck, which would respond with an ID and potentially other data, such as the driver's ID, then the gate would automatically open.

There was a realization on how RFID technologies, specifically electronic vehicle identification (EVI), could change transportation. This was evidenced in the transportation efforts included work at Los Alamos and by the International Bridge and Turnpike and Tunnel Association (IBTTA) and the U.S. Federal Highway Administration. Unfortunately, the IBTTA and U.S. Federal Highway Administration held a conference in 1973 and concluded that there is no national interest in developing a standard for EVI. The late 1970s is when companies realized the potential commercial aspects of RFID, such companies as Identronix, a spin-off from Los Alamos Scientific Lab, Amtech that later became part of Intermec and Transcore were developed.

Later in the mid-1980s, this type of system was commercialized when former Los Alamos scientists left and formed companies that developed automated toll payment systems. These systems have become widely used on roads, bridges, and tunnels around the world. Organizations such as the Port Authority of New York and New Jersey tested electronic toll collection systems built by General Electric, Westinghouse, Philips, and Glenayre.

Also in the 1970s, animal tracking efforts were initially investigated using microwave systems at Los Alamos and using inductive technologies in Europe. Animal ID was pursued in Europe by Alfa Laval, Nedeap, and others.

Other forward-moving occurrences in the 1970s included the use of modulated backscatter. In 1975, Alfred Koelle, Steven Depp, and Robert Freyman introduced the transcript "Short-range radio-telemetry for electronic identification using modulated backscatter," which is the foundation for current RFID passive tags. Other events include the development of the Raytheon's "Raytag" along with other events from RCA and Fairchild in RFID development. Richard Klensch of RCA developed the "Electronic identification system" in 1975, and Sterzer of RCA developing an

"Electronic license plate for motor vehicles" in 1977. Thomas Meyers and Ashley Leigh of the Fairchild organization developed "Passive encoding microwave transponder" in 1978.

In the 1980s, RFID history documents many commercial implementations. The most common implementations in the United States were for transportation, personnel, and animals. In Europe, interests were in short-range systems for animals, industrial and business applications. Toll roads in Italy, France, Spain, Portugal, and Norway were equipped with RFID. The Association of American Railroads and the Container Handling Cooperative Program was active with RFID initiatives. Though testing of RFID tags for collecting tolls had been going on for many years, the first commercial application began in Europe in 1987 in Norway. This was followed quickly in the United States by the Dallas North Turnpike in 1989. Port Authority of New York and New Jersey began commercial operation of RFID for buses going through the Lincoln Tunnel. RFID was finding a home with electronic toll collection.

Also, Los Alamos was requested by the agricultural department to develop passive RFID tags to track cows. The goal was to facilitate the tracking of the amount of hormones and medicines that were administered to cows when they were ill. The challenge of ensuring that each cow received the correct dosage was having multiple economic factors. Los Alamos came up with a passive RFID system that used 125 kHz radio waves. A transponder encapsulated in glass is injected under the cow's skin or is attached to an identification tag. The transponder draws energy from the reader and reflects back a modulated signal to the reader using a technique known as backscatter. This system is still used in cows around the world today (Figure 4.3).

These low-frequency transponders were also put in cards and used to control the access to buildings.

Over time, companies commercialized 125 kHz systems. Later, other companies developed systems that operate on higher radio spectrum to high frequency (13.56 MHz). This frequency was chosen because it was unregulated and unused in most parts of the world. This frequency offered greater range and faster data transfer rates. Companies in Europe began using it to track reusable containers and other assets. The 13.56 MHz frequency RFID systems are used for access control, payment systems (mobile speed pass), contactless smart cards, and as an antitheft device in cars. The cars have a reader in the steering column that reads the passive RFID tag in the plastic housing around the key. The car is rendered disabled if the ID number it is programmed to look for is not found.

In the early 1990s, IBM engineers developed and patented a UHF RFID system that offers longer read ranges—20 ft under good conditions—and faster data transfer. IBM did some early pilots

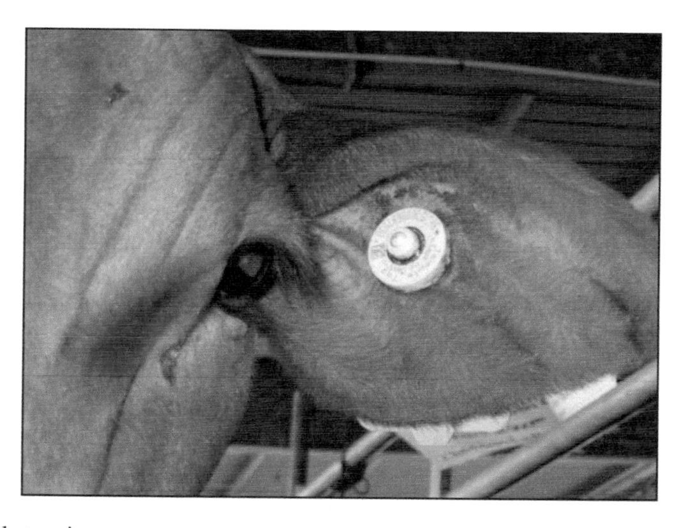

FIGURE 4.3 Cattle tagging.

FIGURE 4.4 Harris County Toll Road Authority E-ZPass.

with Walmart, but never commercialized this technology, but when IBM ran into financial trouble in the mid-1990s, it sold its patents to Intermec, a material handling systems provider. Intermec has installed the system in numerous different applications, from warehouse tracking to farming. Intermec invested in this future of this technology due to the fact that at the time it was expensive because of the low volume of sales and lack of international standards.

Also, in 1990s, electronic toll collection using RFID technologies expanded to wide-scale deployment of electronic toll collection in the United States. Open highway electronic tolling system opened in Oklahoma in 1991 cars in which cars pass scanning points at highway speeds (no need for camera's or barriers). The world's first combined toll collection and traffic management system was installed in the Houston area by the Harris County Toll Road Authority in 1992. Both of the authors personally were able to witness the construction of the toll road in Houston (Figure 4.4).

Later, the Kansas turnpike used a system based on the Title 21 Standard, which allows usage by other states such as Georgia, which also used the same standard. The Title 21 Standard was designed to have a multi-protocol capability in electronic toll collection applications. Also, the Northeastern United States, seven regional toll agencies formed the E-ZPass Interagency Group in 1990 to develop a regionally compatible electronic toll collection system. Also, toll tags were integrated for multiuse like parking garages, toll booths, gated communities, and business campuses, an example would be the Dallas Toll Tag. Texas Instruments began developing the TRIS used for starting cars.

The development of computer engineering technology, which allowed microwave Schottky diodes fabricated on a regular CMOS IC, permitted the construction of the microwave RFID tags. These tags contained a single circuit, which previously had been limited to inductively coupled RFID transponders. Thus, allowing for cheaper active tags and readers.

Federal Communications Commission allocated spectrum in 5.9 GHz band for expansion of intelligent transportation systems, which will spur more RFID development and applications. RFID systems have been installed in numerous different applications, from warehouse tracking to farming. But the technology was expensive at the time due to the low volume of sales and the lack of open, international standards.

4.4.4 Development of Cost-Effective Protocol

In the early 1999, when the UCC, EAN International, Proctor & Gamble, and Gillette established the auto-ID center at the Massachusetts Institute of Technology (MIT), two research professors, David Brock and Sanjay Sarma initiated the idea of integrating low-cost RFID tags on to products in order to track them through the supply chain. Their idea of transmitting a unique number from the RFID tag in order to keep the cost of the technology cost effective was novel. The idea of using a simple microchip that stored very little information as opposed to using a more complex chip that may require batteries, and require more memory allowed for the cost-effective implementation of the idea. Data associated with the serial number on the tag would be stored in a database that would be accessible over the Internet.

Sarma and Brock changed the way people used RFID in the supply chain. Previously, RFID tags were considered mobile databases that contained information about the product, case, pallet, or container on which they were attached. Sarma and Brock promoted the idea of RFID as an associating networking technology that linked objects to databases through the Internet through the tag. This was an important change to businesses because this enabled the idea of visibility. For example, a manufacturer could automatically let a business partner know when a shipment was leaving the dock at a manufacturing facility or warehouse, and a retailer could automatically let the manufacturer know when the goods arrived.

Between 1999 and 2003, the auto-ID center gained industry acceptance of the passive RFID tagging system with the support of more than 100 large end-user companies, the U.S. Department of Defense (DOD), and RFID vendors. Auto-ID research labs were opened in Australia, the United Kingdom, Switzerland, Japan, and China. The auto-ID center is credited with developing two air interface protocols (Class 1 and Class 0), the EPC numbering scheme, and a network architecture for associating data on an RFID tag. The technology was licensed to the UCC in 2003. The UCC created EPC Global organization as a joint venture between the auto-ID center and EAN International in order to commercialize EPC technology. The auto-ID center closed its doors in October 2003, and its research responsibilities were passed on to auto-ID labs.

The industry support is evidenced in the fact that some of the biggest retailers in the world—Albertsons, Metro, Target, Tesco, Walmart—and the U.S. DOD have initiated plans to use EPC technology to track goods in their supply chain. The pharmaceutical, tire, defense, and other industries are also moving to adopt the technology. EPCglobal ratified a second-generation standard in December 2004 in order to compensate for some of the shortcomings of the first-generation technologies, improving challenges such as read distance and better integration between vendor products.

4.4.5 Overview of Passive and Active Radio Frequency Identification Technologies

An RFID system consists of a reader, tags, and an air interface. The reader, also known as an interrogator, sends out a signal through an antenna. This signal is usually in the form of an electromagnetic wave. Because the signal is in the form of an electromagnetic wave, a direct LOS is not needed to read the information on the tag. This is a major advantage of RFID. The signal is received by the tag and a response signal is sent back to the reader. This response signal contains a unique identifier associated with tag. The response signal can be powered in two ways corresponding to the type of tag. Passive tags utilize the energy of the original signal to send a response signal back to the reader. Passive tags have a limited amount of energy to power the response signal. Therefore, the amount of information transmitted by a passive tag is fairly small, quite similar to the information carried in a bar code.

Active and semi-active tags use energy from an attached battery to power the response signal. The use of the embedded battery allows the response signal to contain more information and travel farther. The reader receives the response signal, decodes it, and sends that information to a database. Often the information in the response signal is connected to additional information in the database.

RFID technology can be used throughout the supply chain in order to promote visibility (Figure 4.5). This visibility helps coordinate actions between entities in the supply chain. Figure 4.1 shows the relationships within the supply chain that can be affected by the implementation of the RFID technologies. An example of RFID implementation is the use of active tags for detecting, tampering, and monitoring security of maritime containers (Figure 4.6). These types of tags also have the tracking advantages of RFID and can be used to improve operations management. Those tags can be seen in Figure 4.2.

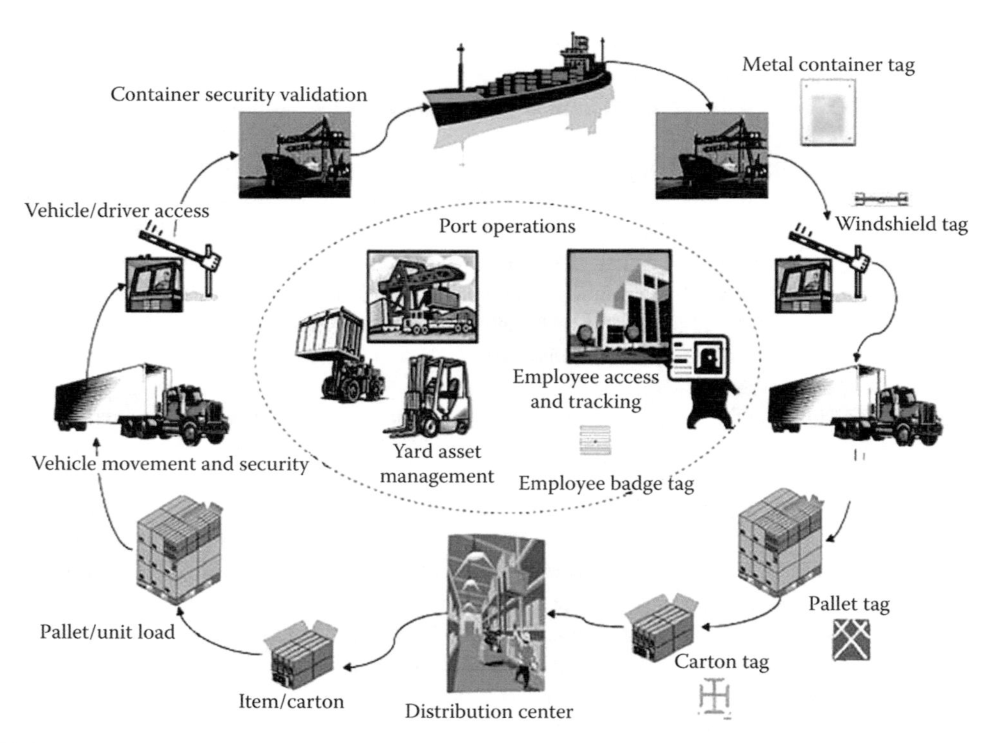

FIGURE 4.5 Integrated supply chain with RFID. (Courtesy of Savi Technologies, a division of Lockheed Martin, Lawrenceville, GA.)

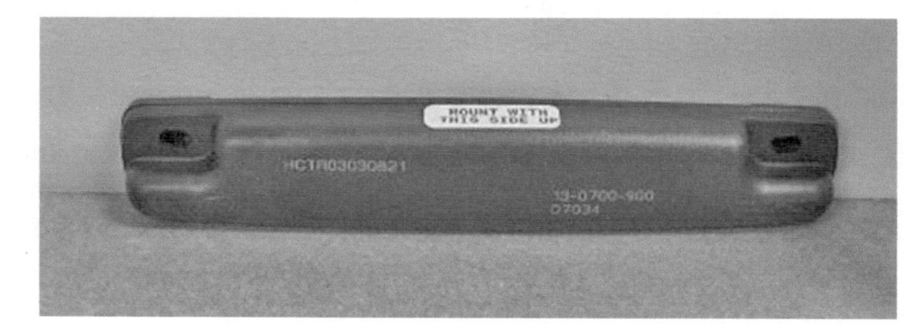

FIGURE 4.6 RFID container seal. (Courtesy of Savi Technologies, a division of Lockheed Martin, Lawrenceville, GA.)

4.5 GLOBAL POSITIONING SYSTEMS

The Navistar Global Positioning System also referred to as GPS has three components:

1. A constellation of satellites orbiting approximately 20,000 km above the earth. There are approximately 27 satellites transmitting ranging signals on frequencies in the microwave ranges in the radio spectrum. The transmission frequencies range from 1 to 10 GHz. Most point-to-point services offered today use the frequency bandwidths of (1) 5.92–6.425 GHz for transmission from earth to satellite called uplink, (2) 3.7–4.2 GHz for transmission from satellite to earth called downlink.
2. Ground control monitor stations that control uplink to the satellites.
3. Receivers for the users. There are both civil and military users (Stalling, 2006).

4.5.1 Integration of Real-Time Technologies and GPS

RFID and GPS are emerging technologies that will allow for real-time data collection to assist with decision support in supply chain management (SCM). RFID has a wide variety of applications. Some examples of RFID uses are library checkout stations, automatic car toll tags, animal identification tags, and inventory systems. Real-time data collected using RFID allows a supply chain to synchronize reorder points and other data. Real-time information can also be used to design and operate logistical systems on a real-time basis. GPS is currently used solely as a means to locate equipment and derive navigation directions.

An RFID system consists of a reader, tags, and an air interface. The reader, also known as an interrogator, sends out a signal through an antenna. This signal is usually in the form of an electromagnetic wave. Because the signal is in the form of an electromagnetic wave, a direct LOS is not needed to read the information on the tag. This is a major advantage of RFID. The signal is received by the tag and a response signal is sent back to the reader. This response signal contains a unique identifier associated with tag. The response signal can be powered in two ways corresponding to the type of tag. Passive tags utilize the energy of the original signal to send a response signal back to the reader. Passive tags have a limited amount of energy to power the response signal. Therefore, the amount of information transmitted by a passive tag is fairly small: quite similar to the information carried in a bar code. Active and semi-active tags use energy from an attached battery to power the response signal. The use of the embedded battery allows the response signal to contain more information and travel farther. The reader receives the response signal, decodes it, and sends that information to a database. Often, the information in the response signal is connected to additional information in the database.

RFID and GPS are radio wave-based technologies that are currently used by many organizations. RFID is primarily used in inventory and material handling processes. Tags are placed on items. When these items pass by checkpoints where readers are located, the tag is read and the appropriate action can be taken. Real-time inventory can be kept by monitoring tag reads at strategic points like loading docks. RFID can also be useful in material handling. Items on a conveyor can be diverted at the appropriate times based on the information received from the RFID tag. GPS is primarily used to track assets such as vehicles and other expensive equipment. For example, if a truck breaks down, it is possible to locate the truck and get the shipment moving again in the fraction of the time it would take with a GPS receiver.

Current applications of RFID and GPS systems have allowed for more effective tracking of inventory and assets. These technologies can be used in conjunction, but, the data have to be captured and written to a database to be correlated to other tags or receivers. If these technologies can be combined to produce hybrid systems, greater gains can be achieved. One focus of research is the nesting of GPS receivers and various RFID tag types. If tags and receivers were able to communicate with one another, even more accurate real-time data collection could be achieved during transportation. This would also reduce equipment costs, because fewer readers would be required. The nesting would follow the form in Figure 4.7.

If these technologies can be nested, it would allow the information in a bar code or a passive RFID tag to be collected by an active tag. This information could then be combined with the information contained within the active tag and transferred to a GPS receiver. The GPS receiver could then send not only its location but all of the information about the cargo being shipped (Reade and Lindsay, 2007). A possible application of this nested technology approach would be in the railroad industry. Currently, there two passive RFID tags attached to the sides of all rail cars in the United States. In addition, most railroads are using GPS receivers to track locomotives. If nesting became possible, implementation would be easy in this case. Active tags could be used to capture the information correlated to the cargo in all of the rail cars and transmit it to the GPS receiver and thus to the inventory databases.

In addition to nesting technologies, more advanced tags can be developed to allow more detailed data collection. Tags that utilize sensors to capture and write data to the tag are being developed.

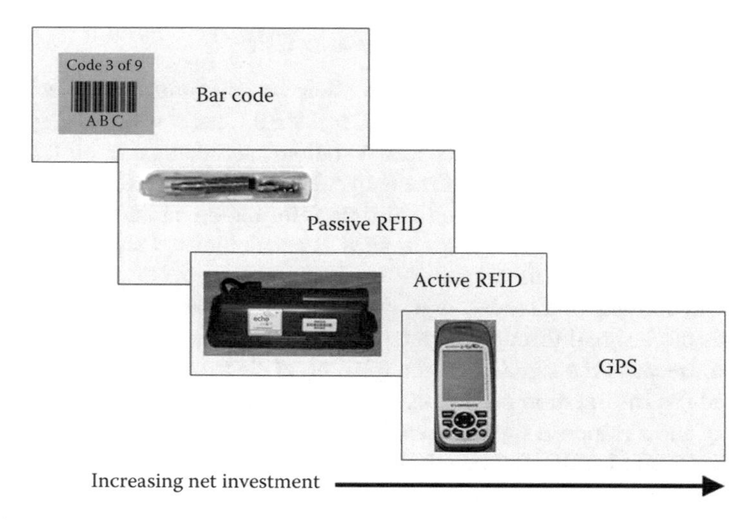

FIGURE 4.7 Nesting diagram. (Courtesy of Savi Technologies, a division of Lockheed Martin, Lawrenceville, GA.)

Some tags have been developed, but are still very unreliable. These sensor tags could be used to monitor physical parameters, like temperature and humidity, as well as security parameters. The main problem faced by these passive sensor tags is the limited power supply. The sensor cannot use any energy while outside the range of the reader. Also, the amount of energy available while in read range is very small. This limits possible measurement techniques (Want, 2007). With these sensor tags, perishable goods could be monitored to guard against possible safety issues. This could include *Salmonella* outbreaks caused by frozen chicken reaching too high of temperatures for too long and medications being held at temperatures that reduce potency.

4.5.2 Conclusion

Technologies are being used to allow real-time data collection. This allows for more dynamic SCM systems that are able to adjust to varying market and environmental conditions. RFID and GPS facilitate this dynamic SCM. RFID allows for up-to-date inventory levels, and when combined with GPS, can provide a means of tracking inventory as it moves from supplier to customer through the supply chain. New technologies are being developed to further the amount of information to decision support systems for SCM.

4.6 REAL-TIME LOCATION SYSTEMS

Current RTLS systems are typically active systems (using battery-operated tags) to detect presence and location within a 2D coordinate system (XY position only, not height). The license-free frequency ranges are most popular (300–433 MHz) and in particular, the 2.45 GHz. At this frequency range, time of flight (TOF) is efficient and many RTLS are based on WiFi, Bluetooth, or ZigBee, which occupy this frequency band. Almost all these current RTLS systems rely on the signal strength as an indicator for distance approximation. Other systems use active RFID, infrared, ultrasound, ultra-wide band (UWB), or a combination of them to perform the localization.

Below are a subset of the RTLS systems that are generally based on the aforementioned procedure.

WiFi RTLS: The Ekahau company is currently incorporating an RTLS using an existing WiFi (802.11) network. The system works by using RSSI between multiple access points throughout the tracking area. This provides accuracy indoors of 3–9 ft. The system does not cause any interference to the existing network traffic because the tag communicates only about 60 bytes of data per location update. The tags are active and require a 4–6 V power source. The tags have built-in accelerometers and can be configured to report the location any time it is moved (Ekahau rtls, 2007). Many other companies have previously employed versions of a WiFi RTLS system in operations, which has created a great amount of interest due to the fact that these location systems can be incorporated into existing networks.

Active RFID RTLS: RF Code has combined active RFID and infrared in an RTLS implementation.

The system has a read range of up to 10 m (Rf code, 2007).

Ultra-wide band: Multispectral Solutions, Inc. incorporated UWB into an RTLS system based on triangulation ranging. These active tags use TOF measurements and can achieve location accuracy within 10–30 cm. The read ranges are also around 200 m (LOS) and about 50 m indoors through obstructions (Mssi, Inc., 2007).

Passive RFID RTLS: An approach was used, which was able to achieve an accuracy of 0.6 m using a Bayesian statistical method on readings collected from multiple RFID readers. This was accomplished without exploiting any prior information about the location, orientation, or power delivered to the tag (Rf code, 2007).

4.7 DIFFERENCES IN USING RFID, RTLS, AND GPS

RTLS is generally considered similar to GPS or vice versa in that they both use localization algorithms to triangulate positions. Technically, GPS can be considered a satellite-based RTLS system. The main difference for non-satellite-based RTLS is that they mainly operate indoors. On the contrary, GPS is traditionally used for outside position location. Also, some RTLS are considered more accurate than GPS in outdoor conditions but the main limitation is the need for infrastructure.

4.8 TREND TO INTEGRATED AIT APPLICATIONS

Recently, there has been some confusion by supply chain managers between the objectives for using RFID technologies and GPS. To add to the confusion, the marketing of the technologies by different vendors has made it difficult for many organizations to decide which technology is best for their operations. Though each technology has different operational objectives, the association of these technologies as ADC technologies has led many to believe the benefits of the technologies are interchangeable. Recently, some manufacturers have created technologies that have integrated both these technologies so that their benefits can be leveraged.

Current applications of RFID and GPS systems have allowed for more effective tracking of inventory and assets. These technologies can be used in conjunction, but, the data have to be captured and written to a database to be correlated to other tags or receivers. If these technologies can be combined to produce hybrid systems, greater gains can be achieved.

One focus of research is the nesting of GPS receivers and various RFID tag types. If tags and receivers were able to communicate with one another, even more accurate real-time data collection could be achieved during transportation. This would also reduce equipment costs, because fewer readers would be required. The concept of requirement layers for determining the required technology is demonstrated in Figure 4.1. These layers were developed by Bob Kenney at Savi Technologies (Lockheed Martin) and based off of the following ideas (Figure 4.8):

- ISO has grouped supply chain goods into logistic unit hierarchy
- Objects in each layer are handled differently and have different requirements for each part of the supply chain
- Necessitates different requirements for each layer

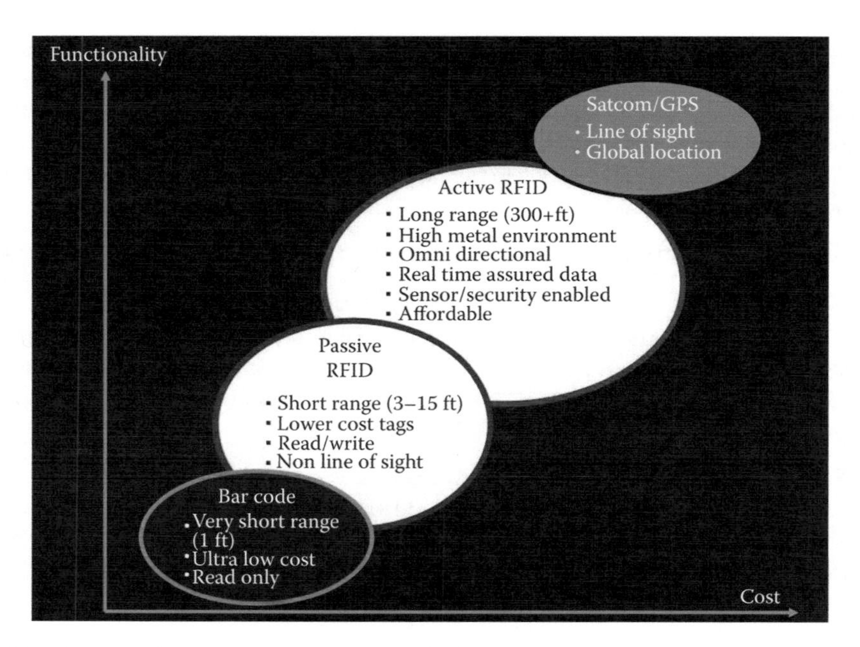

FIGURE 4.8 Functionality layers. (Courtesy of Savi Technologies, a division of Lockheed Martin, Lawrenceville, GA.)

If these technologies can be nested, it would allow the information in a bar code or a passive RFID tag to be collected by an active tag. This information could then be combined with the information contained within the active tag and transferred to a GPS receiver. The GPS receiver could then send not only its location but all of the information about the cargo being shipped (Reade and Lindsay, 2007). This concept was demonstrated earlier and shown in Figure 4.7.

A possible application of this nested technology approach would be in the railroad industry. Currently, there are two passive RFID tags attached to the sides of all rail cars in the United States. In addition, most railroads are using GPS receivers to track locomotives. If nesting became possible, implementation would be easy in this case. Active tags could be used to capture the information correlated to the cargo in all of the rail cars and transmit it to the GPS receiver and thus to the inventory databases.

An application that has recently been developed is demonstrated by the Dow and Chemtrec Company (O'Connor, 2007). This application was created to incorporate a tag that uses GPS, RFID, and sensors to track hazardous materials on railcars. The goal of the application was for constant monitoring of hazardous material conditions for railcars. The technology was based on RFID tags required by the American Association of Railroad RFID initiative (Automatic Equipment Identification) since the 1970s.

In addition to nesting technologies, more advanced tags can be developed to allow more detailed data collection. Tags that utilize sensors to capture and write data to the tag are being developed. Some tags have been developed, but are still very unreliable. These sensor tags could be used to monitor physical parameters, like temperature and humidity, as well as security parameters. The main problem faced by these passive sensor tags is the limited power supply. The sensor cannot use any energy while outside the range of the reader. Also, the amount of energy available while in read range is very small. This limits possible measurement techniques (Want, 2007). With these sensor tags, perishable goods could be monitored to guard against possible safety issues. This could include salmonella outbreaks caused by frozen chicken reaching too high of temperatures for too long and medications being held at temperatures that reduce potency.

An example of a tag that captures sensor data and transmits using GPS technologies is the Hammer tag system (O'Connor, 2007). This system provides for the creation of a PDA that can write both image and geospatial location data to active RFID tags that can be later captured and integrated into a geographical information system. The main use was to capture information for geological digs.

Other applications include nesting the RFID and GPS technologies for use with maritime container applications to offset the shortcomings of the different technologies.

The combination of GPS and active RFID tags on marine cargo containers allows for location of maritime containers using RFID-based RTLS that utilizes wireless access points. In the event the RFID wireless access points are not available, the device switches to a GPS-based RTLS. This system design was created to improve the challenges and system limitations of the RFID/RTLS system, which are caused by the operational conditions of marine ports (O'Connor, 2007).

Other scenarios for this type of nesting technologies include military use (Bacheldor, 2008).

A satellite-based RFID service is described by an OrbitOne application. This GPS tag utilizes a battery-powered active RFID tag to capture information at predetermined times to capture information in challenging military environments. The information capture through RFID interrogation is later transmitted via low earth orbit (LEO) based satellites (Bacheldor, 2008).

Another application utilizes the nesting concept of RFID, GPMS, and GPS for tracking gas distribution in gas tankers (Wessel, 2007). This RFID-based system creates a device that contains an active RFID reader that captures active tag information from tanker gas valves. As the gas valve is opened, the active tag associated with the valve transmits a signal to the active RFID tag reader or interrogator. Later, the RFID tag reader transmits information to a general packet radio service (GPRS) connection located computer system onboard the tanker's trailer cab (Wessel, 2007). Finally, this GPRS transmits information to operations through a GPS. Another example of this nested GPS and active RFID system includes the Unipart system, which is used for traditional supply chain operations as a means to tracking inventory on tractor trailers (Bacheldor, 2007).

4.9 SUMMARY

Technologies are being used to allow real-time data collection. This allows for more dynamic SCM systems that are able to adjust to varying market and environmental conditions. RFID and GPS facilitate this dynamic SCM. RFID allows for up-to-date inventory levels, and when combined with GPS, can provide a means of tracking inventory as it moves from supplier to customer through the supply chain. New technologies are being developed to further the amount of information to decision support systems for SCM.

Some technologies can be used to make real-time adjustments to the supply chain. Those adjustments could be due to many events such as manpower shortages or equipment breakdowns. For example, if a problem occurs to a truck or the road conditions change due to weather, the system, supplied with this updated information, should be able to make the necessary corrections to the transportation routes of other trucks to compensate for the truck failure. This system would be very useful during natural disasters. With real-time information, the system would reallocate transportation and production to a place that would be the optimum solution. This kind of modeling would reduce the response time for such events from months or weeks to days or hours. This system can also be expanded to urban transportation within a city or long distances between two cities.

5 Basic Introduction to Common RFID Components

The design and implementation of a successful radio frequency identification (RFID) system is a complex process. One of the most important sets of decisions that the RFID engineer must make involves the selection of individual system components. Making informed decisions with respect to this process will increase the probability of producing a fully functional reliable system. Conversely, less-than-informed decisions can easily result in a poorly performing system or even worse, having to redesign the system from scratch. A poor implementation of any new technology such as RFID can increase the resistance to other attempts at using technology to increase productivity.

With these thoughts in mind, the objective of this chapter is to introduce the basic components that are required for a functioning RFID system. For now, we will begin by describing the general nature of each component and then follow up with more detailed information. Each of these specific individual components, as well as their implementation, is discussed in greater depth throughout the remainder of this book.

5.1 GENERAL COMPONENT OVERVIEW

We will begin with a generic discussion of general RFID system components. At this high-level viewpoint in our discussion, the exact nature of the system process we are attempting to apply RFID is not of paramount importance. As long as it is a system that requires inventory to be tracked, the following discussion is applicable.

As with many other types of automatic identification systems, an RFID system requires a number of interrelated components. From a very high-level perspective, an RFID system must contain a set of tags, one or more antennas, and a reader. Figure 5.1 illustrates how these components work in conjunction with each other.

5.1.1 Tags

Tags are the devices attached to the items or materials that the RFID system is intended to track. The tags may be placed directly on individual items such as in the case of consumer goods or on shipping containers or pallets, which hold multiple items. Tags come in all sorts of shapes and sizes.

The primary function of the tag is to transmit data to the rest of the RFID system. Tags generally contain three basic parts. These include

- The electronic integrated circuit (IC)
- A miniature antenna
- A substrate to hold the IC and the antenna together and to the inventory item

Tags are also classified depending on how they are powered. These classifications include

- Passive
- Active
- Semi-passive

FIGURE 5.1 Basic RFID system components.

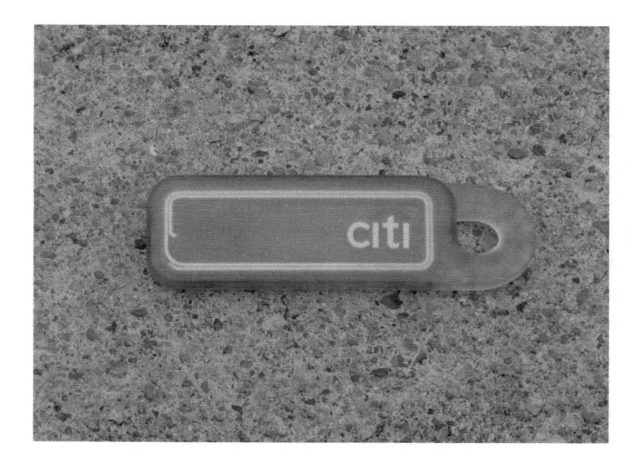

FIGURE 5.2 Common RFID tag.

Figure 5.2 illustrates just one of many types of common RFID tags. In this figure, the E-ZPass passive RFID tag is illustrated. This tag is used in tollway systems across the country. In the subsequent tag section of this chapter, we will describe and illustrate a number of different types of tags.

5.1.2 Reader

The RFID reader is a device that creates an electromagnetic (EM) signal that is transmitted to the RFID tags through one or more antennas. Under normal operation, the reader is continuously transmitting the EM signal in search of one or more RFID tags. The RFID reader also performs a second function of monitoring for EM signals from the RFID tags via the same antenna. Figure 5.3 illustrates a common type of RFID reader. This particular reader is manufactured by the Symbol Corporation.

FIGURE 5.3 Common type of RFID reader.

5.1.3 Antennas

The function of the antenna is to both transmit and receive EM signals between the tags and the reader. The effective EM field that the antenna transmits is in RFID terms known as the interrogation zone. That is the antenna creates a three-dimensional space that is used to communicate with the RFID tags. In order to obtain successful communication, the tags must be within range of the antenna or in the interrogation zone. Figure 5.4 illustrates a common type of RFID antenna. This antenna is manufactured by the Symbol Corporation.

FIGURE 5.4 Common RFID antenna.

5.1.4 Host

The host is a computer system that communicates with the RFID reader. It is the host that actually makes sense of the input from the reader. The host will typically have a number of software applications to support the RFID system. One application is commonly known as RFID middleware. This software is used by the user to setup and control the reader. Another needed software application involves data management. This includes functions that perform database and inventory tracking functions.

Now that you are familiar with the basic purpose of each of the major RFID system components, we will examine each individual component in greater detail. In addition, we will also identify some of the major manufacturers of each type of component.

5.2 TAGS

RFID tags come in an incredible array of shapes, sizes, and capabilities. While we cannot describe every possible type of tag, we will describe RFID tags in more detail. This will include describing the differences and advantages and disadvantages of the different types of tags that are commercially available. These differences between tags will be examined with respect to

- Power sources
- Frequencies
- Writing capabilities
- Tag components
- Tag generations
- Tag costs

5.2.1 Power Sources

All RFID tags must receive power in some form or another. Power is required by the tag in order to communicate information to the reader via the antenna. There are at least three different means currently available to power RFID tags. These include passive tags, active tags, and semi-passive tags. Each type of tag has its own advantages and disadvantages, which must be carefully considered when designing an RFID system. The type of tag required can have the tendency to drive the selection of the rest of the components of the system.

5.2.1.1 Passive Tags

Passive tags do not contain a power source. To power the tag's circuitry, the tag relies on EM power obtained from the RFID systems antenna. Since passive tags do not contain their own power sources, the designs can be simpler and less expensive. They can also have an unlimited shelf life in comparison to active tags. This has made the passive tag the focus of most government and commercial RFID mandates.

The downside of all passive tags is their extremely limited range. Since passive tags depend on power from the reader and antenna, with the current technology, passive tags must be in close proximity to the reader and antenna in order to obtain sufficient power to transmit a signal (Figure 5.5).

Many RFID experts believe that passive tags are the future of RFID. In the last few years, the unit price of passive RFID tags have steadily gone down in cost. This is a result of at least increase scales of production. Some industry analysts believe that when the cost of individual tags reaches 5 cents, significant acceptance will be achieved. At that point, RFID tags may be placed on many consumer consumables. Should this happen, the vision of consumers by passing the checkout counter may soon follow.

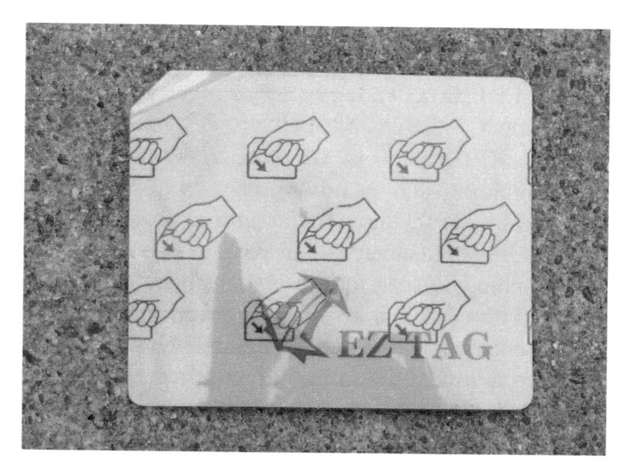

FIGURE 5.5 Common passive tag.

5.2.1.2 Active Tags

In contrast to passive tags, active tags contain an onboard power source. This is usually in the form of a small battery. The battery powers both the tag's internal circuitry and the onboard antenna. The additional circuitry required by the battery, as well as the battery itself requires that active tags be larger and more expensive than passive tags. Many active tags, for example, have plastic housings. These cannot simply be adhered to high-volume inventory in the same manner as a film or mylar-based passive tag. Because of this, specific consideration must be made to affixing the active tag to the inventory or pallet being tracked.

As a result of the additional power offered by the battery, the range of active tags is generally far superior to that of passive tags. Active tags can have transmission ranges measured in hundreds or even thousands of feet instead of just a few feet as is normal in the case of passive tags (Figure 5.6).

Active tags conserve battery power by normally existing in a sleep mode. The tag is woken up or activated by entrance into an RFID system interrogation zone. The powered tag then provides data to the RFID system as requested. The ability to normally exist in a sleep mode greatly lengthens the operational life of an active tag. The minimal power consumption in the normal sleep mode enables many tags to remain operational for several years. The actual length of the battery life will be dependent on the number of times that the tag is activated. Thus, the RFID engineer will have to design or set up an active RFID system so that in the event that tagged material is stored within an interrogation zone, the tags will not be continuously activated until their batteries are exhausted.

RFID active tag batteries come in many shapes and sizes. Many RFID-specific batteries superficially resemble normal commercial equivalents. However, the RFID batteries are likely to function at a higher voltage of 3.6 versus 1.5 for smaller cells as in the case with many defense-related RFID tags.

FIGURE 5.6 Common active tag.

To avoid potential damage to both RFID and other conventional battery equipment, it is imperative that control be maintained over both the storage and replacement of RFID-specific batteries.

The high power demands of RFID active tag batteries may also require different battery chemistry than conventional equipment. Whereas most electronics utilize alkaline, nickel-cadmium, or nickel metal hydride batteries, RFID batteries are more likely to be based on substantially more costly, advanced battery technology such as lithium chemistry. The higher cost associated with lithium batteries may also lead the RFID tag manufacturer to produce rechargeable systems.

Active tags can also be more sophisticated than passive tags. In some cases, active tags can be interfaced with other technologies such as the global positioning system (GPS) and/or satellite communication systems. The GPS is a set of U.S. government satellites orbiting around the earth. A GPS receiver communicates with the GPS satellites. Nowadays, receivers can be found in cell phones, radios, and wristwatches. If the minimum required number of satellite signals can be acquired by the GPS receiver, the system can determine its location within as little as 10 m. This means that GPS-interfaced tags can be placed on large shipping containers or tractor trailer rigs to determine both the identification and location of product.

The larger size and greater expense of active tags does prevent their use on smaller, cheaper types of products, particularly those with high volume. That means that it is unlikely that active tags will ever be used at the individual consumer product level. As you would expect, expensive active tags are obviously not considered disposable and may be intended to be recycled according to operating policy.

5.2.1.3 Semi-Active Tags

Tags can also be designed with features found in both passive and active tags. These are attempts to retain the advantages while eliminating the disadvantages of each type. Semi-active tags typically use an internal battery to power circuitry that is internal to the tag itself. Typically, circuitry on semi-active tags includes sensors for monitoring environmental conditions such as temperature and humidity. Sensors can also be powered to detect vibration or movement. These are typically used to monitor the possibility of damage or unauthorized movement during transport or storage.

However, in contrast to active tags, the semi-active tag does not use its internal power source to communicate with the antenna. For communications functions, the semi-passive tag relies on EM field power received from the system's antenna. By conserving internal power in this manner, the internal battery life can be greatly extended.

5.2.1.4 SAW RFID Tags

Surface acoustic wave (SAW) tags are passive RFID tags, which operate in a fundamentally different way from typical RFID tags. Typical RFID tags are based on semiconductor physics to provide power for transmission of their identification data (ID) number. SAW tags convert an incoming wave from the interrogator into nanoscale SAW on the surface of the chip. The wave travels past a set of acoustic wave reflectors that encode the wave into a unique pulse train. This pulse train is converted to a radio wave to be sent back to the reader.

These tags operate at a globally legal 2.45 GHz frequency. Due to the nature of the tags, there is no need for any DC power to operate the tag, allowing for greater read range. They have sufficient data storage capacity to comply with EPC or other global standards. In addition, the nature of the tag allows the distance between reader and tag to be calculated, as well as provides for direct measurement of tag temperature. The tags have an operating range of –100°C to 200°C. They can also survive high-energy x-rays and gamma-ray sterilization.

5.2.2 Tag Frequencies

Tags primarily operate at either high frequency (HF) or ultrahigh frequencies (UHF). HF is most often 13.56 MHz, while UHF can range from 902 to 928 in the United States. The 2400–2500 MHz range may also be used. Some active tags for specialized applications may utilize microwave frequencies. The use of either HF or UHF tags for more normal applications is dependent on the range required and the materials present in the system.

HF tags are generally limited in ranges measured in inches. This lends HF tags to inventory applications where the items are in close proximity to each other and in close proximity to a reader. UHF tags operating between 902 and 928 MHz, on the other hand, can be utilized out to several feet or even yards. The greater range of UHF tags makes them more applicable to shipping dock-type applications. 2400–2500 MHz tags may have a range between 1 and 4 ft.

Both the packaging material and the material itself is a significant RFID system issue as some materials are known as radio frequency (RF) absorbing while others are RF reflecting. Examples of RF reflecting materials are metallic items or containers. The RF reflecting characteristic of metals can prevent the tag antenna from absorbing sufficient RF energy to be powered by deflecting the RF wave.

Examples of RF absorbing materials are liquids. Liquids reduce the effectiveness of the RF wave by absorbing the energy. The reduced-strength RF signal then does not have sufficient power to activate the tag.

5.2.3 Writing Capabilities

When the tag enters the interrogation zone, the data stored in the tag are transmitted to the RFID reader antenna. The data can be ASCII, hex characters, or decimal characters. The data that are stored in the tag is dependent on the tag's writing capability. The three general types of writing capabilities are

- Read only
- Write once read many
- Read–write

5.2.3.1 Read Only

Read-only tags are tags where the ID are entered by the tag's manufacturer. Thus, these types of tags must be specified by either the manufacturer and accepted by the purchaser or specified by the purchaser. In many cases, the ID are used by a number of different organizations. Therefore, it is actually easier to control if the ID are assigned by the manufacturer. A typical example is the E-ZPass tollway system. A vehicle is assigned a tag with a specific number regardless of which tollway system the tag is purchased through. Since the tag number is controlled, it may be used on other tollway systems in the tag consortium.

5.2.3.2 Write Once Read Many

Write once read many (WORM) tags are not programmed by the manufacturer. The purchaser is given the opportunity to write the ID to the tag. However, with the WORM type of tag, these ID cannot be erased. This means that once the data are written, they cannot be changed. However, in some cases, if additional memory space is available, additional ID can be added. Generally, in the event that incorrect data are written to the tag, the tag must be discarded.

5.2.3.3 Read–Write

As with WORM tags, read–write tags are not programmed by the manufacturer. It is the purchaser that programs the tags. The advantage of the read–write tag is that the purchaser can reprogram the ID held by the tag. Thus, any ID writing errors can be corrected. Read–Write tags are generally the most sophisticated type of the three types of tags. Often, additional information may be used to store additional information. It is also possible to lock certain areas of the tags memory so that it cannot be erased.

5.2.4 Tag Components

Passive, active, and semi-passive tags all contain a minimum of

- Integrated circuitry or chip
- An antenna
- A Substrate or tag housing

The components are illustrated in Figure 5.7.

Of course, more sophisticated active tags will also contain the battery and specialized integrated components such as GPS circuitry or monitors for humidity, temperature vibration, or movement.

5.2.4.1 Tag Integrated Circuitry

The tag IC or chip is that part of the tag, which contains the data that are transmitted. It also contains the logic to decode the RF signal from the reader and code the data recorded on the chip for subsequent transmission by the tag's antenna. Passive tags commonly have the capability to transmit 96 bits of data. Active tags, on the other hand, are limited only by the other system components integrated with the RFID tag.

5.2.4.2 Tag Antennas

Not to be confused with the system reader antenna, the tag antenna is an integral component on the actual RFID tag. The tag antenna is used to both receive and send RF waves. In the case of passive tags, the tag antenna receives the RF energy and passes the energy onto the tag's integrated circuitry. The integrated circuitry's response to the RF energy is then transmitted back out through the tag's antenna.

The configuration of the tag's antenna is dependent on what type of frequency is used by the tag. HF tag antennas will most likely be shaped as coils. UHF antennas will be more linear in shape (Figure 5.8).

Another tag antenna issue is the size of the antenna itself. This is a critical issue as the size of the antenna is related to its ability to both absorb RF energy and transmit RF energy. Since greater range is generally desired regardless of the type of tag that is used, most tag antenna designs attempt to maximize the size of the antenna. Thus, a larger part of any tag is going to consist of antenna.

A significant antenna issue revolves around the potential placement of the tag on the intended object. The effectiveness of the antenna may depend on how it is oriented with respect to the product, which in turn is dependent on how it is positioned with respect to the reader's antenna. To address this issue, some RFID tags contain multiple antennas or antennas that have different branches. All of

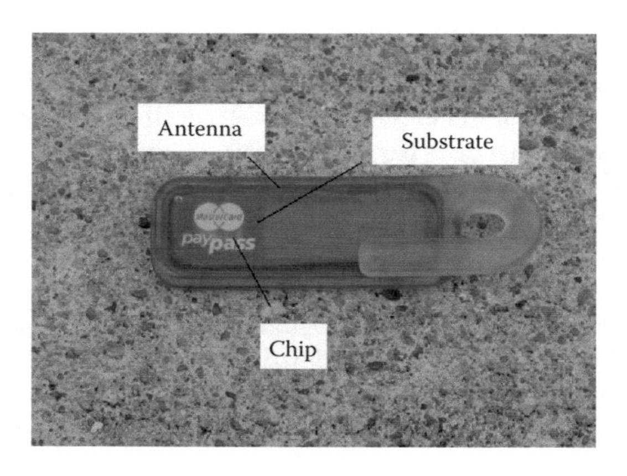

FIGURE 5.7 Tag components.

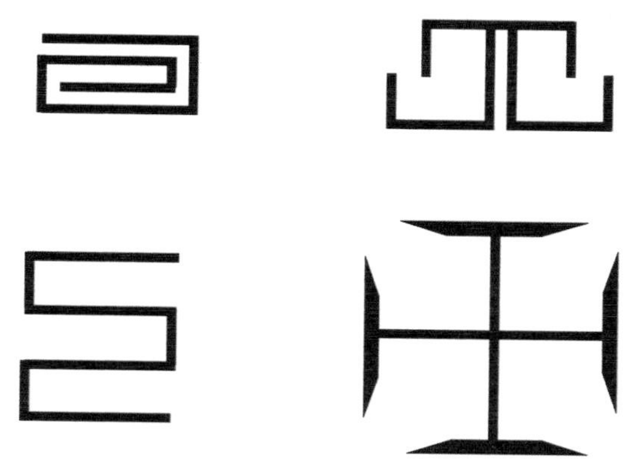

FIGURE 5.8 Tag antenna designs.

these attempts are aimed at increasing the RFID system's ability to obtain more reliable communication. The successful placement of tags on objects requires that the RFID engineers conduct some experimentation in order to determine the optimum position. The subject of tag placement testing will be covered in greater depth in a later section of this book.

5.2.4.3 Tag Substrate or Tag Housing

The tag substrate performs two primary functions. First, on the front surface, it provides a surface or housing for the other tag components. Second, on the back surface, it can provide means for attaching the adhesive or tape for positioning the tag to the item. A variety of materials can be used for the tag substrate. Common materials are thin plastics including mylar.

The following table summarizes the major differences between the three major types of RFID tags. SAW tags have not been included in this table.

	Passive	Active	Semi-Passive
Power source	External electromagnetic antenna field	Onboard battery	Onboard battery for internal circuitry external electromagnetic field for transmission
Range	Measured in feet	Up to thousands of feet	Measured in feet
Size	Smaller	Larger	Larger
Data storage	Less	More	More
Cost	Less	More	More

5.2.5 Tag Generations

One of the vastly superior characteristics inherent in RFID tags over bar codes is the ability to change data that are transmitted to the antenna and reader. Bar codes once printed can only provide a fixed set of data. The most current RFID tags can be reprogrammed. However, to reach this point, RFID tags have undergone evolution.

Earlier generation of RFID tags referred to as Generation 1 Class 0 are programmed by the tag manufacturer. With these class tags, there is no way for the local end user to modify the data. This limitation means that a lot of coordination is needed between the manufacturer and the end user. The end user also has to be extremely careful not to make any tag placement errors. An incorrectly affixed tag could result in a wasted tag. This lack of flexibility somewhat reduces the benefits of the entire RFID system.

Newer Generation 1 Class 1 tags are known as WORM. With the proper equipment, the end user can program blank tags. This produces less waste, as the tags are programmed as the need arises.

The latest tag protocol is known as Generation 2 Class 1. These are also WORM-type tags. The major difference between Generation 1 Class 1 and Generation 2 Class 1 tags is that the Generation 2 protocol is intended to have increased global acceptance.

5.3 SCANNERS AND READERS

As previously discussed, RFID tags receive an EM signal that is transmitted through the system's antennas. The RFID tags then subsequently return a signal to the system. For these functions, either simple scanner or complex readers may be utilized.

5.3.1 Scanners

In simple systems, the EM signal may be transmitted from a simple scanner. Simple scanners may be handheld or mounted to mobile equipment such as a forklift. This means that scanners are primarily used for situations where you go to the material rather than the material coming to you. Scanners are also commonly used for situations where RFID data must be verified. Both of these requirements imply a relatively low rate of reads is required. An application that is well suited to scanner use is order fulfillment. Here, the individual processing the order may travel throughout a warehouse picking individual items from stock. As each item is picked, it is scanned and removed from the on hand inventory database.

Typically, the data gathered by scanners are usually downloaded or linked to other information systems components. Figure 5.9 illustrates a typical handheld RFID scanner.

5.3.2 Readers

In more complex or rapid moving systems, the EM signal is transmitted by a smart reader. Smart readers may be commonly seen in conveyor or loading dock portal applications. With smart readers, the characteristics of the RF field are generated by the reader that is controlled by the RFID middleware installed on the host computer. These include the size of the interrogation zone field and the RF that the interrogation zone field. The reader also monitors the interrogation zone for tag transmissions. Figure 5.10 illustrates a common type of RFID readers.

5.3.3 Reader Frequencies

Readers may use one of two common types of RF. These include HF and UHF. As previously described in the tag section of this chapter, HF is most often 13.56 MHz while UHF can range from 902 to 928 in the United States. It goes without saying that the reader and the tags must operate on the same frequency in order to properly function.

5.3.4 Reader Interrogation Modes

Most readers will have a setup protocol including the mode in which the reader is to interrogate tags within the interrogation zone. With passive tags, this is not so much an issue. However, with active tags, the RFID engineer must decide whether to set up the system with continuous or intermittent interrogation. If any tagged material is present in the interrogation zone and the RFID system is set up for continuous interrogation, the active tags' batteries may be quickly depleted.

One method of sidestepping this problem is to reduce the interrogation zones size. This, however, will prevent the RFID system from being able to maintain any kind of vigilance over potentially expensive tagged material. On the other hand, a large interrogation zone can be maintained and

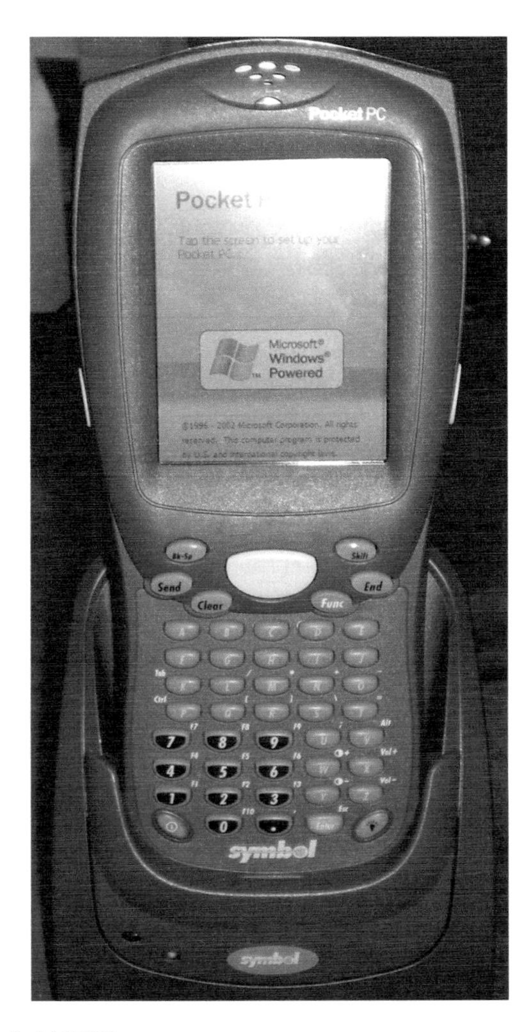

FIGURE 5.9 Typical handheld RFID scanner.

FIGURE 5.10 Common type of RFID reader.

the reader can only activate the interrogation zone on an intermittent basis. The downside to this approach is that all of the material can be monitored, but only at specific intervals. Between interrogation intervals, the presence or location of the tagged material is unknown.

5.4 ANTENNAS

As previously discussed, an RFID antenna is used to transmit the RF signal from the reader to the tags. An RFID antenna is also used to receive the RF signal from the tag for subsequent processing by the RFID reader.

In situations where the orientation of the tag with respect to the reader will not change, it is possible to have a single antenna. This can work in manufacturing applications where a product is undergoing a process. However, in more complex situations where the orientation of the tag is not guaranteed, it is normal for RFID systems to utilize more than one antenna. Generally speaking, for a given sized interrogation zone for an antenna, the greater the number of antennas, the greater the probability of a successful read by the system.

RFID antennas are commonly contained in an outer rectangular-shaped plastic housing. The housing protects the antenna and associated electronic components from damage. The housing also protects the antenna from minor environmental hazards such as dust. Thus, many RFID antennas have little resemblance to the type of antenna that you may be used to seeing. The plastic housing also provides a means of attaching the antenna in position.

The positioning of the antennas is also an important issue. As previously discussed, both the packaging material and the item to be tracked can affect the ability of the RFID system to conduct a successful read. In the classical loading dock application, it would be standard to include an antenna on each side of the portal. In a forklift-type application, the antennas may be positioned above the driver's safety cage. Lastly, in a shrinkwrap application, the antennas would be positioned in locations around the turn table that would still allow access to the table.

Another antenna placement issue is the height of the antenna. In some applications, the material will not necessarily be passing through a specific portal. In many cases, the range of the antenna

FIGURE 5.11 Large common RFID system antenna.

and the size of the interrogation zone can be increased by raising the antenna above ground level. Some experimentation will be necessary in order to determine the optimal sized zone. The subject of effectively testing and positioning antennas is covered in depth later in this book.

Yet another possible antenna issue is a situation where the antenna cannot be allowed to interfere with the surface movement of the material. In situations like this, it is possible to mount the antenna suspended from the ceiling with the field oriented downward. This method works particularly well when the tags can be placed on the topmost horizontal surface. In this case, the system has an unobstructed view of the tags and the successful read rate is likely to be very high (Figure 5.11).

5.5 HOSTS

As previously described, the host is a computer system with application-specific software. The software can include RFID middleware to setup and control the reader and some type of database software to control the information received from the reader.

5.5.1 Communication Protocols

The reader may have the capability of communicating to the host via more than one type of communication protocol. The choice of communications protocol can depend on the distance between the reader and the host, the required data transfer rate, and the system budget. Common types of protocols include RS-232, RS-485, and Ethernet-based systems.

5.5.1.1 RS-232

RS-232 is the simplest of the protocols likely to be available with an RFID system. RS stands for Recommended Standard by the Electronics Industry Association (EIA). This serial protocol is commonly utilized in industry to communicate directly between a host computer and one or more devices via individual dedicated cables. RS-232 ports have either 9 or 25 pins. While reliable, RS-232 suffers from slow data transfer rates of 20 K and limited transmission cable lengths of 50 ft. While some transmission distance issues may be overcome, other issues may make another protocol more attractive. For example, RS-232 cable is more expensive than other alternatives such as Ethernet. Despite many industry announcements of the death of RS-232, the protocol continues to be utilized in the industrial environment (Figure 5.12).

5.5.1.2 RS-485

RS-485 is a more capable serial communication protocol than RS-232. Advancements include increased data transfer rates and increased transmission distances. The RS-422 specifications

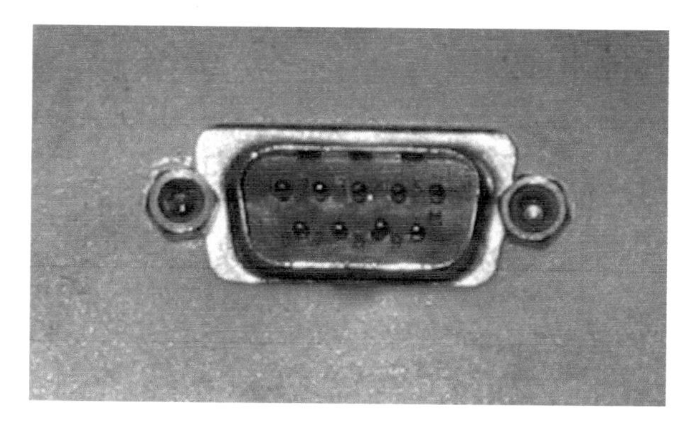

FIGURE 5.12 RS-232 port.

FIGURE 5.13 Ethernet port.

include up to a 100K data transfer rate and a 4000ft transmission distance. It also permits the communication to more than one device via a bus cable architecture. Up to 10 receiving devices can be utilized. RS-485 also has the improved capability over RS-232 of being differential. Among other things, this means that the system is less susceptible to problems associated with voltage shifts and electrical noise. Like RS-232 cable, RS-485 cable is more expensive than Ethernet cable.

5.5.1.3 Ethernet

Another possibility for communication between the RFID reader and the host computer is Ethernet. The attraction of an Ethernet-based system is that most facilities probably already have an Ethernet network in place. Ethernet also has a much higher data transfer rate than either RS-232 or RS-422. Newer Ethernet systems have the capability to transfer data at 100M. If one does not already exist, Ethernet cable can be more inexpensively obtained and installed than either RS-232 or RS-485 cable (Figure 5.13).

5.6 SUMMARY

The objective of this chapter was to provide a high level, easy to understand introduction to the various components required to implement an RFID system. These components include tags, antennas, readers, scanners, and hosts. Tags are classified as either passive or active depending on the source of power. Antennas come in many shapes and sizes depending on the application. Readers send power through the antenna to active tags. Scanners are less sophisticated handheld versions of readers. Hosts are the computer systems that control the entire RFID system.

With a basic understanding of these various components, it will now be possible to discuss specific commercially available RFID components, RFID system configurations, and RFID applications.

QUESTIONS

5.1 What are the four basic components of any RFID system?
5.2 What are the three main components present in a passive RFID tag?
5.3 What is the fundamental difference between passive and active tags and systems?
5.4 Under what conditions would you want to use HF versus UHF for an RFID system?
5.5 What techniques are used in tag design to reduce the tag positioning sensitivity?
5.6 What technique can be used to conserve the battery life of active tagged material while still maintaining some level of control?
5.7 What type of communications protocol would you normally want to use for an RFID system?

6 Passive RFID System Components

6.1 INTRODUCTION

In this chapter, we provide a more in-depth examination of passive RFID system components than in the common RFID systems components chapter. As with the active RFID system components chapter, for details on how these components are actually used in a system, readers are directed to the applications section of the book.

As discussed in the common RFID systems components chapter, passive RFID systems are distinguished from active RFID systems primarily by the absence of an internal power source in the tag. The lack of an internal power source yields both advantages and disadvantages to the RFID system designer.

6.1.1 Major Advantages to Passive RFID Systems

The major advantages of passive RFID systems include

- Lower expense
- Smaller sizes
- Greater operational life
- Environmental robustness

6.1.1.1 Lower Expense

Passive tags are well known for their lower expense in comparison to active tags. The lower expense is a result of having lower capabilities and no internal battery requirement. The lower expense of passive RFID tags is a great advantage in applications that do not require the greater capabilities of active RFID tags. For example, the use of passive RFID tags on consumer products is rapidly approaching. In this application, the RFID tag will obviously be disposable in nature, as the consumer product will permanently leave the logistical system. With extremely low-cost tags, the small increased expense of the tag becomes negligible. The same cannot be said of the much more expensive active RFID tags.

6.1.1.2 Smaller Sizes

Currently, very small-sized passive tags are in use. The completely operational Trovan passive tag, for example, is approximately the size of a grain of rice. This small size allows great flexibility in applications and mounting positions. Tags of this type are used currently for

- Companion animal tracking
- Individual sports competitions

The small size allows passive RFID tags to be implanted between the shoulder blades of companion or pet-type animals. Neither the animal, nor the owner can physically detect the presence of the tag under normal conditions. It is only when the animal's identity needs to be determined, does the tag have any impact.

The small size of passive tags also means that they are very light. This is important in cases where individual sports competition times need to be recorded. The champion tag, for example, is a passive RFID tag positioned in a housing that can be attached to a runner's shoe. The runner's time is recorded as they run across an RFID antenna mat at the finish line. Runners cannot object to the presence of the tag. The tag is so small and light as to not interfere with the competitor's performance.

6.1.1.3 Greater Operational Life

Greater operational life is achieved through the lack of an internal power source. Since no internal power source is necessary, the tag cannot be nonoperational as a result of battery depletion. Active tags in contrast must have their batteries replaced every 3–5 years depending, of course, on the exact nature of the usage. The operational life can, in fact, be almost infinite as long as the tag is not damaged. A more likely source of obsolescence would be an upgrade in the RFID infrastructure. This would most likely come about in the form of a change in protocol standards.

6.1.1.4 Environmental Robustness

Since passive RFID tags do not need to have a provision for replacing an internal power source, they may be hermetically sealed during the manufacturing process. This makes passive RFID tags inherently environmentally robust. This robustness allows passive RFID tags to be subject to environments where the operation of any non-hermetically sealed tag could be compromised. In the case of the Champion Tag, the tag is subject to rain and perspiration. Since the tag is sealed, these sources of moisture cannot enter the tag.

6.1.2 Major Disadvantages to Passive RFID Systems

The major disadvantages to passive RFID systems include

- Less range
- Less identification capability

6.1.2.1 Less Range

Passive RFID tags generally have much less range than active tags. Part of this is a result of the fact that passive tags must acquire power from the electromagnetic transmissions from the RFID reader antenna. The other part is due to the limited size of the antenna in the passive RFID tag. The end result is that the range of passive RFID tags is measured in a few inches or feet, whereas active RFID tags may transmit 100 m or more.

The lesser range capabilities of passive tags are both a blessing and a curse. From a blessing standpoint, the reduced range is advantageous for privacy reasons. For example, few consumers would be interested in carrying a credit card that could be activated by any reader hundreds of feet away. Imagine being charged for goods as you passed by any number of readers in a department store. Similarly, E-Passports incorporate a passive RFID chip to transmit personal data to the port of entry RFID reader. For both privacy and security purposes, the government would not want this data accessible to unauthorized individuals. Even with the short-range passive RFID tags, manufacturers are beginning to bring out transmission proof passport cases and wallets. Supposedly, these will help prevent illicit activation of passive tags.

The downside to lesser range is that a passive RFID system is much more sensitive to bad reads. The system must ensure that the tag is in closer proximity to the RFID reader antenna. In many applications, this is not a problem. The tag user deliberately places the passive tag near the scanner. This is the case with credit cards, passports, and companion pets. However, with other applications such as individual sports competitions, the placement of the tag is not as certain. For example, it is recommended that two separate antenna mats be utilized at the finish line in races. If the primary mat does not pick up the runner's tag, hopefully, the backup mat will acquire the signal.

6.1.2.2 Less Identification Capability

A second major disadvantage to passive tags is their general lesser identification capability. In many cases, passive tags can only hold and transmit a single 10 space unique alphanumeric string. In systems of this type, the identification code must be linked to an external database in order for the identification code to become meaningful.

The E-Passport perhaps represents one of the most advanced applications of passive tag identification capability. The passive tag embedded in the back cover of the E-Passport holds the individual's name, date of birth, place of birth, passport number, and even a digital photograph. The data are also encrypted. It is necessary to possess the key in order to translate the encoded data. However, in contrast to the capabilities of active tags, even this amount of data is limited in comparison.

6.2 CHAPTER ORGANIZATION

The remainder of this chapter focuses on a sample of specific passive RFID chips that are readily available for use by RFID system designers and engineers. New tags are being developed every day. Similarly, older tags are being constantly phased out. The inclusion of specific chips does not imply any endorsement, nor does the absence of any specific chips mean that they are not suitable for particular applications. The passive tags presented here are merely representative of the capabilities of passive chips in general.

6.3 TROVAN ELECTRONIC IDENTIFICATION SYSTEMS

Trovan Electronic Identification Systems is a privately held U.K. company that has been in business since 1988. It produces a variety of passive RFID tags. Trovan is the market leader in animal tagging applications. Its products are currently used by commercial organizations including Volkswagon, Diamler Benz, Coca Cola, Nissan, Merck, and many others. Trovan products are also used by over 60 government agencies in 13 countries around the world.

6.3.1 Trovan Passive Tags

The Trovan tags have 39 bits of data available for use. The tags are written with unique identification numbers during manufacture. With this number of bits, there are 550 billion unique code numbers. The Trovan Web site is www.trovan.com.

The intended uses, capabilities, and dimensions of following Trovan passive RFID Tags will be summarized:

- ID 100 Series
- ID 200 and 300 Series
- ID 400 Series
- ID 600 Series
- ID 700 Series
- ID 800 Series
- ID 1000 Series

6.3.1.1 ID 100 Series

The 100 series tags are primarily utilized for implantation. This includes implantation in both animals and humans. The tags are stated as currently being utilized in over 300 zoos worldwide. The tags are encapsulated in glass. They are approximately the size of a grain of rice at 0.45 in. long and 0.08 in. in diameter. The read range depending on the reader is approximately 9–14 in. The basic 100 series tag is illustrated in Figure 6.1.

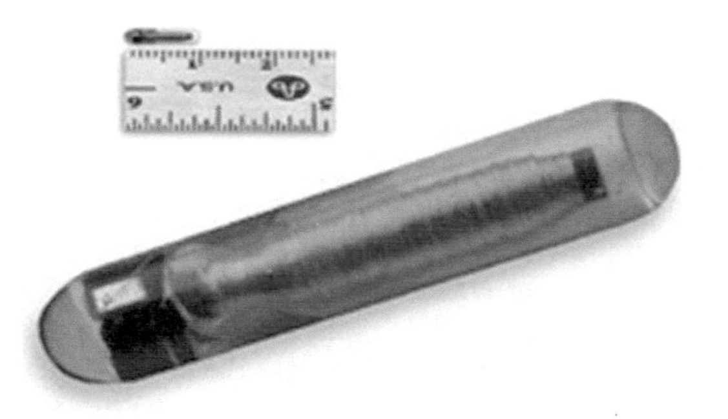

FIGURE 6.1 Trovan 100 series tag.

A ruggedized version of the tag is also available for industrial and garment application use. This tag has a capsule that is approximately twice as thick as the normal 100 series tag. As a result, the tag is slightly larger at 0.51 in. long and 0.18 in. in diameter. The tag is also listed as being able to withstand temperatures as high as 180°. This tag is illustrated in Figure 6.2.

As many of the major applications involve implantation, the 100 series tags have a dedicated implantation tool. This tool is illustrated in Figure 6.3. The RFID tag can be seen near the tip of the tool. The tool has a special gauge to ensure that the tag has been properly inserted.

6.3.1.2 ID 200 and 300 Series

The 200 series circular transponder is a passive tag in the form of donut. The predrilled, countersunk hole allows the tag to be screwed or riveted to a host product. The 200 series tag is 1.02 in. in diameter and 0.18 in. thick. The range of the 200 series is between 13 and 24 in., depending on the

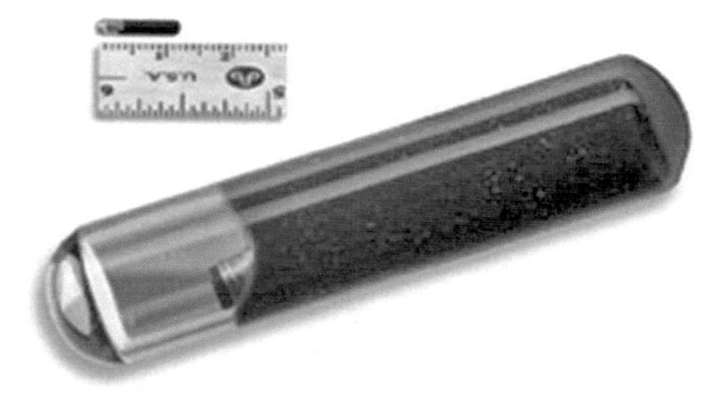

FIGURE 6.2 Trovan hardened 100 series tag.

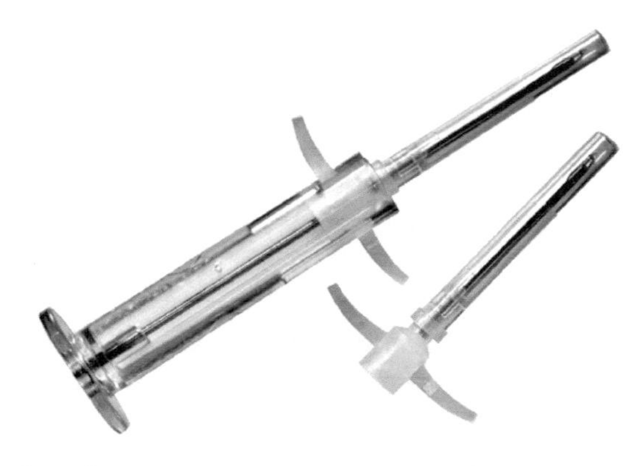

FIGURE 6.3 Trovan implantation tool.

FIGURE 6.4 Trovan 200 and 300 series tags.

FIGURE 6.5 Trovan 400 series card.

reader. The 300 series pellet transponder is similar to the 200 series minus the mounting hole. This also allows the tag to be more compact. It has a diameter of 0.50 in. and a thickness of 0.19 in. The range of the series 300 is between 9 and 18 in. Figure 6.4 illustrates the 200 and 300 series tags.

6.3.1.3 ID 400 Series

The 400 series tag utilizes a card form factor. It is primarily utilized as an identification card. It is suitable for use in security and entertainment applications where the identification number can be linked to an external database. The read range of the card is 20–33 in. It is 3.4 × 2.12 in. The 400 series card is illustrated in Figure 6.5.

6.3.1.4 ID 600 Series

The 600 series tag is designed for use with vehicles. It possesses the characteristic of having a very wide operating temperature range of –20°C to 200°C. Its range is between 18 and 27 in. Its dimensions are 2.4 in. long, 1.34 in. wide, and 0.55 in. thick. The ID 600 is illustrated in Figure 6.6.

6.3.1.5 ID 700 Series

The 700 series metal mount transponder is designed to be mounted on kegs, boxes, or vehicles. It may be mounted by welding or epoxying. It is specifically designed to handle conditions requiring durability and physical ruggedness. It has a read range of approximately 10 in. It is 1.69 in. in diameter and 0.24 in. thick. Figure 6.7 illustrates the ID 700 series tag.

6.3.1.6 ID 800 Series

The 800 series rod transponder is designed for extremely rugged applications including those requiring extended immersion in water. The read range is approximately 29 in. The 800 series is 5 in. long and 1 in. in diameter. The 800 series is illustrated in Figure 6.8.

FIGURE 6.6 Trovan 600 series tag.

FIGURE 6.7 Trovan 700 series tag.

FIGURE 6.8 Trovan 800 series tag.

FIGURE 6.9 Trovan 1000 series tags.

6.3.1.7 ID 1000 Series

The 1000 series tags are similar to the 200 series tags. However, their intended use is for pallet tracking and animal eartags. They are donut in shape with a variety of diameters from 0.5 to 2.0 in. The smaller sizes have 9 in. range while the larger sizes have 20 in. range. Figure 6.9 illustrates a few of the 1000 series tags.

6.3.2 Trovan Portable Readers

Trovan offers a number of portable readers for use with their passive RFID tags. The readers that Trovan offers include

- LID WAPR workabout pro reader
- GR-250 high-performance reader
- LID pocket series readers

6.3.2.1 LID WAPR Workabout Pro Reader

The LID WAPR workabout pro reader is the most sophisticated portable reader offered by Trovan. This reader has the capability to read RFID tags and bar codes. It also possesses a keyboard for data entry. The screen is color and is also touch sensitive. The reader has USB and docking port connectors. The read range depends on the tag being scanned. The maximum range is approximately 5 in. The reader is 8.7 in. long and 3 in. wide. Figure 6.10 illustrates the workabout pro reader.

FIGURE 6.10 Trovan workabout pro reader.

6.3.2.2 GR-250 High-Performance Reader

The GR-250 high-performance reader is designed to maximize the read range and read area. It sacrifices some of the capability of the walkabout pro reader in order to improve performance. The GR-250 does not have a color touch screen, nor a keyboard. The screen is a two-line, 16-character LCD display. The reader can store up to 3072 identification codes. As with the walkabout pro reader, ranges vary with tag type. The maximum range is 29 in. with the Trovan ID 500 tag. The reader is 11.7 in. long, 6.9 in. wide, and 4.5 in. tall. The reader is also sealed for better withstanding environmental conditions. The GR-250 is illustrated in Figure 6.11.

6.3.2.3 LID Pocket Series Readers

Trovan offers a series of pocket readers with different capabilities. The readers are all 12.5 cm long, 7 cm wide, and 2.4 cm thick. All of the readers possess a two-line, 16-character LCD display. The maximum read range of all of the readers is 5 in. The LID 560 basic pocket reader can only store one tag at a time. The LID 571 reader can store up to 1600 codes and has an RS-232 interface. The LID 571 can also provide a date and time stamp. The LID 572 is similar to the LID 571 with the exception of a USB port. The basic pocket reader is illustrated in Figure 6.12.

FIGURE 6.11 GR-250 high-performance reader.

FIGURE 6.12 Trovan LID 560 pocket reader.

6.4 SMARTCODE

SmartCode was founded in 1998. It has headquarters in New York, London, and Hong Kong. SmartCode is a world leader in providing low-cost, high-performance RFID solutions. The company manufactures RFID inlays and passive tags. SmartCode also offers active tags, which will be discussed separately in the active tag chapter. The SmartCode Web site is www.smartcodecorp.com.

6.4.1 SmartCode Inlays

SmartCode inlays are passive RFID tags constructed on a polymer substrate. The tags are typically distributed on reels. Customers can use the inlays between layers of paper or foil to create passive tags for a variety of inexpensive operations. These include product labels, tickets, and tracking tags. The inlays are available in a wide variety of antenna forms for use in particular applications. The tags can be as small as 30 × 30 mm. The largest tags are 100 × 100 mm. The weight varies between 80 and 200 mg. They are write once read many types of tags. They operate on 869/902–928 MHz and have a range up to 25 ft. Up to 256 bits of memory can be programmed on the tags. The tags can operate between −25°C and 80°C. A variety of SmartCode RFID inlays are illustrated in Figure 6.13.

6.4.2 SmartCode Passive Tags

SmartCode also manufactures complete passive RFID tags. Passive tags are available in read only and read/write format. The tags operate on frequencies of 125 kHz/13.56 MHz/915 MHz/2.45 GHz for read/write. Up to 16 K of memory is available. The tags can be as small as 8 mm and weigh between 6 and 54 g. The tags have a range up to 6 m. The operating temperature range is −40°C to 70°C. Figure 6.14 illustrates a variety of SmartCode passive tags.

6.5 SYMBOL TECHNOLOGIES

The New York-based Symbol Technologies Corporation has long been associated with automatic identification systems beginning with bar code technology. Symbol has continued in automatic identification with RFID technology. The company manufactures both passive RFID Gen 1 and Gen 2 inlays and tags. Symbol Technologies Web site is www.symbol.com.

6.5.1 RFX3000 Series Inlays

The RFX3000 series of passive RFID tag inlays come in a variety of shapes and sizes. These tags are Gen 1 Class 0 with read rates up to 400 per s. The tags are general purpose in nature, suitable for boxes, cartons, plastic bottles, and totes. The antennas are either single-dipole designs or dual-dipole designs with omnidirectional polarization. The dual-dipole design antennas incorporate

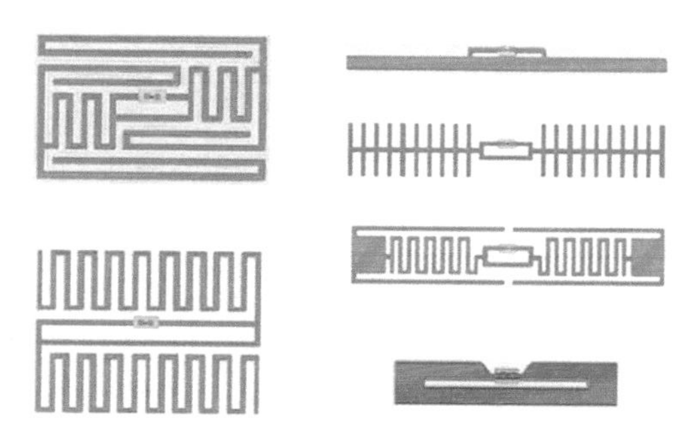

FIGURE 6.13 SmartCode RFID inlays.

FIGURE 6.14 SmartCode passive tags.

two antennas to improve read/write capability at any angle at any direction. Figures 6.15 and 6.16 illustrate some common Symbol Technologies inlays.

6.5.1.1 RFX3000 1 × 1 Read/Write Tag

The RFX3000 1 × 1 tag can be read at up to 5 ft and written to up to 3 ft. It is 1.00 × 1.383 in. It utilizes a single-dipole antenna design. In addition to general use, it is designed for plastic parts, bottles, totes, and containers. The RFX3000 1 × 1 tag is illustrated in Figure 6.15.

6.5.1.2 RFX3000 1 × 6 Read/Write Tag

This tag can be read up to 25 ft away and written to up to 8 ft away. It utilizes a single-dipole antenna design. It is a large tag with dimensions of 6.342 × 0.670 in. Its intended uses are cartons and corrugated cardboard containers. This tag is illustrated in Figure 16.6.

FIGURE 6.15 RFX3000 1 × 1 read/write tag.

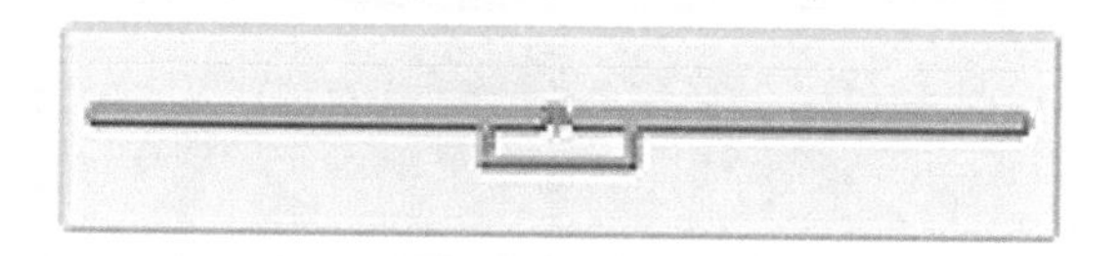

FIGURE 6.16 RFX3000 1 × 6 read/write tag.

FIGURE 6.17 RFX3000 2 × 2 read/write tag.

6.5.1.3 RFX3000 2 × 2 Read/Write Tag

This tag is also designed for plastic parts, bottles, containers, and totes. It differs from the RFX3000 1 × 1 tag by having a dual dipole antenna design. It can read up to 20 ft away and be written to up to 8 ft away. It is 2.234 × 2.234 in. The RFX3000 is illustrated in Figure 6.17.

6.5.1.4 RFX3000 4 × 4 Read/Write Tag

The RFX3000 4 × 4 tag is the largest, most capable of the 3000 series tags. This tag is designed for use with cartons and corrugated cardboard containers. It has a dual dipole antenna design with a read range of 30 ft and a write range of 8 ft. It is 3.692 × 3.692 in. The tag is illustrated in Figure 6.18.

6.5.2 Gen 2 RFX6000 Series Read/Write Inlay

The Gen 2 RFX6000 series tags are UHF EPC compliant tags. They have 96-bit memory for a unique ID w/32-bit access password and 32-bit "kill" password. They have a read rate of up to 400 times per s. The tags have an expected life of up to 10 years with 1000 write cycles. The temperature operating range is −40°C to 65°C. The 6000 series also incorporates both single and dual dipole antenna designs.

6.5.2.1 RFX6000 1 × 1 Series Read/Write Inlay

This tag is intended primarily for use on plastic bottles, parts, containers and trays. It utilizes a single dipole design antenna. It has a read range of up to 5 ft and a write range of up to 3 ft. The tag measures 1.00 × 1.383 in. The RFX6000 1 × 1 is illustrated in Figure 6.19.

6.5.2.2 RFX6000 2 × 4 Series Read/Write Inlay

The RFX6000 2 × 4 tag is intended for use with cartons, boxes, and corrugated cardboard containers. It has a dual dipole antenna design with a read range of up to 15 ft and a write range of up to 6 ft. It measures 1.700 × 3.875 in. The RFX6000 2 × 4 is illustrated in Figure 6.20.

FIGURE 6.18 RFX3000 4 × 4 read/write tag.

FIGURE 6.19 RFX6000 1 × 1 series read/write inlay.

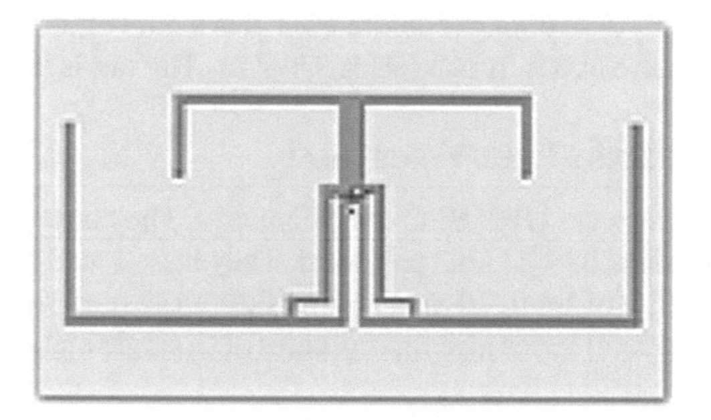

FIGURE 6.20 RFX6000 2 × 4 series read/write inlay.

6.5.2.3 Cargo Tag

The Symbol Technologies EPCglobal Gen 2, Class 1 Cargo Tag is intended for all weather applications. It is designed to be robust enough to withstand vibration, shock, rain, and temperature extremes. The tag has mounting holes, so it can be mounted with screws or rivets or it can be glued. A small area is also available for a conventional bar code. The tag has an aluminum backplate to enhance the tags signal. The tag can be read up to 40 ft. It contains 96-bit memory with

FIGURE 6.21 Symbol cargo tag.

16-bit control for lock and kill features. It operates in UHF at 860–960 MHz. It is 6 × 6 in. with a 4 × 4 in. antenna. It weighs 9.8 oz. It can operate between −20°C and 85°C. The cargo tag is illustrated in Figure 6.21.

6.5.3 Symbol Antennas

Symbol Technologies manufactures a wide variety of antennas. In this section, we will summarize the following:

- AN200 general purpose antenna
- AN400 high-performance area antenna

6.5.3.1 AN200 General Purpose Antenna

The AN200 is designed as a general purpose antenna for use in both indoor and outdoor applications. These antennas are intended for use in pairs with left and right hand polarization. The antenna is relatively small at 11.1 in. wide × 11.1 in. long, and 1.9 in. thick. It weighs approximately 3 lb. The operating temperature range is −40°C to 65°C. It has been tested specifically for rain, humidity, and salt spray. The AN200 is illustrated in Figure 6.22.

FIGURE 6.22 Symbol AN200 general purpose antenna.

FIGURE 6.23 AN400 high-performance area antenna.

6.5.3.2 AN400 High-Performance Area Antenna

The AN400 is designed for long-range and wide-area applications. They can be mounted on both ceilings and walls. The antenna and housing is constructed of aluminum with a polycarbonate cover. The AN400 is 28.3 in. long and 12.5 in. wide and 1.5 in. thick. The antenna can operate between 0°C and 50°C. It is approximately 8 lb in weight. The AN400 high-performance area antenna is illustrated in Figure 6.23.

6.5.4 Symbol Readers

Symbol Technologies manufactures both mobile and fixed readers. The mobile reader is the RD5000. The fixed readers are designated as the XR400, XR440, and XR480. These readers use the same form factor. The difference between the models includes increased sophistication and a greater number of read points. The XR400 has one read point while the other two have four and eight read points, respectively.

6.5.4.1 RD5000

The RD5000 is designed as a mobile reader for mounting on forklifts, mobile carts, and other material handling equipment. It operates in conjunction with a mobile computer such as the Symbol VC5090. It is 7 in. long, 9 in. wide, and 2 in. tall. It weighs 3 lb 10 oz. It is powered by a removable, rechargeable 7.2 V lithium-ion battery. The reader has a special feature that only powers the reader when the platform is in motion. This helps preserve the battery life. The operating temperature is –20°C to 50°C. The RD5000 is illustrated in Figure 6.24.

6.5.4.2 XR400 Series

The XR400 is a multiprotocol fixed reader that operates between 902 and 928 MHz. It has Ethernet capability and USB and RS-232 ports. Feedback is provided with green, yellow, and red LEDs to indicate power, activity, and errors. The XR400 is 11.75 in. long, 8.75 in. wide, and

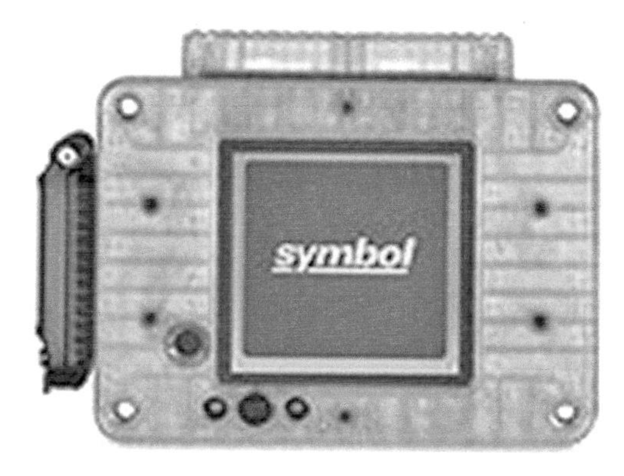

FIGURE 6.24 Symbol RD5000 mobile reader.

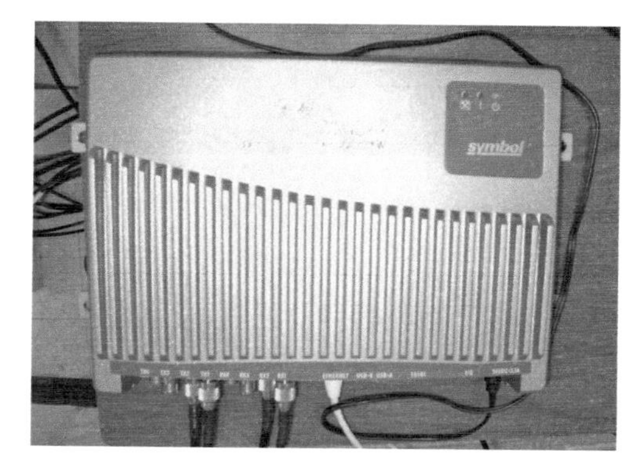

FIGURE 6.25 Symbol XR400 reader.

2.00 in. thick. It weighs 4.85 lb. The reader can operate in temperatures between −20°C and 50°C. Figure 6.25 illustrates the XR400 reader.

6.6 INTERMEC

Intermec's line of passive read/write RFID tags is marketed under the Intellitag name. The tags operate at 915 MHz. Their tags come in a variety of form factors. Applications include container tags, identification cards, and windshield tags.

6.6.1 Intellitag Windshield Tag

This tag is intended to be affixed to vehicle windshields for highway toll and security access applications. The substrate is flexible and has an adhesive backing for mounting. The tag measures 3.1 × 1.81 × 0.051 in. It can operate between −40°C and 85°C. It has a read range of approximately 13 ft. This enables the tag to be read by overhead antenna readers. The Intellitag Winshield tag is illustrated in Figure 6.26.

6.6.2 Intellitag Container Tag

The Intellitag container tag is a general purpose tag designed to be used with pallets, cartons, and containers. It measures 4.13 × 1.28 × 0.125 in. It has a read range of approximately 13 ft. The container tag is also available in other formats specifically suited toward reusable plastic containers. The basic Intellitag container tag is illustrated in Figure 6.27.

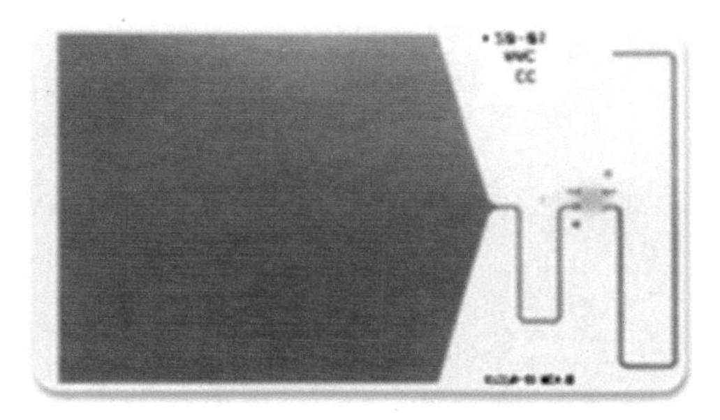

FIGURE 6.26 Intermec Intellitag windshield tag.

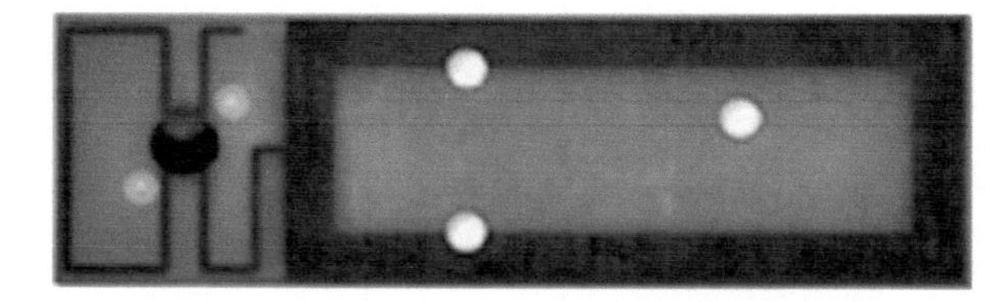

FIGURE 6.27 Intermec Intellitag container tag.

6.6.3 Intellitag ID Card

The Intellitag ID card is listed as being the first credit card format 915 MHz RFID card to have both long-range read and write capabilities. The card is specified as having a read range of 8 ft. It has a total of 110 bytes of memory for use. It can operate between 0°C and 50°C. Applications include border security, customer loyalty, and luggage tags. The card is also available with a magnetic stripe. The Intellitag ID card is illustrated in Figure 6.28.

6.6.4 Intermec Readers

Intermec offers fixed readers, handheld readers, and vehicle mount readers. The designations for these readers are

- IF series of fixed readers
- IP4 handheld reader handle with Intermec 700 series computer
- IV7 vehicle mount reader

FIGURE 6.28 Intermec Intellitag ID card.

FIGURE 6.29 Intermec IF 30 fixed reader.

6.6.4.1 IF Series of Fixed Readers

The IF series of fixed readers includes the IF 30, IF 4, and IF 5. The series is intended for dock doors, portals, and conveyors. The IF is set up for 865, 869, and 915 MHz frequencies.

The IF 30 can read tags from up to 15 ft. It also has the capability to filter out tag information prior to transmission to the host. This prevents redundant information being sent to the host computer. The IF 30 has both Ethernet and RS-232 ports. The dimensions are 12.27 in. long, 8.90 in. wide, and 3.25 in. tall. It weighs 6.75 lb. The IF 30 is illustrated in Figure 6.29.

6.6.4.2 IP4 Handheld Reader Handle with Intermec 700 Series Computer

The IP 4 reader is actually a reader designed to be an accessory handle for the Intermec 700 series of portable computers. This combination allows users to perform exception reading when it would be difficult to perform scans or writes with fixed readers. The system can be interfaced with existing networks for data transmission. Figure 6.30 illustrates the IP 4 reader and Intermec 700 series computer.

FIGURE 6.30 Intermec IP4.

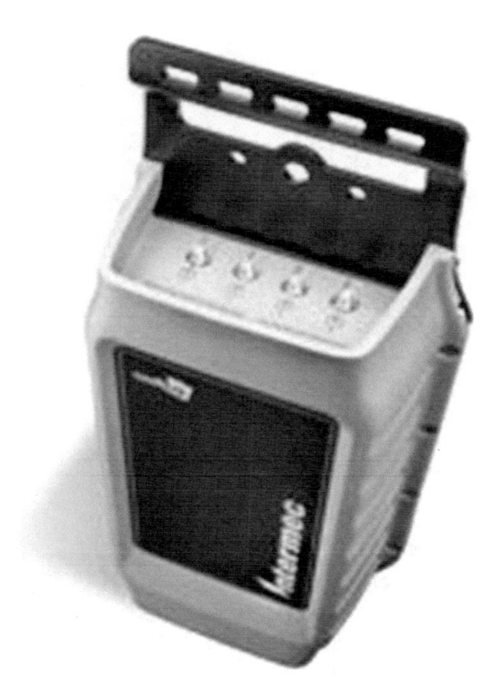

FIGURE 6.31 Intermec IV7 vehicle mount reader.

6.6.4.3 IV7 Vehicle Mount Reader

The IV7 is designed to be mounted directly to a vehicle such as a forklift. It combines the power of a fixed reader with the mobility of a portable reader. The IV7 is particularly useful when the number of dock doors exceeds the number of forklifts. It allows handsfree scans for operators. The IV7 is intended to be interfaced with Intermec's CV60 vehicle mount computer. The IV7 vehicle mount reader is illustrated in Figure 6.31.

6.7 SUMMARY

In this chapter, we presented some representative passive RFID system components. As with active RFID components, there are many more manufacturers than those presented here. Users are encouraged to identify as many different manufacturers as needed to ensure that the equipment best suits their needs.

7 Active RFID System Components

7.1 INTRODUCTION

In this chapter, we provide a more in-depth examination of active RFID system components than in the common RFID system components chapter. For details on how these components are actually used in a system, readers are directed to the applications section of the book.

As discussed in the common RFID system components chapter, active RFID systems are distinguished from passive RFID systems primarily by the presence of an internal power source in the tag. The existence of an internal power source yields both advantages and disadvantages to the RFID system designer.

7.1.1 Major Advantages to Active RFID Systems

The major advantages include

- Greater range
- Greater identification capability

7.1.1.1 Greater Range

The presence of an internal power source enables active tags to possess greater read ranges than passive tags. Active tag ranges are measured in meters rather than inches or feet as in the case with passive tags. This increased range enables an active RFID system to employ fewer reader antennas and read a larger number of tags from a given position than a passive system. The greater range enables active tags to be used for

- Expensive large asset tracking
- Container control

In the case of expensive large asset tracking, the greater range of active tags allows a number of expensive and/or large assets to be inventoried simultaneously with a small number of antennas. A typical example of this would be a vehicle yard. Some active tag systems have a read range as great as 100 m. A large number of vehicles could be near instantaneously inventoried within the 100 m radius of an active RFID system.

Similarly, storage units such as intermodal shipping containers can also be inventoried and tracked with the greater range of active RFID tags. The common storage method of stacking multiple shipping containers makes normal inventory techniques difficult to execute. With the same 100 m range previously discussed, mobile inventory vehicles can easily read the active tags while driving between rows of containers.

7.1.1.2 Greater Identification Capability

The internal power source allows active RFID tags to be far more flexible and capable than passive tags. The internal power also allows greater memory to be utilized. In contrast to 16 or 64 Kbits of memory in passive tags, there may be significantly more memory in active tags. The Savi active

RFID tags, for example, contain 128 kb of user available memory. Active tags may also be interfaced with environmental sensors for humidity and vibration. Even GPS capabilities can be built into active tags. This allows tags to transmit not only identification information, but also the tags physical location.

7.1.2 Major Disadvantages to Active RFID Systems

The major disadvantages of active RFID systems include

- More expensive
- Less operational life
- Larger physical size

7.1.2.1 More Expensive

Active RFID tags are generally far more expensive than passive tags. This is generally a result of the additional circuitry and components required for the increased capabilities of active tags. For this reason, active tags are currently only being utilized in applications where the active tags enhanced capabilities can be taken advantage of and where the tags can be recycled. Unfortunately, recent history indicates that many shipping container-type tags never actually end up being returned for reuse. As the use of active tags becomes more widespread, this disadvantage may become less of an issue.

7.1.2.2 Less Operational Life

Since active RFID tags possess an internal battery, their operational time is limited by the battery life. This is a serious issue as most active tags are far more expensive than passive tags. Many active tags actually have batteries that cannot be replaced by the user in the field. Unless the manufacturer is willing to replace the batteries, these types of tags must be discarded. Fortunately, many sealed active tags of this type can operate up to 10 years if the tag is not subjected to frequent identification collection operations.

Some active tags have the capability to have batteries replaced in the field by the users. However, the battery life is generally shorter than that of the non-replaceable active tags. Still, the battery life can be as long as 3–5 years. However, this is only possible when tag information is collected at a very limited schedule. For example, in some Savi active tags, the 5 year battery life is only achievable with two collections per day.

Several approaches can be taken to minimize this weakness. One approach is to utilize very power-dense battery chemistry. Rather than utilizing conventional alkaline batteries, manufacturers may specify far more expensive lithium-based batteries. Another approach is to design the tag so that the battery can be easily replaced in the field rather than having to return the tag to the manufacturer. Even replacing the battery in the field can be burdensome. If the tag is mounted in an inaccessible position or is protected from theft, battery replacement can become a serious issue when a large number of tags are involved. If frequent tag battery replacement is anticipated, users should ensure that the battery can be replaced while the tag is still mounted.

7.1.2.3 Larger Physical Size

The increased range and data transmission capabilities of active tags naturally require their physical size and weight to be larger in comparison to passive tags. This automatically precludes the use of active RFID tags in any application where these characteristics would be detrimental. Typical examples of this would involve the use with either human beings or animals. The presence of the tag might prevent the host from functioning in a normal manner.

The remainder of this chapter focuses on specific active RFID tags that are available for use by RFID system designers and engineers. As with the passive tag chapter, the active tags presented in this chapter do not constitute any sort of endorsement. The absence of any particular active tags should not be interpreted as being unsuitable for any particular applications.

7.2 SAVI CORPORATION

The California-based Savi Corporation was founded in 1989. It is predominately involved in active RFID systems and is probably the market leader in this area. Savi is heavily involved in defense applications for both the United States and NATO countries. Specific applications include the tracking of expensive assets and containers through logistical systems. Savi was acquired by Lockheed Martin in 2006. The Savi Web site is www.savi.com.

7.2.1 SAVI ACTIVE TAGS

Most Savi tags are based on similar highly developed active RFID technology. The tags generally operate on 433 MHz at 28 kbps. Their range is approximately 200 ft with mobile RFID readers and 300 ft with fixed readers. The difference in tags lies mainly in their intended end application. Their tags are categorized as either asset or data-rich tags. The following Savi tags are reviewed:

- ST-602
- ST-604
- ST-654
- ST-656

7.2.1.1 SaviTag ST-602

The small ST-602 is classified as an asset tag. It is designed to be attached to relatively small inventory items. It is 2.4 × 1.7 × 0.48 in. The tag weighs approximately 1 ounce. The tag is powered by a 3 V, 540 mA non-replaceable, non-rechargeable lithium-ion battery. The battery life is listed as 4 years with 10 reads per day. The tag has 16 bytes of memory for identification and 16 bytes of user memory. The tag returns the tag identification, the tag status, and the reader identification. The ST-602 is illustrated in Figure 7.1.

FIGURE 7.1 Savi ST-602.

FIGURE 7.2 Savi ST-604.

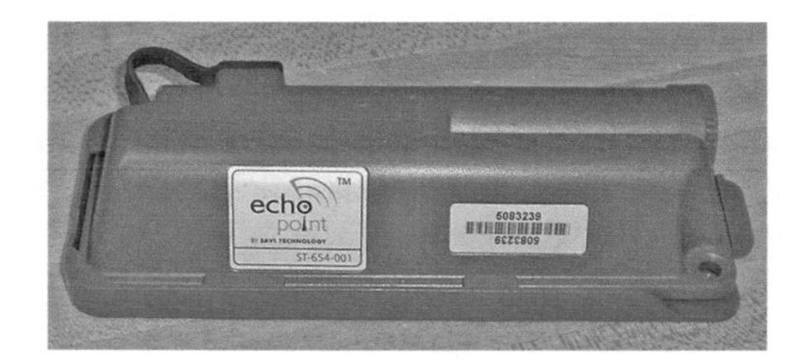

FIGURE 7.3 Savi ST-654.

7.2.1.2 SaviTag ST-604

The ST-604 tag is designed for users needing more battery life than the ST-602. The ST-604 is 6.25 × 1.7 × 1.3 in. It has the same 16-byte identification and user memory as the ST-602. The tag also returns the tag ID, the tag status, and the reader ID. The battery life is much longer than the ST-602 at 10 years, with 10 reads per day. As with the ST-602, the battery is non-replaceable and non-rechargeable. The ST-604 is illustrated in Figure 7.2.

7.2.1.3 SaviTag ST-654

The Savi ST-654 is intended for use on shipping containers. It is classified by Savi as a data-rich tag. The ST-654 has 128 K bytes of memory in comparison to the ST-602's and ST-604's 16 bytes. This allows the user a great deal of flexibility. Aside of the tag ID, the user can also program the contents of the container that the tag is attached to. The tag is similar in size to the ST-604 at 6.25 × 2.12 × 1.125 in. The tag is also more sophisticated than the ST-604 in that the 3.6 V lithium battery may be replaced in the field without special tools. However, the expected battery life is less at 5 years with two collections per day. The tag also has a buzzer for assistance in location. Figure 7.3 illustrates the ST-654.

7.2.1.4 SaviTag ST-656

The ST-656 is designed to be used on large ISO shipping containers. It can be attached to the left door of an ISO-compliant container. It is weatherproofed in order to operate under adverse conditions. The housing has the appearance of a clamp with a front and back case. In position, the tag is inside of the container. The front case is 4.75 × 4.5 × 1.5 in. The back case is 4.6 × 3.2 × 2 in. Like the ST-654, the ST-656 utilizes a replaceable 3.6 V lithium battery. The battery life is also the same at 5 years with two collections per day. The battery is also replaceable without special tools. The tag also has a beeper for location and other alarms. The ST-656 is illustrated in Figure 7.4.

7.2.2 Savi Fixed Readers

Savi offers both fixed and mobile readers. In this section, we will review fixed readers. The Savi fixed readers are intended to be positioned permanently in storage yards, terminals, and warehouses. The readers can be interfaced together through an Ethernet network.

FIGURE 7.4 Savi ST-656.

The following Savi fixed readers will be reviewed:

- SR-650
- Savi signposts

7.2.2.1 SR-650 Fixed Reader

The SR-650 fixed reader is a long-range omnidirectional antenna/reader. Its intended use is for asset inventory. The reader operates on 433 MHz. It is compatible with all Savi active tags. The range of the antenna is 100 m. The antenna/reader is designed to be weatherproof with a polypropylene housing with UV inhibitors. It can operate between –32°C and 60°C. It is circular in shape with a diameter of 12 in. and a height of 5 in. Figure 7.5 illustrates the SR-650 fixed reader.

7.2.2.2 Savi Signpost

The other fixed reader that Savi offers is the signpost. The signposts are designed for use in dock portals and vehicle gates through which tagged material passes. This prevents unintentional cross-reads from occurring, which might be possible with the omnidirectional SR-650. By utilizing a number of signposts, the RFID system can positively determine the location of the tags. The signposts can read tags traveling as fast as 60 miles per h. The signposts are available in short, medium, and long range. The model designations are SP-65X-111, SP-65X-211, and SP-65X-311. The long-range model has a range of up to 12 ft. Figure 7.6 illustrates a Savi signpost.

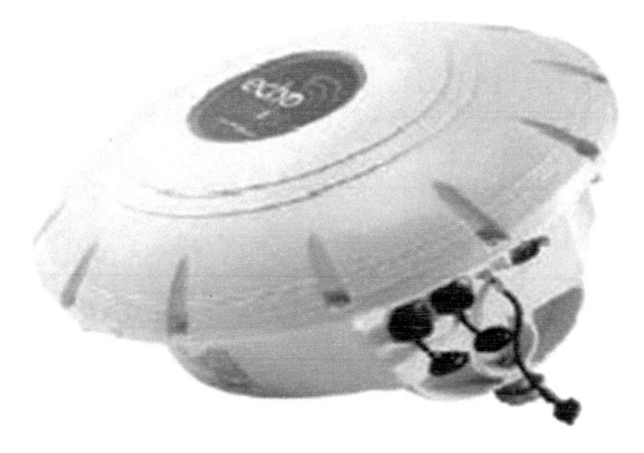

FIGURE 7.5 Savi SR-650 fixed reader.

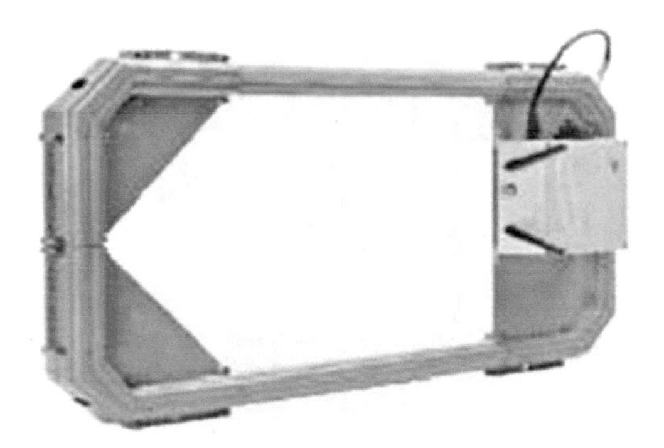

FIGURE 7.6 Savi Signpost.

7.2.3 Savi Mobile Readers

When the permanent nature of fixed readers is less appropriate, Savi offers a mobile solution based on the SMR-650 antenna reader. Readers of this type are particularly useful for commissioning tags. This reader has a range of 200 ft. This unit interfaces with a PC through a DB-9 pin RS-232 port. The dimensions are 6.25 × 1.5 × 1 in. The weight is 6.4 ounces. The unit uses a 3.7 V 420 mAh lithium-ion battery. The battery life is listed as 3–4 days of continuous operation. Figure 7.7 illustrates the SMR-650 antenna/reader.

The unit can be attached to either the Intermec 751 G/A computer or a conventional notebook computer. A SMR-650 attached to the Intermec 751 G/A is illustrated in Figure 7.8.

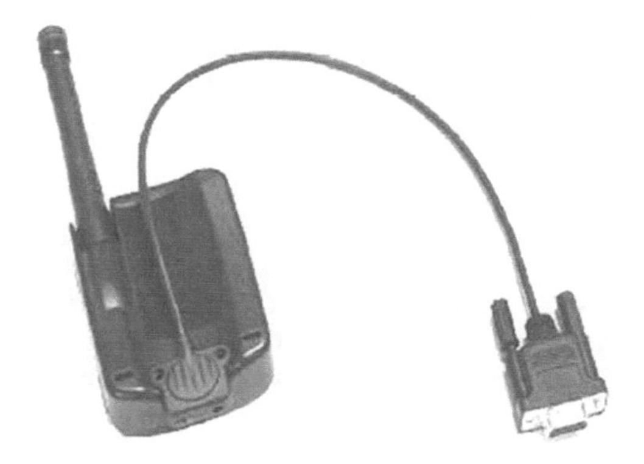

FIGURE 7.7 Savi SMR-650 antenna/reader.

FIGURE 7.8 Savi SMR-650 attached to the Intermec 751 G/A.

7.3 MARK IV INDUSTRIES

Mark IV Industries manufactures active RFID tags primarily for the electronic toll industry. Their RoadCheck Flat Pack Transponders (FPT) line of RFID tags come in a variety of form factors for internal, external, and license plate mount styles. The tags transmit on 915 MHz at 500 kbps. They have 256 bits of memory. The internal lithium battery can operate the tags for up to 10 years. However, most organizations are specifying replacement between 3 and 5 years. The operating temperature range is −40°C to 70°C. The Mark IV industries Web site is www.ivhs.com.

7.3.1 Mark IV Transponders

Mark IV offers a total of five transponders for different vehicle applications. These include

- Internal flat pack transponder
- License plate transponder
- Roof mount transponder
- Fusion transponder
- Ubiquity transponder

7.3.1.1 Mark IV Internal Flat Pack Transponder

The interval version typically mounts behind the rear-view mirror so that it does not interfere with the driver's vision. The basic form factor is 3.5 in. wide, 3.0 in. tall, and 0.6 in. thick. It weighs 2.5 ounces. Figure 7.9 illustrate basic internal form factor.

7.3.1.2 License Plate Transponder

Some vehicles possess antennas internally laminated into the windshield. Vehicles with these types of windshield often experience transmission problems with the normal internal RFID tags. To address this problem, Mark IV offers the license plate transponder form factor with similar performance specifications as the internal flat pack form factor. The license plate tag is illustrated in Figure 7.10.

7.3.1.3 Roof Mount Transponder

Mark IV has also developed an RFID tag for situations in which both a line-of-sight problem may exist and the tag needs to be mounted so that it is out of reach. This may occur with heavy trucks and buses. The roof-mounted transponder fulfills these needs. This tag is illustrated in Figure 7.11.

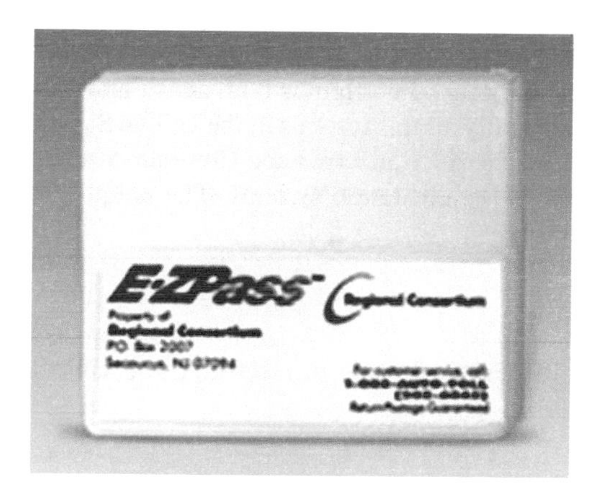

FIGURE 7.9 Mark IV internal flat pack transponder.

FIGURE 7.10 Mark IV license plate transponder.

FIGURE 7.11 Mark IV roof mounted transponder.

7.3.1.4 Fusion Transponder

Mark IV also offers a more sophisticated active RFID tag for commercial use. This tag is known as the fusion transponder. This 915 MHz tag is 3.1 in. wide, 4.3 in. high, and 0.9 in. thick. Its additional capabilities include audiovisual driver capabilities and enhanced read/write abilities. The audiovisual capabilities include LEDs to provide the driver with feedback as to the success of toll transactions. The write capability of the tags means that toll systems can program special data to calculate toll payments. These tags have a lithium battery that is listed as having a life of 5 years. The fusion transponder is illustrated in Figure 7.12.

7.3.1.5 Ubiquity Transponder

The final RFID transponder that Mark IV offers is both active and passive. This tag enables commercial operators to utilize virtually all toll systems in the United States, including the Inter Agency Group EZ-Pass system and the Tier 21 California and Colorado system. In addition, it also has the capability to utilize a variety of weigh station systems. The ubiquity transponder is illustrated in Figure 7.13.

7.3.2 Mark IV Readers

Mark IV offers a number of readers for use in electronic toll collection (ETC) systems. These include

- Badger reader
- MGate reader

FIGURE 7.12 Mark IV fusion transponder.

FIGURE 7.13 Mark IV ubiquity transponder.

7.3.2.1 Badger Reader

The Badger reader is Mark IV's product for general applications. It is capable of interfacing with eight regular lane antennas or four high-speed lane antennas. For larger applications, several Badger readers may be interfaced together. Vehicles traveling as fast as 100 mph can be detected with the Badger. The Badger operates on 915 MHz. Both RS-232 and RS-422 interfaces are available to communicate with the host computer. The Badger's dimensions are 20.95 × 19.00 × 12.85 in. The Badger can operate in temperatures between −40°C and 158°C. The Badger reader is illustrated in Figure 7.14.

7.3.2.2 MGate Reader

The low-cost MGate reader is designed for less demanding applications than the Badger reader. The MGate has also been used as a temporary mobile replacement for permanent facilities under

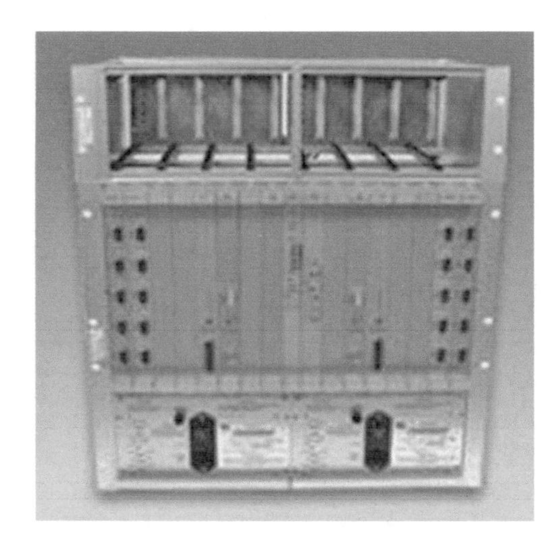

FIGURE 7.14 Mark IV Badger reader.

FIGURE 7.15 Mark IV MGate reader.

FIGURE 7.16 Mark IV lane antenna.

construction or repair. The MGate is 12 × 3.4 × 4.1 in. With these dimensions, the MGate is one of the smallest ETC capable readers. The MGate can operate between −40°C and 70°C. The MGate is illustrated in Figure 7.15.

7.3.3 Mark IV Antennas

Due to the nature of ETC systems, Mark IV offers their antenna separate from their readers. In most applications, a number of antennas will be attached to a single reader. Each of these antennas will be mounted either above or alongside the traffic lane. The lane antenna is 37 × 30 × 4 in. The outer housing is fiberglass. It is hermetically sealed to withstand exposure to the elements. The antenna can operate in temperatures between −40°C and 158°C. Figure 7.16 illustrates one of Mark IV's lane antennas.

7.4 SUMMARY

In this chapter, we have reviewed actual product offerings from two manufacturers in the active RFID system market. For logistics chain-type applications, we examined tags and antenna/reader systems from one of the market leaders Savi Technology. For ETC operations, we examined the tags, readers, and antennas from the E-ZPass manufacturer, Mark IV Industries. These are obviously not the only manufacturers that supply active RFID system components to those markets. Likewise, many other markets also exist that utilize active RFID systems. Readers are encouraged to explore the offerings from as many manufacturers as necessary to meet their needs.

Part III

Overview of Equipment Interoperability Protocols and Standards

8 Important RFID Mandates

8.1 INTRODUCTION

Mandates by large suppliers such as Walmart and the Department of Defense (DOD) are driving development, use, and price of RFID passive technologies in the supply chain. Due to these mandates becoming a main driver for RFID technologies, we provide an overview of the most important organizations mandating that suppliers use RFID technologies. We will discuss DOD, Walmart, and other companies that are considered leaders in supporting RFID.

8.2 DEPARTMENT OF DEFENSE MANDATE

The DOD has developed a plan for passive RFID tagging with the warfighter as their customer. The mandate and dates are mainly laid out in the implementation plan. RFID technology will be implemented through a phased approach, applied both to supplier requirements and DOD sites. The implementation approach that DOD follows according to U.S. DOD Suppliers' Passive RFID Information Guide Version 1.0 7 is as follows.

8.2.1 Commencing January 1, 2005

RFID tagging will be required for all DOD manufacturers and suppliers who have new contracts issued on or after October 1, 2004 according to the following implementation guidelines: The following classes of supply will require RFID tags to be placed on all individual cases, all cases packaged within palletized unit loads, and all palletized unit loads:

- Class I Subclass—Packaged operational rations
- Class II—Clothing, individual equipment, and tools
- Class VI—Personal demand items
- Class IX—Weapon systems repair parts and components

When these commodities are being shipped to the following locations:

- Defense Distribution Depot, Susquehanna, Pennsylvania (DDSP)
- Defense Distribution Depot, San Joaquin, California (DDJC)

8.2.2 Commencing January 1, 2006

In addition to the requirements above, RFID tagging will be required for all DOD manufacturers and suppliers who have new contracts issued on or after October 1, 2004 according to the following implementation guidelines.

The following classes of supply will require RFID tags to be placed on all individual cases, all cases packaged within palletized unit loads, and all palletized unit loads (pending appropriate safety certification):

- Class I—Subsistence and comfort items
- Class III—Packaged petroleum, lubricants, oils, preservatives, chemicals and additives
- Class IV—Construction and barrier equipment
- Class V—Ammunition of all types
- Class VII—Major end items
- Class VIII—Pharmaceuticals and medical materials

8.2.3 Commencing January 1, 2007

RFID tagging will be required for all DOD manufacturers and suppliers who have new contracts issued after October 1, 2004 according to the following implementation guidelines:

- All classes of supply will require RFID tags on all individual cases, all cases packaged within palletized unit loads, all pallets, and all unit packs for unique identification (UID) items.
- RFID tagging will be required on commodities shipped to any DOD location.

8.2.4 Guidelines and Requirements

The cost of implementing and operating RFID technology is considered a normal cost of business. If DOD customers desire the inclusion of a passive RFID tag on shipments for these type purchases, this requirement must be specifically requested of the shipping supplier/vendor and the shipment must be accompanied by an appropriate advanced shipment notification (ASN) containing the shipment information associated to the appropriate RFID tag.

All solicitations awarded on or after October 1, 2004 for delivery of material on or after January 1, 2005 will require that passive RFID tags be affixed at the case, pallet, and UID item packaging level for material delivered to the DOD, in accordance with the implementation plan, which is mentioned above under the section entitled "Implementation Approach." The plan can also be found at http://www.dodrfid.org/supplierimplementationplan.htm.

8.3 WALMART MANDATE

In recent years, Walmart has been working with the auto-ID center to develop and test RFID technology, which will allow companies to track goods using a universal electronic product code (EPC). Their long-term goal is to use smart shelves at the retailer to monitor how many items are on each shelf. When inventory is low, software would signal the management that products such as Gillette razors need to be brought from the storeroom. Readers in the storeroom would monitor inventory and alert the distribution center when more product is needed and automatically send a replenishment order, and this continues as a pull system throughout the supply chain.

But Walmart and other sponsors of the auto-ID center have always envisioned that it might take as long as 10 years before RFID tags would become inexpensive enough to put on individual items in stores. To date at Walmart, RFID has been used successfully in closed-loop supply chains. Generally, the mandate requires that suppliers provide pallet-level RFID tag that can be scanned at 10 ft wide dock door at all times. The RFID EPC tag, which is based on the pallet bar code that uses EAN.UCC GTIN or SSCC, is the required tag. Walmart suppliers have been communicated the requirement and all updates to the standards through Walmart's retail link, which is their online communications link to suppliers. The Walmart RFID mandate means its top 100 suppliers not only have to put tags on pallets and cases, they must also install RFID readers in their manufacturing facilities, warehouses, and distribution centers. They, in turn, can require their suppliers to tag shipments and so on through the supply chain.

Walmart is unlikely to back off its requirement because the retailer is convinced the benefits are justifiable. Sanford C. Bernstein & Co., a New York investment research house, estimates that Walmart could save nearly $8.4 billion per year when RFID is fully deployed throughout its supply chain and in stores. Walmart has been studying the potential of RFID for more than a decade before rolling out the standard. Walmart communicates to their suppliers what they need to do to fulfill the retailer's requirements, but after that, they are held accountable.

8.4 OTHER ORGANIZATIONS

Other companies that have similar mandates as Walmart include Metro Corporation in Germany. Metro is the fifth-largest retailer in the world and Germany's largest retailer. Their mandate is very similar to the Walmart standard, which is important to RFID pundits due to the fact that both Walmart and Metro use many of the same suppliers. Tesco and Marks & Spencer, larger retailers in the United Kingdom, also have initiated RFID mandates, providing further influence for suppliers to use RFID technologies.

Many pundits expect RFID use at the pallet and case level to take off rapidly because of something economists call the "network effect," which basically says that the more people use a physical network (say, the Internet) or shared service (Google), the more valuable it becomes. That encourages even more people to use the network, creating exponential growth, a tipping point, or a groupthink type of mentality.

9 Standards Organizations and RFID Standards

9.1 INTRODUCTION

Standards describe data content, air interface protocol, conformance testing, and applications usage. Data content describes how content is stored and formatted. The air interface describes how the tags will talk to each other. The conformance standards describe how to test for acceptable performance. Application standards describe how various devices such as shipping labels are used.

9.2 INTERNATIONAL STANDARDS ORGANIZATION STANDARDS

The International Standards Organization or ISO is one of the key organizations in the world for standardizing equipment and operations for over 20 years. RFID standards at the item level are described for ISO under the ISO–IEC Automatic Identification and Data Capture (AIDC) standard (Figure 9.1). The section is labeled JTC-1 SC31/WG4. This is often referred to as the EPC global standard.

The ISO is an international association of national standards bodies of 148 countries with one member per country. It was founded in 1947 with the headquarters in Geneva, Switzerland. ISO produces guidelines, procedures, and policies on a wide range of issues and applications. Standards produced by ISO provide a template for member bodies to develop their own standards. Regulators may adopt these ISO standards unchanged, or modify them to suit local conditions or requirements. The result is standards that are internationally compatible, consistent, and clear.

9.3 ISO STANDARDS AND RFID

Decisions about RFID ISO standards are made by two groups. ISO and International Electro-Technical Commission (IEC) jointly sponsor Joint Technical Committee number one, JTC 1, to address subjects of interest to both organizations. JTC 1 has several subcommittees to address specific issues, including SC31. SC31, AIDC techniques oversee standardization of data formats, data syntax, data structures, data encoding, and technologies for the process of AIDC. There are seven work groups from this subcommittee: data carriers (WG1), data syntax (WG2), conformance (WG3), and RFID (WG4), real-time locating systems or RTLS (WG5), mobile item identification and management or MIIM (WG6), and the newly formed file management and security for item management (WG7).

RFID standards at the item level are described for ISO under the ISO–IEC AIDC standard. See the attached chart. The section is labeled JTC-1 SC31/WG4. This is often referred to as the EPC global standard for (1) Class 0 Gen 1, (2) Class 1 Gen 1, and (3) Gen 2.

The SC31 organization structure for RFID is carried out in SG1, SG3, SG5, and SG6. The SG 1 is responsible for data content and includes the ISO/IEC 15961 data protocol—application interface description, and the ISO/IEC 15962 data protocol—data encoding rules. SG3 is responsible for

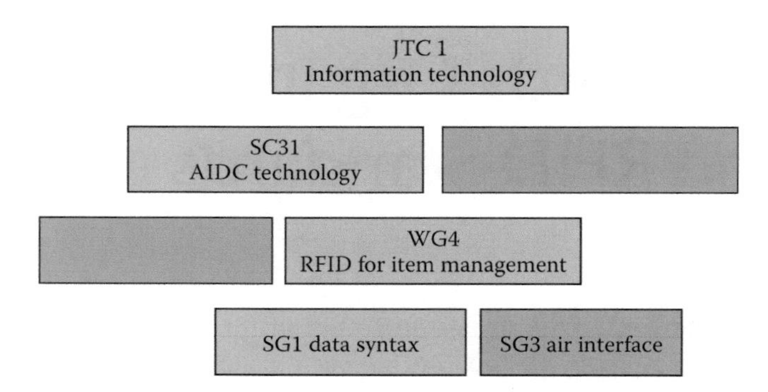

FIGURE 9.1 ISO/IEC standard structure.

ISO/IEC 18000 series air interface standards at <135 kHz; 13.56 MHz; 433 MHz; 860–960 MHz; 2.45 GHz. The ISO/IEC 18000, RFID for item management—air interface is described in the following parts.

Part 1: Air Interface Communication for Globally Accepted Frequencies
Part 2: Air Interface Communication below 135 kHz
Part 3: Air Interface Communication at 13.56 MHz
Part 4: Air Interface Communication at 2.45 GHz
Part 5: Air Interface Communication at 5.8 GHz withdrawn
Part 6: Air Interface at 860–930 MHz
Part 7: Air Interface at 433.92 MHz

9.3.1 18000-1 Part 1: Generic Parameters for the Air Interface for Globally Accepted Frequencies

The scope of this standard is to describe the reference architecture for RFID for item management and to establish the parameters that shall be determined in any standardized air interface definition in the ISO 18000 series. The subsequent parts of this standard providing specific values for air interface definition parameters shall then, once approved, provide the frequency-specific values and value ranges from which compliance to (or noncompliance with) this standard can be established.

This standard limits its scope to transactions and data exchanges across the air interface at reference point delta. The means of generating and managing such transactions, other than a requirement to achieve the transactional performance determined within this standard, are outside the scope of this standard, as is the definition or specification of any supporting hardware, firmware, software, or associated equipments.

This standard is an enabling standard, which supports and promotes several RFID implementations without making conclusions about the relative technical merits of any available option for any possible application.

9.3.2 18000-2 Part 2: Parameters for Air Interface Communications below 135 kHz

This document specifies the physical layer that shall be used for communication between the interrogator and the tag. The interrogator shall be capable to communicate with tags of both Type A (FDX) and Type B (HDX).

9.3.2.1 Protocol and the Commands

The method to detect and communicate with one tag among several tags ("anti-collision").

9.3.2.1.1 Tag Types

This standard specifies two types of tags: Type A (FDX) and Type B (HDX).

These two types differ only by their physical layer. Both types support the same anticollision and protocol.

FDX tags are permanently powered by the interrogator, including during the tag-to-interrogator transmission. They operate at 125 kHz.

HDX tags are powered by the interrogator, except during the tag-to-interrogator transmission. They operate at 134.2 kHz.

An optional anticollision is described in the informative Annex D.

9.3.2.2 Compliance Rules

9.3.2.2.1 Tag

To claim compliance with this standard, a tag shall be of either Type A or B.

9.3.2.2.2 Interrogator

To claim compliance with this standard, an interrogator shall support both Types A and B.

Depending on the application, it may be configured as Type A only, Type B only, or Types A and B. When configured in Types A and B, and when in the inventory phase, the interrogator shall alternate between Type A and Type B interrogation. See Annex C.

9.3.3 18000-3 Part 3: Parameters for Air Interface Communications at 13.56 MHz

This standard is to provide physical layer, collision management system and protocol values for RFID systems for item identification operating at 13.56 MHz in accordance with the requirements of ISO 18000-1. This standard provides parameter value for each MODE determined in the requirements clause below.

In this version of the standard, two non-contending MODES are defined:

- The MODES are NOT interoperable
- The MODES, whilst not interoperable, are non-contending

9.3.3.1 Intellectual Property

Both of the MODES require a license from the owner of the intellectual property (IP), which shall be available on terms in accordance with ISO Policy (RAND) (reasonable and nondiscriminatory). Details of IP are shown at the end of Clause 7. Neither of the MODES in this version of the standard may be used without a license.

9.3.4 18000-4 Part 4: Parameters for Air Interface Communications at 2.45 GHz

9.3.4.1 Frequency

This standard is intended to address RFID devices operating in the 2450 MHz industrial, scientific, and medical (ISM) frequency band.

9.3.4.2 Interface Definitions

This standard supports a standard API (ISO/IEC 18000-1) and standard air interface implementations for wireless, noncontact information system equipment for item management applications. Typical applications operate at ranges greater than 1 m.

There are two MODES. MODE 1 is a passive tag, MODE 2 is a battery assisted, tag talks first tag.

9.3.4.2.1 MODE 1: Passive Backscatter RFID System

The FHSS backscatter option or the narrow band operation RFID system shall include an interrogator that runs the FHSS backscatter option 1 RFID protocol or in narrow band operation, as well as one or more tags within the interrogation zone.

9.3.4.2.2 MODE 2: Long-Range High Data-Rate RFID System

This clause describes a RFID system, offering a gross data rate up to 384 kbps at the air interface in case of read/write (R/W) tag. In case of read only (R/O) tag, the data rate is 76.8 kbps. The tag is battery by back scattering. Battery-powered tags allow for a system is well designed for long-range RFID applications. This air interface description does not explicit claim for battery assistance in the tag, also real passive tags or tags for mixed operation are conceivably.

9.3.5 18000-5 Part 5: Parameters for Air Interface Communications at 5.8 GHz

This standard is to provide physical layer, anticollision system and protocol values for RFID systems for item identification operating at 5.8–5.9 GHz in accordance with the requirements of ISO 18000-1. This standard provides parameter value for each MODE determined in the requirements section below.

In this version of the standard, two non-contending MODES are defined:

- None of the MODES are interoperable
- All of the MODES whilst not interoperable are non-contending

9.3.5.1 Intellectual Property

Some of the MODES require a license from the owner of the IP, which shall be available on terms in accordance with ISO policy. Details of IP are shown at the end of each MODE.

MODE 1 in this version of the standard may be used without a license.

This part has been withdrawn due to lack of global acceptance. The physical, anticollision, and transmission protocols determined in this MODE are consistent with the approach taken in ISO/IEC 15693. This section provides the normative part of MODE 1 by reference.

The physical layer for the MODE 1 air interface at 5.8 GHz shall be consistent and compliant to CEN 12253. The data link and MAC layers shall be compliant to CEN 12795.

The anti-collision system and protocols for the MODE 1 air interface at 5.8 GHz shall be consistent and compliant to CEN 12834.

9.3.6 18000-6 Part 6: Parameters for Air Interface Communications at 860–930 MHz

This standard describes the following:

- The physical interactions between the interrogator and the tag
- The protocols and the commands
- The collision arbitration schemes

There are four types of protocols; Type A and Type B, Type C and Type D. For the forward link, Type A uses pulse interval encoding, Type B uses bi-phase modulation and Manchester encoding. For the collision arbitration, Type A uses an Aloha-based mechanism; Type B uses an adaptive binary tree mechanism. Both types use the same bi-phase space FM0 return link encoding. Type C is based on the EPC global Gen 2 protocol and Type D will be defining tag only talks after listen (TOTAL) interrogator and tag behavior.

9.3.7 18000-7 Part 7: Parameters for Air Interface Communications at 433 MHz

Information technology AIDC techniques—RFID for item management—air interface, Part 7—parameters for an active RFID air interface communications at 433 MHz. This standard is intended to address RFID devices operating in the 433 MHz frequency band.

At the time of publication, ISO/IEC 18000-7 was the only RFID air interface protocol standard operating with a single mode (or type).

9.3.7.1 DASH7 Alliance

The DASH7 Alliance (www.dash7.org) was formed in 2009 to advance the use of DASH7 wireless data technology by developing extensions to the ISO 18000-7 standard, ensuring interoperability among devices, and educating the market about DASH7 technology. This consortium is continuously growing: manufacturers, systems integrators, developers, regulators, academia, and end users are all working together to promote the use of DASH7 technology in a wide array of industries and applications. We should expect many new developments for this ISO active RFID air interface protocol standard in the coming years as adoption accelerates and end users around the world embrace a proven technology used in different environments from human friendly to harsh and chaotic.

Finally, the WG-3 defines standard conformance.

9.4 WORK GROUP ON RFID FOR ITEM MANAGEMENT (WG 4)

This work group's purpose is to provide standards for interoperability of wireless, noncontact omnidirectional RFID devices capable of receiving, storing, and transmitting data while operating at power levels that are in freely available international frequency bands in the area of item level identification and management across the supply chain such as finished good asset management, raw material asset management, material traceability, inventory control, electronic article surveillance, warranty data, production control/robotics, and facilities management. The proposed RFID item management work would align without duplicating and co-exist with the approved work of other international standards committees. It is their intent to utilize the prevailing standards, by normative reference, where appropriate.

WG 4's subgroup (SG) 1 is responsible for the following:

- ISO/IEC 15961 data protocol—application interface
- ISO/IEC 15962 data protocol—data encoding rules
- ISO/IEC 15963: Unique identification for RF tags
- ISO/IEC 24791 Software system infrastructure(six parts)
- ISO/IEC 29167 Air interface for file management and security services for RFID
- ISO/IEC 24753: Application protocol: encoding and processing rules for sensors and batteries

WG 4's SG3 is responsible for

- ISO/IEC 18000 series air interface standards at <135 kHz; 13.56 MHz; 433 MHz; 860–960 MHz; 2.45 GHz.

Application standards illustrate how products are to be used, such as where to place a label.

Conformance standards provide instructions on how a specific device is to be evaluated to ensure it complies with a standard.

9.4.1 ISO Standards Summary

The ISO standards related to RFID include

- ISO 11784, 11785, 14223: RFID standards for animal tracking. 14223 is the air interface standard
- ISO 10536, 14443, 15693: RFID standards for smart cards
- ISO 10374: RFID for rail and ship freight containers
- ISO 15961, 15962, 15963: RFID for item management
- ISO 18000 series: Cover both active and passive RFID technologies
  - 18000-1 Part 1: Generic Parameters for the Air Interface for Globally Accepted Frequencies
  - 18000-2 Part 2: Parameters for Air Interface Communications below 135 kHz
  - 18000-3 Part 3: Parameters for Air Interface Communications at 13.56 MHz
  - 18000-4 Part 4: Parameters for Air Interface Communications at 2.45 GHz
  - 18000-5 Part 5: Parameters for Air Interface Communications at 5.8 GHz (Withdrawn)
  - 18000-6 Part 6: Parameters for Air Interface Communications at 860–930 MHz
  - 18000-7 Part 7: Parameters for Air Interface Communications at 433 MHz
- ISO 15418, 15434, 15459, 24721, 15961, 15962: Data content
- ISO 18046, 18047 (RFID device conformance tests methods for active/passive)
- Application standards: ISO 10374, ISO 18185, 11785
  - Recently, ISO TC104 (freight containers committee) has witnessed increased activity with a clear interest from Asian national standards bodies, in particular, to improve the status quo on e-seals (ISO 18185) and on container security and sensing devices
- ISO/IEC TR 24710: RFID for item management
- ISO/IEC TR 18047-7: Information technology—RFID device conformance test methods—Part 7: Test methods for active air interface communications at 433 MHz
- ISO/IEC 15434: Information technology—Syntax for high-capacity automatic data capture (ADC) media
- New work items are also in progress in the following committees:
  - SC31/WG6 (mobile item identification management). The scope of this committee is standardization of automatic identification and data collection techniques that are anticipated to be connected to wired or wireless networks, including sensor specifications, combining RFID with mobile telephony, and combining optically readable media with mobile telephony.
  - SC31/WG7 (security for item management). The scope of this committee is to provide standards and a framework for security of AIDC systems, particularly the air interface and other SC31 wireless communications components. This group will further define appropriate secure file management techniques for various memory sizes and configurations. These devices would be employed in the area of item level identification and management across the scope of SC31. Covering risk identification, management, and mitigation, the work group will identify risks and potential controls and deliver a suite of solutions that enable the implementation of various tiers of security for item management.

9.5 EPC GLOBAL STANDARDS

The foundation for the EPC global RFID passive tag is the identifier that is called the electronic product code (EPC). EPC is a joint venture between GS1 and GS1 US. It is an organization set up to achieve global adoption of the EPC RFID standard. We provide a brief history of the EPC global, Auto-ID Center, GS1, and GS2 organizations.

Electronic Product Code Type 1			
01–0000A89-00016F–000169DC0			
Header	EPC manager	Object class	Serial number
8 bits	28 bits	24 bits	36 bits

FIGURE 9.2 EPC tag data content.

EPC global was formed in October, 2003 from the confines of the Auto-ID Center.

The Auto-ID Center was founded by David Brock and Sanjay Sarma and supported financially by companies such as Procter and Gamble, and Gillette. The Uniform Code Council (UCC) was also a main supporter of the Auto-ID Center. The center created a global passive RFID-based item identification system that may eventually replace the bar codes. This code that was developed is called the EPC (Figure 9.2).

9.6 GS1 AND GS1 US

The global standards organization, which in includes GS1 and GS1 US, seek to provide member organization a common data structure for information collected and manipulated using ADC technologies. GS1, previously known as EAN International, managed the European Article Number product identification structure, which was commonly used as a supply chain identifier in Europe. GS1 US™, previously known as the Uniform Code Council, Inc. (UCC), managed the UCC identification structure, mainly in the United States. The bar code was often referred to as the UCC-128 bar code. The protocols have been used for over a decade to enhance data identification throughout the supply chain.

The combining of these structures and management of the protocol allow more global coordination. GS1 groups these activities into the GS1 system. They support the following areas:

- Bar codes: Numbering and bar coding
- eCom— Electronic Data Interchange (EDI)
- Global data synchronization network (GDSN)—data synchronization
- EPC global—RFID

They continue to support this standardization effort through their global standards management process (GSMP). GS1 is governed by a management board composed of key leaders from multinational firms, retailers, manufacturers, and other GS1 member organizations. EPC global Inc. and GS1 GDSN Inc. have separate management boards.

Using RFID, a tag "communicates" its number to a reader. The reader then passes the number to a computer or local application system, known as the object name service (ONS). ONS tells the computer systems where to locate information on the network about the object carrying an EPC, such as when the item was produced. Commonly, the physical markup language (PML) is used as a common language in the EPC global network to define data on physical objects. Initially, the Savant software technology was envisioned to act as the central nervous system of the EPC global network. Savant was designed to manage and move information in a way that does not overload existing corporate and public networks.

Currently, this function can be handled by edgeware or middleware software commonly provided by hardware vendors and distributors.

The EPC air interface protocol is intended to describe the following elements, including (1) air interface (waveforms of different symbols), (2) command set, and (3) operating procedure (how to use command set to identify/modify tags).

The air interface protocol describes how the reader talks with the tag, start-up signals, tree traversal negotiation, and command communication. The start-up signal sends radio frequency (RF)

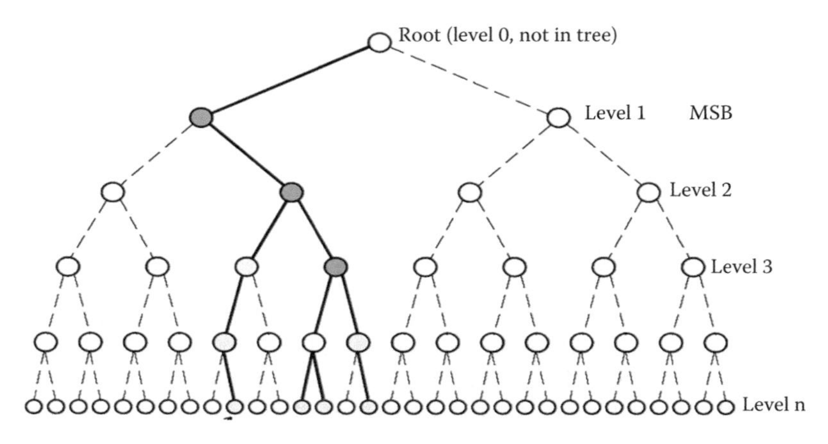

FIGURE 9.3 Air interface protocol.

signals to the tag population. The tree traversal negotiation is the process in which the tag backscatters or reflects the data and how the reader acknowledges that data map the path through the population. The command communication describes commands to retrieve data after confirmation that the tag has been identified (Figure 9.3).

9.7 EPC/GTIN INTEGRATION

Due to the fact that EPC global was a joint venture between EAN International and the Uniform Code Council in order to provide the standard for passive RFID tag nomenclature, the nomenclature is based on a family of bar codes. The GS1 family of bar codes is known as the global trade item number (GTIN), a worldwide system of supply chain identification. The basic UPC bar code extends to include country, classification, and product information (Figure 9.4).

9.8 EPC GENERATION 2

The EPC Generation 2 tag, which is now becoming the most commonly used tag, was developed and widely placed in use in mid-2006. It focused on supply chain customer requirements. The standard had some unique development items, which include that it was not based on a system from any

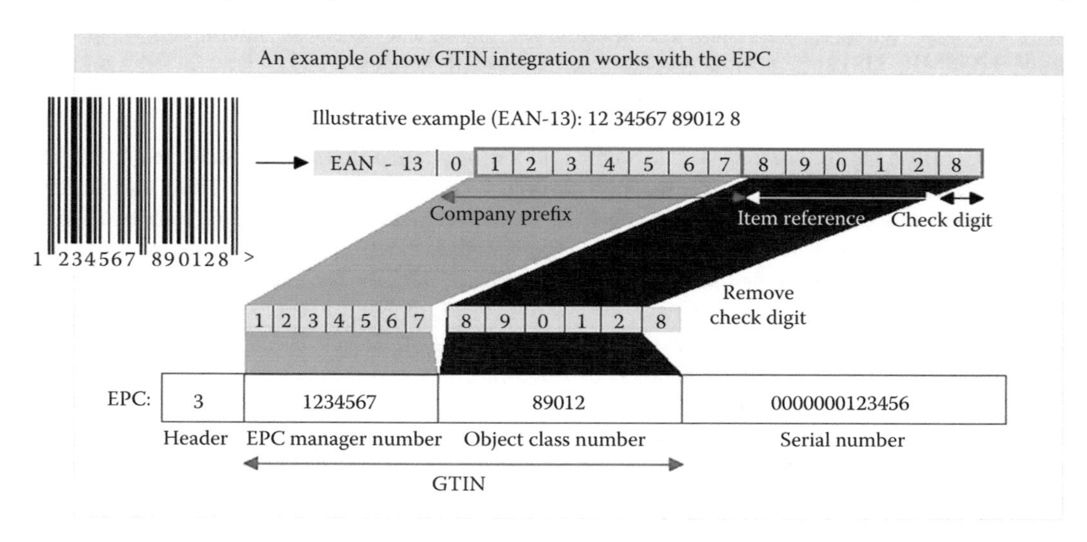

FIGURE 9.4 GTIN and EPC.

TABLE 9.1
EPC Tag Classes

EPC	Description	Functionality	Remarks
0	Read only	Passive tags	Data can be written only once during tab manufacturing and read many times
1	Write once and read only	Passive tags	Data can be written only once by tag manufacturer or user and read many times
2	Read/write	Passive tags	User can read/write data many times
3	Read/write	Semi-passive tags	Can be coupled with onboard sensors for capturing parameters like temperatures, pressure, etc.
4	Read/write	Active tags	Can be coupled with onboard sensors and act as radio wave transmitter to communicate with reader

one technology provider and it was developed by a broad consensus of chip, reader, and tag manufacturers. The improvements that the "Gen 2" tag was based on standardizing tags the goal was a common standard—not multiple competing standards (Table 9.1). The Gen 2 tags advantages over the first-generation passive RFID tags were

1. Open standard—available from multiple sources
2. Interoperable equipment
3. System performance exceeds requirements
4. High reliability—always detects items in the field
5. Global operation
6. Standard enables low-cost tags/readers
7. Protocol detects late arriving tags or intermittently powered tags
8. Tags can be permanently killed by a reader

Up until recently, EPC global has focused on passive RFID Gen 2 derived data. The massive amount of information collected from passive tags led to the development of the EPC global framework, an architecture that could efficiently support EPC data identification, capture, and exchange.

EPC global has also been evaluating existing active RFID protocols and attempting to define a standards strategy that meets the requirements of both the aerospace and defense as well as transportation and logistics services industry action groups. An adhoc committee was set up in 2009 in the HAG to perform this analysis and provide a recommendation to the EPC architecture and technical steering committees.

9.9 OTHER STANDARDS

It is envisioned the passive RFID standards will be integrated with other standards such as the U.S. Department of Defense's unique identifier (UID) standard. The UID is based on identification requirements for military products over $5000. Other standards are target to be integrated with passive RFID standards including other countries' EANs, animal identification standards, and smart cards, to name a few.

9.10 ELECTRONIC PRODUCT CODE DETAILS

An electronic data signal provides the unique identification for a passive tag that is received by an RFID reader. This signal is formatted into a consistent code referred to as the EPC. This code is similar to the standardized framework provided in a universal product code or UPC used in most bar-coding systems. The EPC structure consists of header, manager number, object class, and a serial number. The header identifies the length, type, and structure for the specific EPC standards version number. The manager number identifies the company or enterprise using the EPC number. The object class refers to the class or type of product similar to a stock keeping unit or SKU. The serial number is the specific instance of the object class being tagged. The 96 bit tag has additional fields that allow for more information and additional traceability.

The EPC identifier is a meta-coding scheme designed to support the needs of various industries by accommodating both existing coding schemes where possible and defining new schemes where necessary. The various coding schemes are referred to as domain identifiers, to indicate that they provide object identification within certain domains such as a particular industry or group of industries. As such, the EPC represents a family of coding schemes (or namespaces) and a means to make them unique across all possible EPC-compliant tags.

To better understand the overall framework of the EPC tag data standards, it is helpful to distinguish between three levels of identification: pure identity layer, encoding layer, and physical realization layer. The pure identity layer is associated with a physical or logical item. The encoding layer is the EPC specification. The physical realization is the RFID tag.

Four types of EPCs have been defined as follows:

- 96 bit
- 64 bit, type I
- 64 bit, type II
- 64 bit, type III

A 64 bit EPC may uniquely identify 16,000 companies with a range of 9–1 million classes of products and 33 million serial numbers within a class.

The 96 bit EPC may uniquely identify 268 million companies with 16 million classes of products and 68 billion serial numbers in each class.

Organization by effectively using software can associate RFID tags with other identification schemas. The schemas include the following:

- EAN.UCC Serialized Global Trade Item Number (SGTIN)
- EAN.UCC Serial Shipping Container Code (SSCC)
- EAN.UCC Global Location Number (GLAN)
- EAN.UCC Global Returnable Asset Identifier (GRAI)
- EAN.UCC Global Individual Asset Identifier (GIAI)
- A General Identifier (GID)

9.10.1 Serialized Global Trade Item Number

The SGTIN is a new identity-type based on the EAN.UCC GTIN. A GTIN by itself does not represent a pure EPC identity because it does not uniquely identify a single physical object. Instead, a GTIN identifies a particular class of object, such as a particular kind of product or SKU. The coding structure for the 96 bit tag is given in Table 9.2.

Unlike the GTIN, the SSCC is already intended for assignment to individual objects and therefore does not require any additional fields to serve as an EPC pure identity. Its specific coding structure for the 96 bit version is given in Table 9.3. A lower capacity, 64 bit version is also defined.

9.10.2 Serialized Global Location Number

A serialized GLN can represent either a discrete, unique physical location, such as a dock door or a warehouse slot, or an aggregate physical location, such as an entire warehouse. It may also represent

TABLE 9.2
96 Bit Tag Coding Structure

	Header	Filter Value	Partition	Company Prefix	Item Reference	Serial Number
SGTIN-96	8 0011 0000 (binary value)	3 (Refer to Table 5 for values)	3 (Refer to Table 7 for values)	20–40 999,999–999,999, 999,999 (max. decimal range[a])	24–4 9,999,999–9 (max. decimal range[a])	38 274,877,906,943 (max. decimal value)

[a] Max. decimal value range of company prefix and item reference fields vary according to the contents of the partition field.

TABLE 9.3
Serial Shipping Container Code (SSCC)

	Header	Filter Value	Partition	Company Prefix	Serial Reference	Unallocated
SSCC-96	8 0011 0001 (binary value)	3 (Refer to Table 9 for values)	3 (Refer to Table 11 for values)	20–40 999,999–999,999,999, 999 (max. decimal range[a])	38–18 9,999,999,999– 99,999 (max. decimal range[a])	24 (Not used)

[a] Max. decimal value range of company prefix and location reference fields vary according to contents of the partition field.

TABLE 9.4
SGLN 96 Bit Code

	Header	Filter Value	Partition	Company Prefix	Location Reference	Serial Number
SGLN-96	8 0011 0010 (binary value)	3 (Refer to Table 13 for values)	3 (Refer to Table 15 for values)	20–40 999,999–999,999, 999,999 (max. decimal range[a])	21–1 999,999–0 (max. decimal range[a])	41 2,199,023,255, 551 (max. decimal value) (not used)

[a] Max. decimal value range of company prefix and location reference fields vary according to contents of the partition field.

a logical entity, e.g., an organization that performs a business function, such as placing a purchase order. The coding structure for the 96 bit version is given in Table 9.4.

9.10.3 Global Returnable Asset Identifier

Unlike the GTIN, the GRAI is already intended for assignment to individual objects; therefore, it does not require any additional fields to serve as an EPC pure identity. The coding structure is given in Table 9.5.

9.10.4 Global Individual Asset Identifier

Like the GRAI, the GIAI is already intended for assignment to individual objects; therefore, it does not require any additional fields to serve as an EPC pure identity. The coding structure is shown in Table 9.6.

TABLE 9.5
GRAI Code Structure

	Header	Filter Value	Company Prefix Index	Asset Type	Serial Number
GRAI-64	8	3	14	20	19
	0000 1010 (binary value)	(Refer to Table 17 for values)	16,383 (max. decimal value)	999,999–0 (max. decimal range[a])	524,287 (max. decimal value)

[a] Max. decimal value range of asset type field varies with company prefix.

TABLE 9.6
GIAI Code Structure

	Header	Filter Value	Partition	Company Prefix	Individual Asset Reference
GIAI-96	8	3	3	20–40	62–42
	0011 0100 (Binary value)	(Refer to Table 21 for values)	(Refer to Table 23 for values)	999,999–999, 999,999,999 (max. decimal range[a])	4,611,686,018,427,387, 903–4,398,046,511,103 (max. decimal range[a])

[a] Max. decimal value range of company prefix and individual asset reference fields vary according to contents of the partition field.

9.10.5 Global Identifier (GID-96)

GID is a 96 bit coding scheme that does not use a pre-existing standard. The GID is composed of three fields—the general manager number, object class, and serial number. Encoding of the GID includes a fourth field, the header, to guarantee uniqueness in the EPC namespace (Table 9.7).

Serialization, the ability to uniquely identify objects, is the key benefit to using the EPC standards. This serialization allows for item-level tracking and security against counterfeiting.

TABLE 9.7
GID 96 Code Structure

	Header	General Manager Number	Object Class	Serial Number
GID 96	8	28	24	36
	0011 0101 (binary value)	268,435,455 (max. decimal value)	16,777,215 (max. decimal value)	68,719,476,735 (max. decimal value)

TABLE 9.8
Summary of EPC Classes

Class	Frequency	Description
0	900 MHz	Read-only (0) Read/write (0+)
1	13.56 MHz ISM band	Write once read many (WORM) Read/write
1	860–930 MHz (UHF)	WORM Read/write

Anticounterfeiting is one of the main advantages that RFID poses over bar-coding. Another important component of the EPC standard includes the air interface and command set, often called the specification protocol. They are often listed as classes. Table 9.8 is a summary of the EPC classes.

9.11 DEPARTMENT OF DEFENSE UID

The Department of Defense (DOD) vision for RFID is to utilize RFID to facilitate accurate, hands-free data capture in support of business processes in an integrated DOD supply chain enterprise as an integral part of a comprehensive suite of automatic identification technology (AIT). The key to future functionality of the unique item data in the DOD supply chain will be the ability to temporarily associate "conditional state" information about an item.

DOD expects to fully embrace the use of EPC technology as well as approved EPC tag data constructs in a supporting DOD data environment. The efficiencies of RFID provide another valuable component of the suite of AITs. Active RFID has already improved the ability to track and trace material through the supply chain. Combining the passive RFID technology will create greater efficiencies and data accuracy. Leveraging RFID to the fullest extent possible will improve the ability to get the warfighter the right material, at the right place, at the right time, and in the right condition.

DOD definitions apply to passive RFID technology and tags in support of the DOD requirement to mark/tag material shipments to DOD activities in accordance with the DOD RFID policy.

EPC technology: Passive RFID technology (readers, tags, etc.) that is built to the most current published EPC global Class 0 and Class 1 specifications and that meets interoperability test requirements as prescribed by EPC global™. EPC technology will include ultra-high frequency generation 2 (UHF Gen 2) when this specification is approved and published by EPC global™.

UID unit pack: An MIL-STD-129 defined unit pack, specifically, the first tie, wrap, or container applied to a single item, or to a group of items, of a single stock number, preserved or unpreserved, which constitutes a complete or identifiable package.

Bulk commodities: These items shall not be tagged in accordance with passive RFID tagging requirements. Bulk commodities are products carried or shipped in rail tank cars; tanker trucks; other bulk, wheeled conveyances; or pipelines.

In addition, munitions and explosives shall not be tagged until the following certification requirements are met for the passive RFID tag: electromagnetic effects on the environment (E3), hazards of electromagnetic radiation to ordnance (HERO), hazards of electromagnetic radiation to fuel (HERF), and hazards of electromagnetic radiation to personnel (HERP).

Case: Either an exterior container within a palletized unit load or an individual shipping container.

Exterior container: An MIL-STD-129 defined container, bundle, or assembly that is sufficient by reason of material, design, and construction to protect unit packs and intermediate containers and their contents during shipment and storage. It can be a unit pack or a container with a combination of unit packs or intermediate containers. An exterior container may or may not be used as a shipping container.

Shipping container: An MIL-STD-129 defined exterior container that meets carrier regulations and is of sufficient strength, by reason of material, design, and construction, to be shipped safely without further packing (e.g., wooden boxes or crates, fiber and metal drums, and corrugated and solid fiberboard boxes).

Palletized unit load: An MIL-STD-129 defined quantity of items, packed or unpacked, arranged on a pallet in a specified manner and secured, strapped, or fastened on the pallet so that the whole palletized load is handled as a single unit. A palletized or skidded load is not considered to be a shipping container.

RFID technology will be implemented through a phased approach, applied both to supplier requirements and DOD sites. Shipments of goods and materials will be phased in by procurement methods, classes/commodities, location and layers of packaging for passive RFID.

TABLE 9.9
Acceptable RFID Tag Data Constructs for Supplier Originated Shipments

Class	User Memory Size (Bits)	Tag Data Construct	Controlling Organization	Requires EPC Global Membership to Use Construct?
0	64	SGTIN-64 GRAI-64 GIAI-64 SSCC-64	EPC	Yes
0	64	DOD-64	DOD	No
1	64	SGTIN-64 GRAI-64 GIAI-64 SSCC-64	EPC	Yes
1	64	DOD-64	DOD	No
0	96	SGTIN-64 GRAI-64 GIAI-64 SSCC-64	EPC	Yes
0	96	DOD-96	DOD	No
1	96	SGTIN-64 GRAI-64 GIAI-64 SSCC-64	EPC	Yes
1	96	DOD-96	DOD	No

For DOD suppliers, RFID can be implemented using two options:

- EPC global tag data construct
- DOD tag data construct

Suppliers that are EPC global™ subscribers and possess a unique EPC manager number may choose to use an EPC tag data construct to encode tags per the rules that follow. Suppliers that choose to employ the DOD tag data construct will use the commercial and government entity (CAGE) code previously assigned to them and encode the tags per the rules that follow. Regardless of the selected encoding scheme, suppliers are responsible for ensuring that each tag contains a UID.

Table 9.9 indicates the acceptable tag data constructs and the relationships between the various combinations of tag class, size, data construct, and the organization that controls the data construct.

Based on your membership in EPC global, select either an EPC global tag data construct option or a DOD tag data construct option and proceed to the corresponding following section for details on how to encode RFID tags using the selected option.

9.12 EPC GLOBAL TAG DATA CONSTRUCT OPTION

This option should be selected by a DOD supplier that is

- Already a member of EPC global and has an assigned company prefix
- Intends to join EPC global and obtain a company prefix

This company prefix is required for encoding of all RFID tag classes and sizes. Table 9.10 summarizes the selection of an encoding scheme for either 64 or 96 bit tags based on the type of object

TABLE 9.10
Selecting the Proper Tag Data Construct

Tag Requirement	EPC Data Construct	When Used
UID unit pack	SGTIN-64SGTIN-96	On item packaging for items meeting the DOD criteria for assignment of UID where a serial number is used to augment a GTIN, which is used for the unique identification of trade items worldwide within the EAN.UCC system
	GRAI-64GRAI-96	On item packaging for items meeting the DOD criteria for assignment of UID (reusable package or transport equipment of specific or certain value)
	GIAI-64GIAI-96	On item packaging for items meeting the DOD criteria for assignment of UID (used to uniquely identify an entity that is part of the fixed inventory of a company—GIAI can be used to identify any fixed asset of an organization)
Case, transport package, palletized unit load	SGTIN-64SGTIN-96	Items shipped as either pure case or pallet (see above)
	SSCC-64SSCC-96	Items shipped as either pure or mixed case, or pallet. (SSCC can be used by all parties in the supply chain as a reference number to the relevant information held in computer database or file)

TABLE 9.11
FCC Limits for Maximum Permissible Exposure

Frequency Range (MHz)	Electric Field Strength (E) (V/m)	Magnetic Field Strength (H) (A/m)	Power Density (S) (mW/cm^2)	Averaging Time $\lvert E \rvert^2$, $\lvert H \rvert^2$ or S (min)
(a) *Limits for occupational/controlled exposure*				
0.3–3.0	614	1.63	(100)[a]	6
3.0–30	1,842/f	4.89/f	$(900/f^2)$[a]	6
30–300	61.4	0.163	1.0	6
300–1,500	—	—	f/300	6
1,500–100,000	—	—	5	6
(b) *Limit for general population/uncontrolled exposure*				
0.3–1.34	614	1.63	(100)[a]	30
1.34–30	842/f	2.19/f	$(180/f^2)$[a]	30
30–300	27.5	0.073	0.2	30
300–1,500	—	—	f/1,500	30
1,500–100,000	—	—	1.0	30

[a] Plane-wave equivalent power density.
Note: f, frequency in MHz.

TABLE 9.12
FCC Limits for Localized (Partial Body) Exposure

Specific Absorption Rate (SAR)	
Occupational/Controlled Exposure (100 kHz–6 GHz)	General Uncontrolled/ Exposure (100 kHz–6 GHz)
<0.4 W/kg whole-body	<0.08 W/kg whole-body
≤8 W/kg partial-body	≤1.6 W/kg partial-body

being tagged and its usage. In general, the DOD is integrating the RFID passive technology with the UID standard that exists for item-level tracking.

9.13 FCC PART 15 RADIATION REGULATION

Though standards are important, the regulation of the RFID technologies has been more localized at the state level and focused on privacy issues, e.g., the California anti-RFID legislation. A safety regulation that relates to RFID and human safety is given by the FCC as part 15. Part 15 of the FCC's rules for low-powered devices affects RFID technologies. RFID devices are referred to as "intentional radiators." These low-powered devices do not raise a serious threat of interference with other devices and hence can be operated without a license. All the same, RFID devices have to meet the RF emissions limitations and power restrictions as laid down by the FCC.

> Such intentional radiators as RFID need to obtain certification from the FCC. Obtaining the certification requires an application containing legal information about the device and the filing party; a technical report that includes RF test results; a block diagram of the instrument; and an explanation of the manner in which the instrument complies with FCC regulations. Intentional radiators operating at different frequencies are governed by different rules as per part 15. RFID products using the UHF 902–928 MHz band has to comply with rules in Section 15.247. The section stipulates that the systems should employ a frequency-hopping spread spectrum modulation technique so as to derive the maximum reader transmitted power allowances. UHF readers are allowed to operate at a maximum power of 1 watt and can go up to 4 watts if they have a directional antenna and hop across at least 50 channels (*RFID gazette*, 2005).

The FCC has rules for human exposure to RF energy, as given forth in several of its bulletins. Bulletins 56 and 65 outline the amount of power a person can absorb in both a magnitude and a rate. It also provides limits for maximum electric and magnetic field exposure with limits on average exposure during any given time. Since RFID devices are low power, it is highly unlikely that any of these limits will be reached in a zone, except when a person may be standing in front of an antenna that is radiating at full power. The power output can be measured for each individual antenna to ensure that no excessive exposure will occur. The FCC limits are summarized in Tables 9.11 and 9.12 and found at http://www.fcc.gov /Bureaus /Engineering Technology/Documents/bulletins/oet56/oet56e4.pdf.

10 UID and RFID Standards for Military

10.1 INTRODUCTION

The Department of Defense (DOD) regards unique identification (UID) and radio frequency identification (RFID) as tools that focus on connections, marking, tracking, tracing, and managing inventory (DODRFID summit). While these two programs serve the same purposes, they are distinct programs. This chapter will explore UID, RFID, and the implementation of these programs.

10.2 UID

In 1998, the General Accounting Office (GAO) documented concerns with the DOD's management of its inventory of equipment. GAO found that the DOD's inventory exceeded its war reserve or current operating requirements but lacked key spare parts based on inadequate accountability on material shipments and ineffective monitoring of defective spare parts.

The DOD needed a way to identify tangible assets individually, which would be globally unique and unambiguous, have the ability to ensure data integrity and data quality throughout life, and support multifaceted business applications and users. This approach is called UID (http://uidforum.com/overview.html).

UID became mandatory for all solicitations for material issued on or after January 1, 2004. Further refinements in the UID policy established the requirement to apply UID to all existing personal property items in inventory, operational use, and all government-furnished property.

UID requires the placement of an ISO/IEC 16022 compliant two-dimensional (2D) data matrix bar code with data encoded in ISO/IEC 15434 format on every item the DOD acquires as an end item costing over $5000.00, as well as those embedded items, components, or subassemblies that are serially managed by the DOD; critical items; and items that are spared/repaired by the department (http://uidforum.com/overview.html).

UID is envisioned as a program that enables improved access to information about DOD equipment to make acquisition, repair, and deployment of items faster and more efficient. It also aims to achieve

1. Reduce lower life cycle cost of item management
2. Provide item visibility regardless of platform or "owner," supply item data necessary for top-level logistics and engineering analysis
3. Provide an accurate source for property and equipment valuation/accountability
4. Improve access to historical data for use during systems design and throughout the life of an item
5. Provide better item intelligence for the warfighter for operational planning
6. Reduce workforce burden through increased productivity and efficiency

These many benefits are what started the DOD on the path to UID.

However, there are several difficulties involved with the implementation of item unique identification (IUID), which the department required. The initial serialization schemes for the DOD's inventory were concurrent, or otherwise resulted in several items having the same serialized number. In general, the new items had to follow the updated rule whereas the legacy items had to be retrofitted as discussed earlier. Contract receipt and acceptance item identification data did not always match maintenance data, updates to the IUID registry using maintenance data keys could not be linked to UII records in some cases, and physically marked data sometimes differed from system keys used and may not have been captured in existing systems. Life cycle management traceability could not be achieved until these issues were resolved (RFID Summit Lord UID.pdf).

10.3 RFID

RFID includes technologies that use radio frequency identification and their associated standardized technologies to identify items automatically. Using this technology, the DOD will have full automated visibility of supplies. Items will be automatically entered into the inventory upon receipt, with no person needed to scan bar codes or to input data. It will move the supply chain toward optimization and provide the asset visibility support (DCMAC-JP RFID DOD UIDTask8).

10.3.1 Passive RFID

DOD requirements specify in detail the placement of RFID labels upon items that require RFID labels in their MIL-STD-129 document (current version P change 4). The DOD requires EPCglobal Generation 2, Class 1 RFID tags, a.k.a. Gen 2, which comply with ISO/IEC 18000-6C specifications for all requirements. EPCglobal is an industry organization responsible for setting standards for RFID use. EPCglobal is a wholly owned subsidiary of GS1. The previous efforts of GS1 are now a visible part of our everyday life—the UPC bar code. Suppliers that are EPCglobal subscribers and possess a unique GS1 company prefix number may choose to use the EPC tag identity type (formerly data construct) to encode tags.

The DOD specifies that it is the responsibility of the supplier, with whom the department holds the contract, to ensure that every RFID tag the supplier ships to the department is encoded with a globally unique identifier regardless of the selected tag encoding scheme. These unique EPC data labels are then linked through one of three methods (paper form, entering directly into the database, or through a preferred method for file submission) to an IUID with the DOD. This allows the DOD to use their IUID in all cases, even though the 96-bit EPC standard tags cannot always store the full IUID.

The "serial number" required in the passive RFID tag ID identifier does not refer to the serial number of the product being shipped. The "serial number" in the passive RFID tag ID is merely a unique number assigned by the supplier. This tag ID is associated with a specific case or pallet. This "serial number," combined with the supplier's government-managed identifier (or CAGE code), and header values comprise the RFID tag ID. Identifiers are comprised of four parts: header, filter, government-managed identifier, and serial number.

The 96-bit EPC standard data structure is also specified as the DOD-96 identifier, and is detailed as follows (from the DoD_suppliers_passive_RFID_guide_V13.pdf).

10.3.1.1 DOD-96 Identifier

DOD's expiration date ("sunset date") of February 28, 2007 for EPC Gen 1 Class 0 and Class 1 Specification tags has passed. Therefore, all 64-bit encodings (including all encodings that used 2-bit headers) are obsolete, and must not be used in new applications. The 96-bit tag is comprised of four fields as indicated in Table 10.1.

The details of what information to encode into these fields are explained below. After all, the field values have been determined; the entire contents of the tag can be viewed as a single unique number used to identify a shipment to the DOD.

10.3.1.2 Data Fields

The header specifies that the tag data are encoded as a DOD 96-bit tag identity type (use "2F" encoded in binary as 0010 1111). Detailed in Table 10.2 are the only headers accepted by the DOD, along with their corresponding binary codes.

TABLE 10.1
DOD-96 Identity Type Format

Header	Filter	Government-Managed Identifier	Serial Number
8 bits	4 bits	48 bits	36 bits

TABLE 10.2
Accepted Headers

Hexadecimal Header	8-Bit Binary Header	Identifier
2F	00101111	DOD-96
30	00110000	SGTIN-96
31	00110001	SSCC-96
33	00110011	GRAI-96
34	00110100	GIAI-96
35	00110101	GID-96
36	00110110	SGTIN-198
37	00110111	GRAI-170
38	00111000	GIAI-202

The most recent version of the EPC tag data standard at http://www.epcglobalinc.org/standards/tds/ also details more information about acceptable headers.

The filter identifies a pallet, case, or UID item associated with a tag (0000 = pallet, 0001 = case, 0010 = UID item, all other combinations = reserved for future use). This value is not part of the EPC code, so no two codes are allowed with only the filters being different.

The government-managed identifier, DOD automatic address code/commercial activity government entity (DODAAC/CAGE) identifies the supplier, ensures uniqueness of serial number across all suppliers, and is represented in ASCII format. This ensured that the prefix for GS1, which supports the EPC tag data structure, and the DODAAC, and the CAGE prefixes would be the same length. For CAGE codes, an ASCII, a space character must be placed in front of the CAGE to make a total of six ASCII characters. The serial number uniquely identifies up to 2^{36} = 68,719,476,736 tagged items, represented in binary format.

Table 10.3 shows an example for the four data fields. Notice that this operates as a unique serial number, and is not the specific IUID that will represent the items in the DOD's database. This information and all other information about what is included in a shipment must be sent to the department separately with an advance shipment notice.

RFID numbers are reported and recorded in a hexadecimal format. (Numeric base 16, with characters 0,1,2,3,4,5,6,7,8,9,0,A,B,C,D,E,F.) RFID software converts the 96-bit binary code into a

TABLE 10.3
DOD-96 Identity Example

Header (DOD identity type)	0010 1111
Filter (pallet)	0000
CAGE (2S194)	0010 0000 0011 0010 0101 0011 0011 0001 0011 1001 0011 0100
Serial number (12,345,678,901)	0010 1101 1111 1101 1100 0001 1100 0011 0101

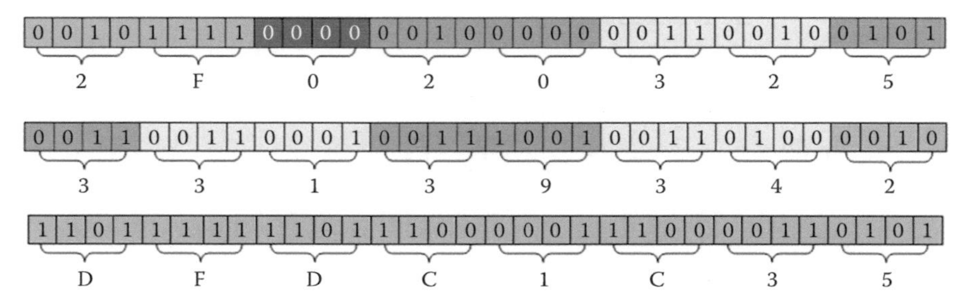

FIGURE 10.1 Binary to hexadecimal conversion.

hexadecimal code to aid in the speed of entering data. The process is demonstrated for our sample code in Figure 10.1.

The result is a unique hexadecimal code that can be written to the tag: 2F02032533139342DFDC1C35. The binary serial number generated by the computer must be converted to a hexadecimal code to be entered into wide area work flow (WAWF). This conversion is normally made by the software when the RFID tag is read. The hexadecimal RFID tag code is also the value displayed on RFID readers and scanners.

10.3.2 Active RFID

The DOD has been using active RFID tags to track shipments since 1994. Currently, the DOD's in-transit visibility RFID network is the world's largest active RFID cargo tracking system, with participation at over 1500 installations—airports, seaports, consolidation ports, and railheads. Active RFID tags on conveyances provide near real-time visibility in over 45 countries, with 37,000 conveyances tracked daily.

The DOD policy for active RFID tags also includes any contractor shipments directly to overseas destinations, but it is currently the responsibility of the procuring activity to arrange for the vendor to apply the active tags (DCMAC-JP RFID DOD UIDTask8).

10.4 IMPLEMENTATION

The cost of implementing and operating RFID technology is considered a normal cost of business. If DOD customers desire the inclusion of a passive RFID tag on shipments for these type purchases, this requirement must be specifically requested of the shipping supplier/vendor and the shipment must be accompanied by an appropriate advanced shipment notification (ASN) containing the shipment information associated to the appropriate RFID tag.

All solicitations awarded on or after October 1, 2004 for delivery of material on or after January 1, 2005 will require that passive RFID tags be affixed at the case, pallet, and UID item packaging level for material delivered to the DOD, in accordance with the implementation plan, which is located above under the section entitled: "Implementation Approach." The plan can also be found at http://www.dodrfid.org/supplierimplementationplan.htm

The DOD is implementing RFID in three phases, and is in Phase 2 as of July, 2007. Passive RFID tags on cases and pallets shipped to specified DOD receiving points, including all the defense distribution depots, for the following item supply classes:

Class I: Subsistence limited to packaged operational rations (meals ready to eat [MREs].) used to be called C-Rations or C-rats

Class II: Clothing, individual equipment, tentage, organizational tool kits, hand tools, and administrative and housekeeping supplies and equipment

Class III: Packaged petroleum fuels, lubricants, hydraulic and insulating oils, preservatives, liquid and gas, bulk chemical products, coolants, deicer and antifreeze compounds, components and additives of petroleum and chemical products and coal

Class IV: Construction material including installed equipment and all fortification and barrier material

Class VI: Personal demand items such as snack foods, beverages, cigarettes, soap, toothpaste, writing material, cameras, batteries, and other nonmilitary sale items

Class VIII: Medical materials (except pharmaceuticals, biologicals, and reagents)

Class IX: Repair parts and components including kits, assemblies and subassemblies, reparable and consumable items required for maintenance support of all equipment, excluding medical-peculiar repair parts

Pharmaceuticals, biologicals, and reagents are not required to have RFID tags at this time. The DOD will follow the FDA RFID rules for these materials when they are finalized. *Class V* (munitions/explosives) is also not required to have passive RFID tags at this time until compatibility tests have been conducted.

The DOD RFID Web site www.dodrfid.org has a Class of Supply Lookup Tool, which can be used to determine an item's supply class by using the national stock number (NSN).

Figure 10.2 shows the RFID tag placement as required in Phase 2. For the pallet tag, the requirements in MIL-STD-129P specify the tag to be placed 32–48 in. from the bottom of the pallet. It should not be placed directly on the pallet, since this may be too low for tag readability (DCMAC-JP RFID DOD UIDTask8).

The third phase of the RFID implementation plan is to include tags on the unit packs of items that fall under the UID requirements. The RFID requirements will then apply to all commodities shipped to all locations except for bulk commodities and explosives. Bulk commodities are defined as those that are shipped in rail tank cars, tanker trucks, trailers, other bulk wheeled conveyances or pipelines. These products include sand, gravel, bulk liquids (water, chemicals, or petroleum products), ready-mix concrete or similar construction materials, coal or combustibles such as firewood, and agricultural products such as seeds, grains, and animal feeds.

The two major requirements for suppliers are (1) passive tagging at the case, pallet, and case within pallet load; and (2) ASN. The RFID electronic data requirements are for a standard Ship Notice/Manifest EDI Transaction Set 856. This will enable the sender to describe the contents and configuration of a shipment in a format that can be recognized at the receiving end or by other interested parties with appropriate access to the data. Three sets of information are included: contract

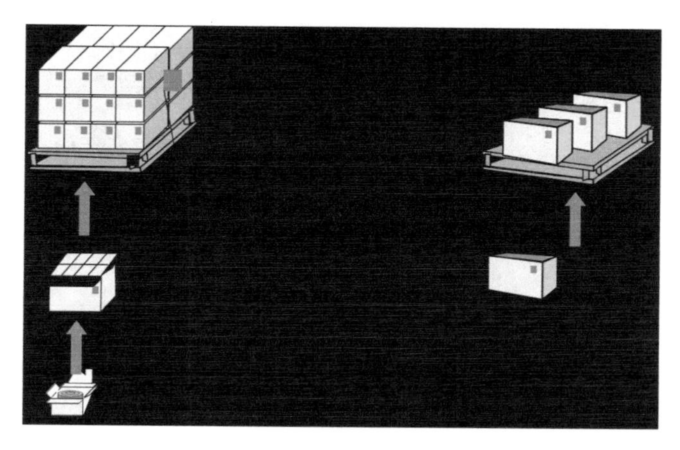

FIGURE 10.2 UID labeling diagram.

information, product description, and RFID tag data. The contract information includes the contract number, shipment number, prime contractor, and shipment date. The product description includes contract line item number, NSN, item description, and quantity. The RFID tag data include the RFID tag number, line item number, and quantity.

MIL-STD-129P, Change Notice 4, page 70, paragraph 4.9.4—Electronic data interchange (EDI) transactions are used to link the passive RFID tag to the content level detail information associated with each of the container types. Consignors are required to transmit these EDI transactions to consignees in advance of the shipment (DCMAC-JP RFID DOD UIDTask8).

Contractors must submit the ASN via the WAWF program. Most of the data required by an ASN are already being submitted to the DOD via WAWF. Contractors must be registered in the central contractor registry (CCR) to become registered to use WAWF.

Once the contractor enters and submits the RFID tag data into WAWF, they will be transmitted to the receiving depot's database, awaiting arrival of the shipment. When the shipment arrives and the RFID tags are read, the shipment will automatically be updated as received by the depot.

RFID antennas are mounted on either side of the truck dock area.

As shipping containers or pallet loads of boxes are moved past the antennas via forklift, RFID tags are automatically scanned and RFID tag data are submitted to the depot's computer system.

The DOD has also determined a nesting scheme to use the capabilities of the currently used active RFID tags with the mandated passive tags as described in Figure 10.3.

The grey rectangular box depicts a "freight container" that will be identified with an active RFID tag. The freight container could be a sea van or other enclosed intermodal container or a 463L pallet. The pallet loads that are consolidated into the freight container will have passive RFID tags that will link to content information for the containers that make up the pallet load.

Figure 10.4 shows the various stages of the transportation process, and the RFID interface they will use as part of the nested design. Manufacturers and suppliers use passive RFID, which then go to distribution centers and depots where shipments are organized and grouped with active RFID, which is used at both the port of entry (POE) and port of departure (POD). These shipments are

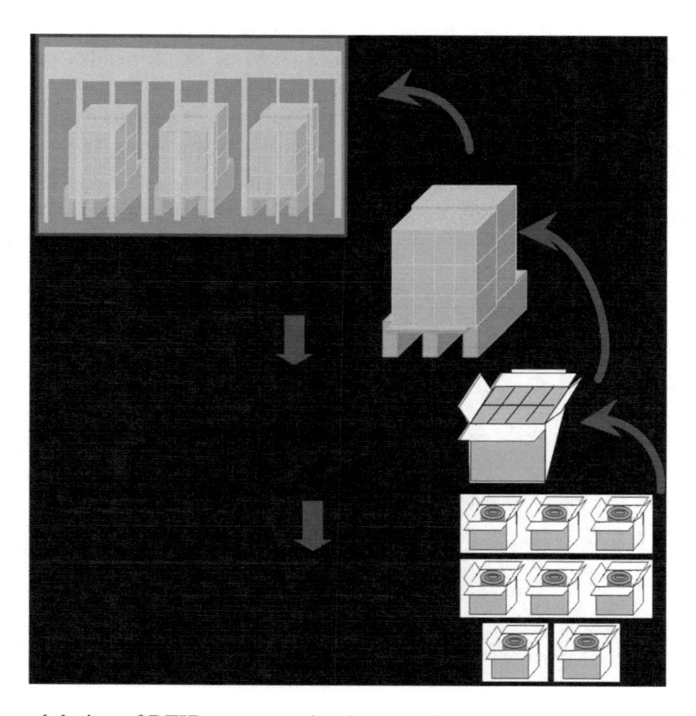

FIGURE 10.3 Nested design of RFID prototype implementations.

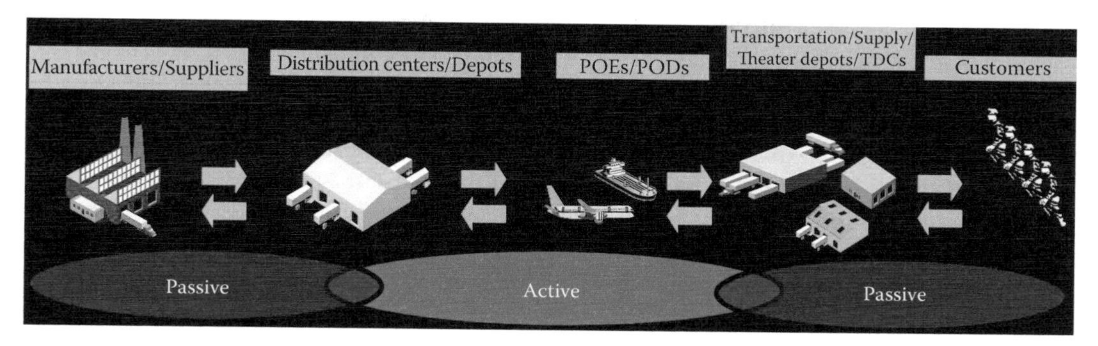

FIGURE 10.4 Diagram of active/passive utilization.

received at the theater or depot, and then broken down to the point where the passive RFID technologies are again in use (DCMAC-JP RFID DOD UIDTask8).

At the manufacturer/supplier locations and the depots, the item can be tracked to the shipping container level and pallet level using passive RFID. Active RFID is applied on large shipping containers to track the consolidated shipments during overseas transit. Passive RFID at the delivery point/customer location tracks the item from the individual container level to its destination.

10.4.1 Implementation Strategies

Various strategies may be employed to meet the DOD RFID requirements. Each supplier's best course of action will depend upon the amount of business with DOD or other customers requiring RFID. Suppliers will have a variety of options when they are looking for ways that they can meet the DOD RFID requirements. Suppliers may use third-party logistics (3PL) provider, purchase programmed tags and apply to cases/pallets, purchase equipment (i.e., printer or reader) to program tags, or incorporate a full RFID infrastructure throughout their business process.

Third-party logistics vendors are responsible for applying RFID-enabled tags, submitting ASN via WAWF, and maintaining uniqueness of the RFID number. 3PL vendors may also provide additional services such as packaging of material to meet contract requirements, transportation management, finished goods storage, and order fulfillment. Advantages of using 3PL vendors is that they should have the expertise to meet the contractual requirements, obtain quotes, and add costs to your unit price to the DOD. However, this may end up costing more than an internal capability in the long run. Companies also do not benefit from the RFID technology within company processes (DCMAC-JP RFID DOD UIDTask8).

Purchasing programmed tags and applying to cases/pallets is also called "Slap and ship." Vendors of these tags must certify tag data, uniqueness, and readability. Labels may contain the MIL-STD-129 markings (MSL) or they may be blank. This is recommended for businesses that are not shipping a significant number of cases to the DOD per year and do not intend to utilize RFID technology in-house. The advantage is a minimal cost to meet the mandate. All that is required of the supplier is to attach labels and provide a corresponding ASN. Costs are low at approximately $80–$5 per pre-programmed tag, and sources appear to be readily available by doing an Internet search using "DOD RFID." Tags verified as readable at the time of purchase may be used at a later date for shipments, and RFID tags may be purchased for several contracts as the RFID serial number is not contract specific. However, suppliers do not have their own quality check if the tags being applied are readable as the DFARS Clause mandates. WAWF data entry can also be time consuming for multiple tags. Some tag suppliers add linear bar codes to the tags, so that the contractor can scan the label with a wedge reader and input the hexadecimal tag serial number automatically, rather than keying in each number; although this will cost more, it is more accurate (DCMAC-JP RFID DOD UIDTask8).

For the case of purchasing equipment to program tags, RFID interrogators have an approximate minimum cost of $2500. Part of this price is that each interrogator is a small radio transmitter, which must be licensed by the Federal Communications Commission. Vendors often offer bundled packages (software and hardware) to meet the DOD requirements. Suppliers may be able to upgrade existing hardware (label printers) to program RFID tags. By printing their own labels, they may be able to obtain discounts on tags, control their own production, and may be able to flow the RFID tag numbers directly to the ASN information without manually inputting the data. However, suppliers do not gain full benefits of the RFID technology if they are simply meeting customer requirements. Minimal cost for an RFID-enabled printer appears to be at least $5000; non-programmed tags can be obtained for 30–50 cents per tag, and software for data exchange to supplier's computer system—$300–$1000, depending on the system. Using their own equipment also requires trained personnel, and requires that the supplier maintain uniqueness of tag data.

The remaining option is to incorporate a full RFID infrastructure throughout business process. This could include implementing a new full RFID from inbound receiving to outbound shipment or obtaining or modifying existing software and hardware to apply RFID technology. Suppliers that take this option are in a position to receive all the benefits of the RFID technology; increased production visibility, control, and accuracy. Automated build of shipment information and generation of ASN will also decrease the inefficiencies seen in the other options. However, this option is a significant investment in hardware and software (DCMAC-JP RFID DOD UIDTask8).

10.4.2 Replacement for Bar Codes

Traditional bar codes (Linear UPC) will remain the dominant auto-ID technology in most mainstream applications for the foreseeable future as it has the lowest cost, broadest applicability, and a huge infrastructure investment. 2D bar codes (PDF 417) have been adopted for value-added applications, portable data files, supplementary retail coding, etc.

RFID active, passive, and semi-passive systems will be increasingly adopted where non-line of sight, read/write, and multiple detection requirements are needed. Will bar codes still be required on cases/cartons? Yes. The requirements of human-readable, linear bar codes, and 2D military shipping labels have not been altered by the RFID initiative. These methods of marking must remain as a backup when the RFID can't be read (DCMAC-JP RFID DOD UIDTask8).

10.5 SUMMARY

UID and RFID are frequently confused for each other. Both are systems of identification, newly mandated by the DOD in the past few years. Table 10.4 is a quick guide to identify the differences between the two systems (DCMAC-JP RFID DOD UIDTask8).

This chapter explored UID, RFID, and the implementation of these programs. These distinct programs both serve the same purpose. Managing, tracing, tracking, marking, and connecting inventory are a focus of the DOD, and why they are a driving force behind the implementation of UID and RFID.

TABLE 10.4
UID vs. RFID

Unique Identification	Radio Frequency Identification
Item markings	Packaging markings
DFARS clause 252.211-7003	DFARS clause 252.211-7006
MIL-STD-130	MIL-STD-129
Linear bar code 39	RFID chip embedded in paper label
Data matrix 2D bar code	
Applied to items $5000 and over	Applied based on destination and item supply class
Used to identify items in various databases	Used to track packages in transportation

Part IV

Implementation, Decision Making, and Testing Approaches

11 RFID Project Management

11.1 INTRODUCTION

In many cases, military units will be given a step-by-step procedure for fielding an RFID system. This is usually in the form of a field manual or a service regulation that must be followed. However, in other cases, the RFID technician may find themselves with a broad directive to get the job done in an expedient and effective manner. This is particularly the case when RFID technology has not been previously applied to the logistical system under consideration. When this occurs, it is in the best interest of the adopter to take an organized, documentable approach to what needs to be done. In fact, the successful adoption of RFID technology could become the basis for many other like applications. Here are some considerations.

In adopting an RFID system, the temptation to follow the "slap and ship" approach can be overwhelming. While this approach may undoubtedly be successful for a few organizations, most organizations would do better otherwise. The unique aspects of each organization's process really demands additional RFID implementation project planning to help ensure success. The implementation of RFID also presents additional considerations above and beyond the normal project planning process. The fundamental reason behind the additional difficulties is that RFID systems can represent a dramatic change in how the organization functions. In situations like this, not only must the implementation be properly planned with respect to the project planning process, but the issue of new technology acceptance must also be addressed.

To put all of these issues in perspective, the RFID implementation project process will be divided into two separate sections. In the first section, we will briefly discuss the general principles of the project management process. In the section, we will provide specific guidance on planning the implementation of an RFID system for a number of different types of scenarios. Readers who are familiar with basic project management issues may wish to advance to the RFID project implementation–specific section.

11.2 RFID PROJECT SELECTION

Organizations contemplating implementing RFID projects may be doing so for two basic reasons. The first compelling reason to adopt RFID technology is as a response to a specific mandate. In recent years, both the U.S. Department of Defense and large retailers such as Walmart have mandated the use of RFID technology to enhance their logistical train. In cases such as these, there is but little choice for the organization to implement RFID technology as the mandate requires. However, for other organizations, the choice of implementing RFID technology may begin with the selection of a specific pilot project to gain experience with the technology. Depending on the success and utility of the project, RFID technology may be expanded into other areas. Yet, the question remains as to which specific RFID project should be selected for pilot purposes. To help resolve this question, the following section touches briefly on a few project management concepts associated with project selection.

11.2.1 Project Selection Models and Factors

In order to rationally and consistently select an RFID project, the organization must select what is known as a project selection model. The project selection model is a means by which the organization can rank competing processes for the application of RFID technology. Project selection models generally contain a set of factors. It is through evaluating these individual factors via the model that the organization selects the projects. The choice of factors is unique to the organization, but in general, many organizations utilize factors associated with

- Production issues
- Financial issues
- Personnel issues
- Marketing issues

The project selection models in which these factors are examined can be broadly classified as either nonnumeric or numeric models. Nonnumeric models as the name suggests, do not specifically utilize values to determine the ranking of projects. Numeric models, on the other hand, rely exclusively on values for the ranking of projects.

11.2.2 Nonnumeric Project Selection Models

Nonnumeric project selection models are generally older and simpler than numeric project selection models. However, this is not meant to imply that nonnumeric models are not necessarily useful. They should, however, be closely scrutinized to determine whether or not their use is appropriate or desired. Common nonnumeric project selection models include

- The sacred cow
- Operating necessity
- Competive necessity
- Comparative models

11.2.2.1 Sacred Cow

The basis for the sacred cow is that some high-level management individual has decided that it is appropriate for the organization to apply RFID technology to a particular process. Sacred cow projects are difficult to deal with since challenging ill-thought-out, unsuccessful project may be difficult. Often, the only way to terminate a sacred cow implementation is when the champion either leaves the organization or the champion's interest turns to other application areas.

11.2.2.2 Operating Necessity

The operating necessity model is based on the fact that the organization might have to adopt an RFID project in order to keep the organization functioning on a daily basis. This might occur in the case of internal tracking of manufactured assemblies. In order to prevent products from being assembled incorrectly, the organization may select one RFID project over another.

11.2.2.3 Competitive Necessity

The competitive necessity model is the nonnumeric project selection model that is most likely to be encountered in RFID applications. In fact, this type of model is not all that dissimilar to that associated with mandates. In other words, in order to continue to be competitive in a certain marketplace, the organization selects RFID projects according to what will allow it to survive. This means that an end-user, retail-based RFID application might take precedence over an internal manufacturing RFID project.

11.2.2.4 Comparative Models

Comparative project selection models are typically used to compare RFID projects that do not have directly comparable project selection factors. For example, one RFID project may have great significance to the production process. Another RFID project may be needed to properly fulfill outgoing orders. In order to select among these types of projects, an organization may appoint an evaluating committee. The responsibility of the committee is to progressively break down the various projects with respect to importance of the organization. The top-level screen may lump projects into not important, somewhat important, and very important. Each category is then looked into greater detail and is subsequently rescreened into additional categories. Eventually, only a very few projects are accepted as sufficiently important to be implemented.

11.2.3 Numeric Project Selection Models

In comparison to nonnumeric models, numeric project selection models are new and more complicated. Numeric models can be broadly classified as those that rely on profit-based data and those that require some sort of scoring mechanism. Simple profit-based models include

- Payback time
- Average rate of return

Note that as RFID projects do not necessarily make money, but save money, the following descriptions have been modified to make the general models more applicable to RFID projects.

11.2.3.1 Payback Time

Payback time selection models are based on the amount of time the project takes to recover the amount of capital invested in the project. The measure of performance is in years. This value is obtained simply by dividing the capital invested by the amount of money that the project is expected to save on an annual basis. The determination of the amount of investment is relatively straightforward. Similarly, the organization can determine the amount of money saved by examining savings in labor costs and error resolution. The equation for determining the payback time is

$$\text{Payback} = \frac{\text{Initial investment}}{\text{Savings}}$$

This means for an RFID project, which initially costs \$100,000, the payback period will be 4 years if, on an average, a savings of \$25,000 will be realized each year:

$$4 \text{ year payback} = \frac{\$100{,}000 \text{ Initial investment}}{\$25{,}000 \text{ Annual savings}}$$

With this model, the payback period or time is in the same units as the savings. For example, if the savings are projected on an annual basis, the payback time will be in years. Organizations will generally gravitate toward the selection of projects that have relatively short payback periods. Thus, projects that can be completed and that have relatively low initial investments or high savings will normally be selected.

11.2.3.2 Average Rate of Return

The average rate of return (ARR) model is used to determine which models yield the best investment. Sometimes, the ARR is compared to how much the organization may yield in comparison to investing the project funds outside of the organization. Generally speaking, those projects with the greatest ARR will be selected for implementation. The equation of ARR is

$$\text{ARR} = \frac{\text{Average savings}}{\text{Initial investment}}$$

In this case, our $100,000 RFID project would yield a 25% ARR:

$$25\%\ \text{ARR} = \frac{\$25{,}000 \text{ average savings}}{\$100{,}000 \text{ initial investment}}$$

A distinct disadvantage of both the payback and ARR models, as well as many other financial models is their inability to incorporate nonnumeric project selection considerations. Thus, these models could potentially only favor projects that look good on paper. No consideration is included to take into account issues such as competitive necessity.

To help address this lack of flexibility, numeric project selection models based on scoring methods were developed. As previously discussed, these models require the identification of scoring factors. These can include both nonnumeric and numeric scoring factors. In the RFID, context scoring factors could include, but are not limited to, issues such as the difficulty of the implementation, labor savings, and error reduction as a result of the implementation, and the probability of successful implementation. Common scoring models include

- Unweighted 0–1
- Unweighted scoring
- Weighted scoring
- Constrained weighted scoring

11.2.3.3 Unweighted 0–1

In the unweighted 0–1 scoring model, each candidate RFID application is scored as a 0 or 1 for each factor that is to be considered. The total score for all of the factors is then totaled and compared against all other RFID application candidates. Mathematically, the model appears as

$$\text{Project score} = \sum_{i=1}^{n} \text{Factor value}_i$$

where

i = 1–n factors used to evaluate the project
Factor value = 0 or 1 for each factor i

Note that this model has the distinct disadvantage of being only able to assign a 0 or a 1 to each factor. This model also suffers from the limitation that every factor is considered as equal importance. In reality, some factors may be more important for the organization than others. Due to these limitations, this model is not normally recommended for use.

11.2.3.4 Unweighted Scoring

To overcome the 0–1 limitation of the unweighted 0–1 model, the scoring model may be utilized. This model replaces the 0–1 value with a value on some scale between 0 and a top value. Typically, the scale will be between 0 and 5, 0 and 7, or 0 and 10. The summed value of the score will indicate how well the project candidate fulfills the project selection factor. This is mathematically represented in the following equation:

$$\text{Project score} = \sum_{i=1}^{n} \text{Factor score}_i$$

where
$i = 1\text{–}n$ factors used to evaluate the project
Factor score = 0–x for each factor i

This scoring model overcomes the 0–1 limitation, but it still considers each factor as of equal importance. To overcome this limitation, we have the weighted scoring model.

11.2.3.5 Weighted Scoring

The weighted scoring model includes both the ability to numerically score project selection factors, as well as consider that the project selection factors may be of different levels of importance. To incorporate this consideration, the weighted scoring model includes a factor weighting component. Normally, this component will be assigned a value between 0 and 1. It is multiplied by the project selection factor score and summed. This is illustrated in the following equation:

$$\text{Project score} = \sum_{i=1}^{n} \text{Factor weight}_i \times \text{Factor score}_i$$

where
$i = 1\text{–}n$ factors used to evaluate the project
Factor weight = 0–1 for each factor i
Factor score = 0–x for each factor i

While the weighted scoring model is a great improvement over the previous numeric scoring models, it does still suffer from one limitation. In some cases, it must be necessary for a given project to receive a minimum score on a particular project selection factor. If the project scores below the minimum value, some mechanism must be incorporated to ensure that the final project score is less competitive than projects that are otherwise completely qualified.

11.2.3.6 Constrained Weighted Scoring

The constrained weighted scoring model can be incorporated in two different manners. In the first implementation, only the specific factor for the project is given a 0 value if the factor score is below the limit and a 1 if it is above the limit. This is represented in the following manner:

$$\text{Project score} = \sum_{i=1}^{n} \text{Factor weight}_i \times \text{Factor score}_i \times \text{Factor constraint}_i$$

where
$i = 1\text{–}n$ factors used to evaluate the project
Factor weight = 0–1 for each factor i
Factor score = 0–x for each factor i
Factor constraint = 0 or 1 for each factor i

This implementation will normally yield a nonzero score even if the project is unacceptable in one or more factors. However, it is likely that the final score is significantly lower than other projects that do not have any unacceptable factor scores.

Another method of incorporating constraint into the above project selection model is to score any model that has any unsatisfactory factor scores as a complete 0. This would effectively eliminate any projects that are unsatisfactory in any manner. Mathematically, this could be represented as

$$\text{Project score} = \text{Factor constraint} \times \sum_{i=1}^{n} \text{Factor weight}_i \times \text{Factor score}_i$$

where

i = 1–n factors used to evaluate the project
Factor weight = 0–1 for each factor i
Factor score = 0–x for each factor i
Factor constraint = 0 if any factor i score is unacceptable, otherwise 1

The constrained weighted scoring model represents the most sophisticated of the reasonably easily implemented numeric project selection models. It has the advantage of being able to incorporate both subjective factors such as competitive necessity, as well as financial performance such as payback time and rate of return.

11.3 RFID PROJECT PARAMETERS

Provided that the organization has decided to properly proceed with the RFID project, a comprehensive project management plan must be developed. An RFID project plan will enable the project manager to properly manage the project with respect to time, cost, and technical performance. The time parameter refers to the schedule allocated to the RFID implementation project. The cost parameter is the budget associated with the project. Lastly, the technical performance refers to the ability of the project to meet the required needs.

The three project parameters are frequently depicted as a triangle with each parameter being represented by one side. The significance of modeling the project as a triangle is that any two of the three legs or parameters can seriously impact the outcome of the third. For example, if the project falls behind schedule and the same level of technical performance is required, in order to finish on schedule, the budget must be increased. Similarly, if the project needs to be completed before the scheduled finish and the same level of technical performance is required, the budget must also be increased. In both of these cases, if the budget is not increased, and the schedule must be maintained or finished early, the technical performance of the system must be compromised. This is a particularly dangerous situation, since the failure to achieve the proper level of technical performance may seriously compromise the original purpose of the project.

The same effect is realized with respect to the budget. If the budget for the project is reduced, the technical performance may have to be sacrificed. Alternatively, fewer resources could be assigned to the project resulting in a longer completion time.

As a final comment on the project parameter triangle, the effects of the technical performance leg can be considered. If the technical specifications of the system are constantly revised through engineering change orders, either the completion time or the budget must be increased. In such situations, sometimes both the time and the cost of the project are increased.

11.4 RFID IMPLEMENTATION LIFE CYCLE

Many projects such as RFID implementation projects follow a general project life cycle. In the case of technological implementations such as RFID systems, the cycle follows four phases. These include conceptual, planning, installation, and startup phases. These phases are illustrated in Figure 11.1.

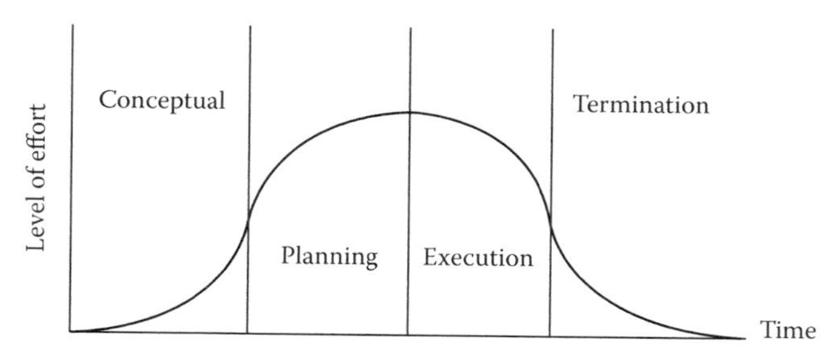

FIGURE 11.1 RFID project life cycle.

11.4.1 Conceptual Phase

In the conceptual phase, the organization is determining the basic objectives of the implementation project. This includes specifying what benefits the organization hopes to achieve by successfully executing the project.

11.4.2 Planning Phase

In the planning phase, the project manager is creating an outline for implementing the project. This includes as a minimum developing a work breakdown structure (WBS), a linear responsibility chart (LRC), and a Gantt chart. The WBS is a division of project tasks into increasingly detailed activities named work packages. The LRC specifies who participates and to what degree for each work package. Lastly, the Gantt chart details the length and relationships between each work package. Each of these planning tools will be discussed in further detail.

11.4.3 Installation Phase

In the installation phase, the organization begins by identifying and acquiring all of the hardware and software necessary to implement the RFID project. The installation phase continues by physically positioning all of the hardware and software in place.

11.4.4 Startup Phase

In the startup phase, the organization is going through the process of testing and debugging the installed hardware and software. The startup phase is completed when the hardware and software is functioning properly. At this point, the project manager completely turns the project over to operational personnel.

11.5 RFID PROJECT MANAGER

The concept of the project manager cannot be discounted. As the primary individual responsible for the project, the project manager is largely responsible for the success or failure of the RFID implementation project. The success or failure of the project may depend on the cooperation of a variety of resources that the project manager may not necessarily have direct authority over.

11.6 RFID PROJECT MANAGER AUTHORITY

Authority is defined as the power to command others to act or not to act. With respect to project management, there are two different types of authority. The first type is de jure authority. This type of authority is that awarded by some official organizational document. De jure authority gives the

RFID project manager the legal power to acquire and utilize organizational resources. This type of authority would typically be held by an engineering or production manager.

The second type of authority is de facto authority. This type of authority is based on an individual's personal knowledge or skills related to a particular task. De facto authority depends on other individuals to comply out of respect for the individual's unique knowledge or skills. This type of authority is typically the kind of authority held by some sort of analysis or engineer.

The RFID project manager may have either de jure, de facto, or both types of authority. However, the most likely scenario is that the project manager is not specifically a manager, but an engineer who must undertake the RFID implementation project. In this case, as an engineer without specific de jure authority, the engineer must take particular care in project management, as he will not have the power to command others to act or not to act. In other words, the engineer is more likely to have to use his de facto authority and interpersonal skills to gain the cooperation of the different individuals involved at different levels and phases of the RFID implementation project.

11.7 RFID PROJECT MANAGER FUNCTIONS

The project manager has five basic functions. These include planning, organizing, motivating, directing, and controlling.

11.7.1 Planning

This RFID project manager function is primarily performed during the planning phase of the implementation. As previously discussed, it involves creating an outline for implementing the project. This includes the WBS and the Gantt chart.

11.7.2 Organizing

Organizing involves the identification and acquisition of project resources. It most specifically includes identifying and arranging for the personnel that will be involved in the system design, specification, installation, and startup processes. The organizing function also includes the development of the LRC.

11.7.3 Motivating

The project manager motivating function involves providing the environment necessary to obtain the desired performance from the project team members. Classical motivation theory includes work by Maslow and Herzberg.

11.7.3.1 Maslow's Theory

According to Maslow's theory, people are motivated by a tendency to full unfulfilled needs according to a hierarchical list. At the bottom of the list are basic physiological needs. These are followed by safety and security. In the modern world, this is analogous to receiving adequate pay and benefits. Next is relatedness and belongingness with respect to coworkers. Fourth is the need for esteem. This includes self-esteem and esteem from others. Last is what is known as self-actualization. This means that the individual is primarily motivated by a need to fulfill themselves by maximizing their potential with respect to individual performance.

The way that this theory works is that individuals start at the lowest level. They progress to the next level in terms of need importance only when the current level has been satisfied. Upper levels beyond the current level hold no motivating value to the individual. For example, if an individual is mired in the safety and security level because their pay is inadequate, offering rewards that generate higher levels of self-esteem are not motivating. Similarly, if the individual is receiving adequate pay, but cannot relate to their coworkers, offering the opportunity to reach their intellectual potential would not have a motivating effect.

What this theory means to the RFID project manager is that they will have to identify what level of the hierarchy each individual on the project team is on. By offering or arranging for rewards or opportunity at that level, the project manager can best motivate their team members.

11.7.3.2 Herzberg's Theory

Herzberg's theory involves classifying motivating factors into intrinsic and extrinsic motivators. Intrinsic motivators are more internal to the specific job. Examples of intrinsic motivators are

- Work deemed as important
- Sense of accomplishment
- Responsibility
- Recognition

Extrinsic motivators are external to the job itself. They are more related to context in which the job is performed. These include

- Pay
- Benefits
- Working conditions

The significance of this motivation theory is that the extrinsic factors relate to the level of job dissatisfaction, while the intrinsic factors relate to the level of job satisfaction. The absence of extrinsic factors produces a worker with high job dissatisfaction, but their presence only results in low job dissatisfaction. However the presence of intrinsic factors results in a worker with high satisfaction. The absence of intrinsic rewards results in low job satisfaction. In order to have a highly motivated team member, it is necessary to have the presence of both extrinsic and intrinsic factors. However, even if strong intrinsic factors are present, you can still have a highly dissatisfied team member. With this theory, the RFID project manager must make an effort to ensure the presence of both extrinsic and intrinsic factors in order to obtain maximum satisfaction and, hopefully, performance from his team members.

11.7.4 Directing

The directing function involves proving leadership to the RFID project implementation team. Leadership is frequently defined as an individual's ability to influence the behavior of others to achieve a specific objective. Directing is a particularly important project manager function as the RFID project manager will not necessarily have de jure authority over the RFID project team members. Many texts have been written on both general leadership and leadership associated with high-technology projects. For the purposes of this book, we will review only a few leadership theories.

11.7.4.1 Situational Leadership Theory

Situational leadership theory is founded on the concept that the project manager's leadership style should be based on the personnel situation surrounding the project. More specifically, the project personnel are evaluated with respect to ability and willingness.

With the first parameter, the personnel are categorized as being either able or unable. Able personnel have both the necessary intelligence and training to perform their responsibilities. Unable personnel are either missing the necessary intelligence or training or perhaps both. With an RFID project, we would have to assume that all of the engineers on the project have sufficient native intelligence to contribute to the project. However, as the RFID technology may be new to the engineers, they might not have had the necessary training.

Similarly with the second parameter, the personnel are categorized as being either willing or unwilling. Willing personnel are motivated to perform their responsibilities. Unwilling personnel are obviously not interested in completing their job duties. Obviously, the RFID project manager would prefer to have mostly willing project team members. However, part of the inherent project management process is the possible assignment of not necessarily willing team members. In some cases, the team members may have been taken from projects in which they had more interest. In other cases, the team members may not be familiar with RFID technology and therefore less interested in an RFID implementation.

The two personnel parameters result in a total of four different possible combinations of project personnel. These are

- Able and willing
- Able, but unwilling
- Unable, but willing
- Unable and unwilling

Most project managers would naturally hope to obtain the able and willing team members. This would present the easiest, but most unlikely of any of the four situations. In the able, but unwilling personnel, the duty of the project manager is to find ways to motivate the project team members. In the unable, but willing personnel situation, the project manager must seek ways of bringing the team members up to the necessary technical standards. If the team members have the intellectual and physical capacity, then just the necessary training is required. However, it is possible that team members simply do not have the ability to perform the job no matter how motivated they are. In the unfortunate unable and unwilling personnel situation, the project manager is in serious trouble. In this case, there is a significant possibility of project failure.

Depending on the individual combination of parameters, the most successful project manager will adapt their leadership style:

- Able and willing, use delegating style
- Able, but unwilling, use participating style
- Unable, but willing, use selling style
- Unable and unwilling, use telling style

With the delegating style, the team members are able and willing. This means that the RFID project manager can take more of a coaching and assisting role with the project. With the participating style, the team members are able, but unwilling. For this situation, the RFID project manager must encourage the sharing of ideas and authority. This will create an atmosphere of ownership, which will help the team members buy into the project. The selling style is used when the team members are unable, but willing. In this case, the RFID project manager must explain in more detail what must be done, since the team members do not have the necessary knowledge in order to perform their responsibilities. Lastly, with the telling style, the team members will be unable and unwilling. As with the selling style, with the telling style, the RFID project manager must explicitly direct the team members to do specific tasks. Since the workers are also unwilling, the RFID project manager must also expend significant effort in following up with respect to project progress and completion.

11.7.5 Controlling

The control function is essential to the successful completion of the project. This function involves the establishment of specific performance standards, the observation of performance, comparing the observed performance to performance standards, and taking corrective action. This process is then repeated as necessary for each significant work package. The control function is most often illustrated as a control cycle as illustrated in Figure 11.2.

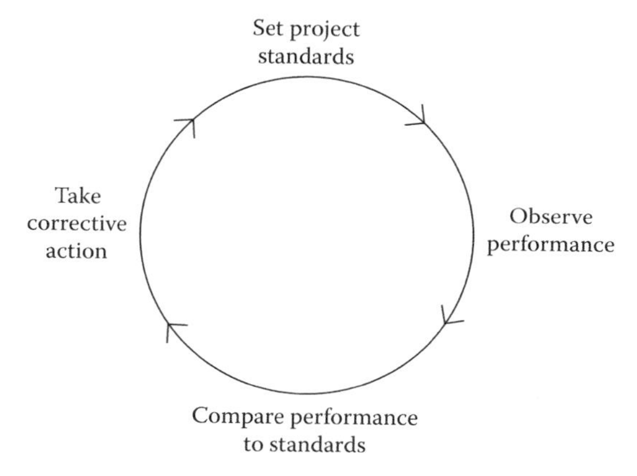

FIGURE 11.2 RFID project control cycle.

11.8 DEVELOPING THE PROJECT PLAN

The project planning process as a minimum consists of developing a WBS, an LRC, and a Gantt chart.

11.8.1 Work Breakdown Structure

A WBS is a successively lower division of project tasks into subtasks. On the first or top-most level, the WBS might consist of only a few of the steps outlined in this book. How many lower levels to use in the project plan is at the discretion of the project manager. However, for the WBS to be meaningful, it is likely to require at least two levels. The following table illustrates a basic two-level WBS for an RFID project.

WBS	Activity
1.0	Problem formulation
1.1	Orientation
1.2	Problem statement
1.3	Objectives
2.0	Project planning
2.1	Work breakdown structure
2.2	Linear responsibility chart
2.3	Gantt chart
3.0	System definition
3.1	Identify RFID components requirements
3.2	Identify layout of RFID system
4.0	Equipment specification
4.1	Identify specific equipment that meets system needs
4.2	Acquire identified equipment
5.0	Pilot implementation
5.1	Installation of pilot system
5.2	Testing and debugging
6.0	Full-Scale implementation
6.1	Installation of full system
6.2	Testing and debugging
6.3	Operator training

If a third level were utilized, it would be annotated with a second digit. For example, a particular type of equipment specification might be represented by the number 4.1.1 identify RFID tags. Another type of equipment specification would be represented by 4.1.2 identify RFID reader.

The level of planning in some projects is so detailed that the division of tasks into subtasks continues until the point at which a single work package is reached. A work package is a discrete unit of work for which authority and responsibility can be assigned.

11.8.2 Linear Responsibility Chart

Most business entities utilize an organizational chart to some extent. Even though a trendy organization may deny the existence of such a chart, there will still be some sort of organizational framework in place. These charts will depict a hierarchical relationship between organizational different divisions and personnel. While these types of charts may be of some value, an RFID project is in need of a type of chart that illustrates the relationship between an activity on the WBS and the different individuals involved in the project.

This type of chart is commonly known as an LRC. The LRC shows who participates in each activity and to what degree. For example, there may be a number of individuals involved in collecting data, but only one individual holds primary responsibility for completing the activity. The LRC can be superimposed on the existing WBS as an x-axis going horizontally across the chart. Across the top, each individual's name or title is listed. There is an intersection between the vertical WBS tasks and the horizontal project personnel designations. At each of these intersections, the project manager can place a code corresponding to the individuals' level of participation. Although the RFID implementation project manager may choose any codes, the following codes are commonly employed:

- P = Primary responsibility
- S = Secondary responsibility
- W = Worker
- A = Approval
- R = Review

Primary responsibility means that the individual has the primary authority and responsibility for completing that work package. Secondary responsibility is assigned as a backup to the individual with primary responsibility. If utilized in the chart, W means that the individual is to assist the primary responsibility individual with completing the task. Approval involves going over the end results of the task and providing feedback to the primary responsibility individual. Review, if utilized, may involve that individual reviewing the results for possible effects on other ongoing activities. Not all tasks are so complex that both a primary and secondary responsibility individuals are required. In addition, if an individual is not involved in a particular task at all, that intersection is simply left blank.

The following table illustrates the use of an LRC with the previously existing WBS.

WBS	Activity	Eng. Mgr.	Eng. 1	Eng. 2	Eng. 3
1.0 Problem formulation					
1.1 Orientation					
1.2 Problem statement					
1.3 Objectives					
2.0 Project planning					
2.1 Work breakdown structure					
2.2 Linear responsibility chart					
2.3 Gantt chart					

(continued)

WBS	Activity	Eng. Mgr.	Eng. 1	Eng. 2	Eng. 3
3.0 System definition					
3.1 Identify RFID components requirements					
3.2 Identify layout of RFID system					
4.0 Equipment specification					
4.1 Identify specific equipment that meets system needs					
4.2 Acquire identified equipment					
5.0 Pilot implementation					
5.1 Installation of pilot system					
5.2 Testing and debugging					
6.0 Full-Scale implementation					
6.1 Installation of full system					
6.2 Testing and debugging					
6.3 Operator training					

Note how in this case, the engineering manager wants to approve everything. Engineer 1 is mostly responsible for planning and control issues. Engineer 2 is good at installing hardware. Engineer 3 must be junior to engineers 1 and 2, as well as a skilled trainer.

11.8.3 Gantt Chart

A Gantt chart illustrates the duration and relationship between different project activities. The duration of individual tasks is represented by horizontal bars. The relationships between activities are illustrated by connecting lines with arrows between the dependent activities. With an RFID project, the Gantt chart can be appended to the right of the LRC or interchanged with the LRC if presentation space is at a premium. This enables the RFID project manager to identify the task, individuals involved, duration, and task relationships at a glance.

Project tasks may have what are known as relationships. These are connections between the tasks, which dictate what sequences must be observed between a preceding task and a succeeding task. Common relationships include

- Finish to start
- Start to start
- Finish to finish

11.8.3.1 Finish-to-Start Relationship

The most common relationship is the finish-to-start relationship. In the finish-to-start relationship, it is necessary for the preceding task to finish before the succeeding task may start. This type of relationship occurs in an RFID project when attempting to test the system. For example, in order to perform a startup test the system, it is necessary to first install both the hardware and the software.

Under most circumstances, the only way to reduce the overall time for a finish-to-start predecessor and a successor is to reduce the individual time of either the predecessor or successor or both. If the tasks could be performed in parallel, then the relationship would not be finish to start in the first place. A finish-to-start relationship is represented by an arrow that leaves the rightmost part of the predecessor and enters the leftmost part of the successor.

11.8.3.2 Start-to-Start Relationship

Less common than the finish-to-start relationship is the start-to-start relationship. The start-to-start relationship means that the predecessor and the successor must start at the same time. This sort of situation may occur when a single previous process splits into two different tasks, which must be worked on simultaneously. In an RFID project, this type of relationship can appear when the RFID hardware components are identified. The RFID tags, readers, and antennas can be ordered simultaneously. It will often be desirable to begin working on a number of tasks as soon as possible, but it is not absolutely positively necessary to have any start-to-start relationships. The start-to-start relationship is represented by two or more arrows feeding into the leading edge of the Gantt chart task bars.

11.8.3.3 Finish-to-Finish Relationship

A finish-to-finish relationship between a predecessor and a successor is found when both tasks are desired to be completed at the same time. This means that if the tasks are of different duration, then the longer duration task must be started before the shorter duration task. The finish-to-finish relationship is most often found in situations where there is some sort of limited life associated with different processes that must be combined in a following task at the same time. This situation is not necessarily going to exist in an RFID project. The finish-to-finish relationship is represented by two arrows exiting the Gantt chart task bars.

11.8.3.4 Lags

A possible variation to the above relationships is the concept of a lag. A lag is a required time period between the predecessor and the successor. For example, it may be desirable to begin testing the functioning of RFID tags as soon as possible after acquiring the tags. In order to begin the testing, it is necessary to actually have the tags to test. However, all of the tags may not necessarily be available for testing at the same time due to shipment issues. In this situation, the project manager may decide that it is more important to wait for all of the tags to arrive before beginning testing. Thus, for project planning purposes, there could be a small lag between these two tasks. Similarly, it would be expected that the tag testing processes finish somewhat behind the tag acquisition process.

11.9 COMPRESSING AND CRASHING PROJECTS

A natural question both during the planning phase and also during the project execution is

> How can the project be compressed?

This question can arise as a result of a requirement to move on to another phase in the overall project or when the project begins to slip behind schedule. An extreme case of compression is known as crashing the projects. This is an analysis of how quickly the project can be complete without respect to cost. In other words, crashing the project optimizes the time parameter of the project at the expense of resource utilization.

Crashing projects is only effective if the activities that are crashed are on what is known as the critical path. The critical path is the link of activities that represent the minimum amount of time that the project can be completed. This means that if the project manager crashes activities not on the critical path that only the length of that activity will be shortened, not the length of the overall project.

Whether the project manager is simply interested in compressing the project time or crashing the project, the approach is similar. The general objective is to attempt to sequence as many tasks as possible in at the same time or in parallel with each other. Since a number of tasks in the RFID project process are in whole or part independent of each other, it is possible to perform significant project compression without affecting the quality of the project. On the other hand, some RFID project tasks simply cannot be significantly compressed because it is necessary to have the predecessor complete before the successor can begin.

As a final warning with respect to crashing RFID projects, the project manager must ensure that the significant increase in effort and expense does not go to waste. When critical activities are crashed, other activities become critical. This means that in order to continue compressing or crashing the project, the project manager must be aware of the entire picture, not just the single activity that they wish to shorten.

Typical RFID project tasks that can be compressed are

- Acquisition of hardware and software
- Testing of RFID tags
- Installation of hardware and software

11.9.1 Compressing the Acquisition of Hardware and Software

RFID hardware and software are not yet in the commodity stage. This means that the acquisition of RFID hardware and software may require the project manager to obtain tags or other components from either the manufacturer or a distributor. As can be imagined, the supply chain for these RFID items themselves may be in need of attention. Significant project time may be hidden in the delivery times of RFID hardware and software. By being willing to incur significantly higher shipment costs, the RFID project manager may compress or crash the time associated with the acquisition of hardware and software. One of the few advantages of compressing the RFID project from this perspective is that there are no additional internal resource costs to expediting the shipment of hardware and software.

11.9.2 Compressing the Testing of RFID Tags

The testing of RFID tags is perhaps the most easily compressed or crashed tasks that the RFID project manager will encounter. Here, it is a simple matter of laying on the necessary resources to assist the original technicians and engineers. Unlike the more complicated processes of installing RFID hardware and software described next, fewer resource problems are likely to arise. Compressing the testing of tags is also beneficial for the entire project as the need for additional or different tags is identified earlier.

11.9.3 Compressing the Installation of Hardware and Software

Compressing the installation of hardware and software will necessitate either an increase in internal or external resources. Since these processes will take a fixed number of man hours, the RFID project manager can either allocate additional internal technical and engineering support or acquire additional outside resources. While this approach may seem initially attractive to the RFID project manager, the concept of the "mythical man month" must be considered. This concept revolves around the phenomenon that adding a particular percentage of additional workers does not necessarily translate into that much more productivity. The reason behind this is that a significant amount of the original workforce may be needed to bring the new team members up to speed and the new team members themselves may not necessarily have the required training to become immediately productive.

11.9.4 RFID Project Tasks That Cannot or Should Not Be Compressed

An RFID implementation project task should never be compressed or crashed when there is a significant possibility that it or a related task would compromise the technical performance of the project. Typical RFID project tasks that should not be compressed are

- Hardware and software selection
- Pilot testing

11.9.4.1 Hardware and Software Selection

The RFID project manager should avoid compressing the hardware and software selection process. This is primarily due to the fact that the success or failure of the project may ride on correct decisions with respect to the specifications and suitability of the hardware and software. Making bad decisions at this point could easily result in problems that are not identifiable until much further into the project. At that point, it may be impossible to take corrective action and still successfully complete the project with respect to time, cost, and technical performance. A typical example would be deciding to acquire a system that is not capable of functioning properly with a specific type of product due to radio frequency transmission issues. By the time this is realized in a pilot implementation, the project manager has already configured the pilot system. Thus, a bad decision made during the hardware and software selection process could easily send the project back to the beginning.

11.9.4.2 Avoid Compressing Pilot Testing

The RFID project manager may decide that it would be beneficial to conduct a pilot test implementation prior to wide operational deployment. If this is the case, the RFID project manager should attempt to avoid compressing or crashing the pilot implementation. The original purpose of the pilot implementation is likely to learn from equipment acquisition, installation, and startup mistakes prior to full implementation. If sufficient time is not allowed to permit this process to be performed, the full-scale implementation may be put in danger. This would essentially be the same as not having a pilot implementation to begin with. There are probably few cases in recorded history where a full-scale implementation was more rapidly and effectively debugged than a smaller pilot implementation.

12 RFID System Design

12.1 SYSTEM DESIGN APPROACH

Methods for conducting a system design analysis. In order to be effective in conducting this type of analysis, the investigative team or engineer should seek to collect information effectively. An approach that is suggested by Foster (2003), a six-step design life cycle for products and processes, is shown in Figure 12.1. They are listed as follows:

1. Idea generation
2. Preliminary design
3. Prototype development
4. Final definition
5. Product design and evaluation
6. Implementation

As we tailor this process for RFID system design cycle, we suggest the following steps (Figure 12.2):

1. Gain ideas through understanding
2. Preliminary design
3. Prototype development
4. Choose an alternative
5. Test and retest the chosen alternative
6. Implement the solution

12.2 STEP 1: GAIN IDEAS THROUGH UNDERSTANDING

In this step, we seek to document the environment that is being investigated. Techniques such as flowcharting and values stream mapping should be utilized to understand the operations. Validation of these types of tools by operations is critical to allow for operational understanding. Also, dialog with operators, managers, and technicians is recommended for keen understanding of common problems that RFID may improve and/or create. The collection of operational data in specific areas such as parts or product selected per hour will provide estimates on how much equipment may be needed for operations. Understanding of the source of the data may provide an evaluation of the credibility of the information. Oftentimes, this type analysis is accomplished by evaluating the real operation by using an on-site analysis. RFID on-site analysis can be demonstrated in three parts: equipment evaluation, environment evaluation, and human factor evaluation.

12.2.1 On-Site Analysis

An on-site analysis is an excellent tool for identifying problem areas prior to deploying an RFID institutive. Consider the three parts of an on-site analysis: equipment evaluation, environmental evaluation, and human factor evaluation.

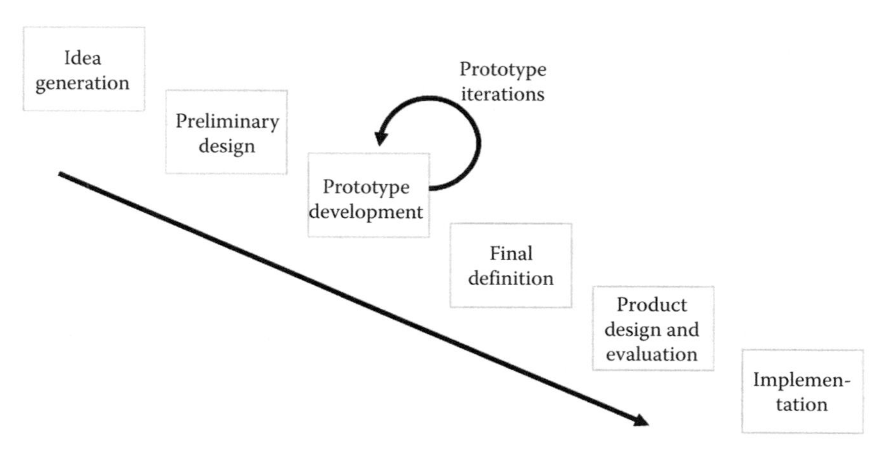

FIGURE 12.1 Design life cycle.

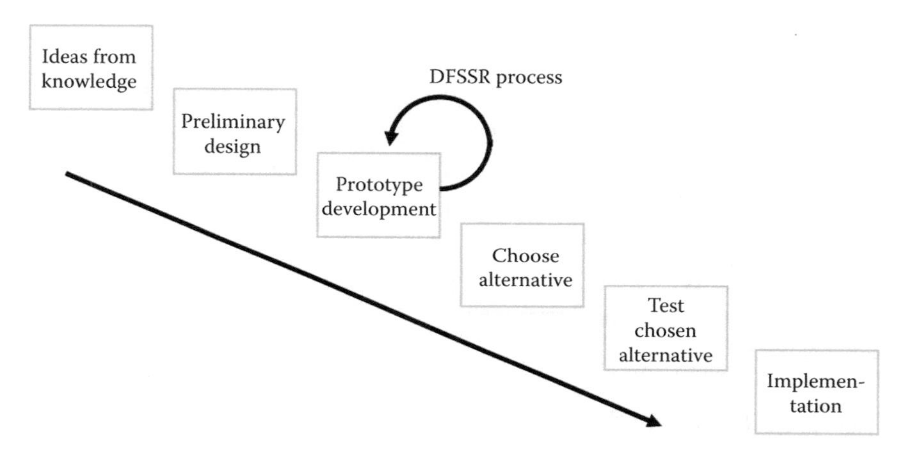

FIGURE 12.2 RFID system design cycle.

12.2.2 Equipment Evaluation

The first part of the analysis should be the equipment evaluation that would include estimating needs, reviewing costs, and training that will be required when using RFID equipment. This would include some of the following tasks:

- Identifying the number of readers required for optimal read response. The number of readers must be determined along with how many antennas will be necessary for boosting the signal in the given environment.
- Identifying the number of tags required for tagging new items, retagging of bad labels, and tagging of miscellaneous items such as returns.
- What type of products, which will be tagged, will have physical problems with the RFID technologies? For example, products encased with metal tubing may have problems with Gen 1 passive RFID tags.

In Appendix 12.A.1, a sample RFID dock door survey shows typical areas that need to be assessed.

12.2.3 Environmental Evaluation

Environmental evaluation allows for the identification and creation of physical and logical read zones such as portals that will not hinder operation. If readers and/or physical enhancements to facilities will be necessary for integrating RFID into the environment, what violation of leases, rental agreement, or building codes may originate. Also, process flows for material handling traffic, replenishment of max/min volume, and other environmental interference must be evaluated. Most pundits recommend that you walk the process and observe operations. Often, some operational understanding can only be achieved by a substantial amount of keen observation. Appendix 12.A.2 provides an RFID environmental evaluation.

12.2.3.1 Radio Frequency Interference Testing

Another form of environmental testing includes the identification of sources of electromagnetic interference (EMI) in the environment. Often, after production operations are clearly understood, it becomes important to understand the radio frequency (RF) environment. First, checking for documented wireless protocol is important to make sure you understand which frequencies are being used and the potential conflicts with RFID frequencies. Next, we suggest utilizing equipment to evaluate frequencies. A spectrum analyzer can be utilized to evaluate potential frequency problems in an environment. A sample exercise on using a spectrum analyzer to do an EMI survey is shown in Appendix 12.A.3.

12.2.4 Human Factor Evaluation

Often, the human element of the process and operations is overlooked when performing an assessment. It is important to consider not only what is being done but who is doing the work. Consider not only the people who will be doing the job tasks; we suggest that you also evaluate who will be performing the implementation. Oftentimes, job satisfaction is overlooked and a technical initiative such as system implementation is sabotaged by disgruntled workers. We suggest that both work measurement and time study analysis be performed, along with a job satisfaction analysis for technical personnel. We provide an overview of work measurement in Appendix 12.A.4 and cognitive turnover job satisfaction survey and case study in Appendix 12.A.5.

12.3 STEP 2: CREATE PRELIMINARY DESIGNS

After collecting relevant information, RFID technologies can be evaluated and selected based on the solution that will best fit the operation. After the understanding of the environment has been made clear, the knowledge of RFID technologies will allow for a better analysis of relevant technologies for the operation.

12.4 STEP 3: PROTOTYPE DEVELOPMENT

Best practices suggest that you create an initial scenario or prototyped environment to evaluate the chosen solution. We suggest using the DFSSR methodology when designing operational prototypes. This is covered in detail in the RFID implementation chapter.

12.5 STEP 4: CHOOSE AN ALTERNATIVE

After testing the solutions for performance, cost justification and return on investment analysis should be evaluated. Choosing the best-valued RFID solution is important to satisfy both short-term needs such as mandate compliance and long-term needs such as operational efficiencies. A 5 year cost analysis technique should be implored for this type of evaluation. We suggest using the RFID ROI chapter as a guide for cost evaluation details. Make sure there is corporate and workplace buy-in in the chosen solutions. Some suggest that involving both decision makers and personnel who utilize the solution most in the process as a means for attaining buy-in.

12.6 STEP 5: TEST AND RETEST CHOSEN SYSTEM

Testing protocols and minimum specification should be set for operational acceptance. Testing should be categorized into subsystem testing and full system testing procedures. It is critical that end users of the system are involved in the testing to identify the problem prior to roll out.

12.7 STEP 6: IMPLEMENT THE SOLUTION

Effective project management is the key to implementing an RFID solution. Coordinating efforts and identifying key milestone using techniques such as PERT and CPM are critical. More details are provided in the project management chapter.

APPENDIX 12.A

12.A.1 Sample RFID Dock Door Survey

A dock door survey may allow users to evaluate performance expectations/specifications to identify and/or model potential RF, EMI, and other interference sources. These elements hinder RFID system of its performance, reliability, and introduce uncertainty regarding the overall system throughput and integrity.

This sample may allow understanding on how to measure RF signal strength, stray signal, multipath, bounce, RF signal absorption/reflection characteristics, harmonic effects, shading, emitted and conducted RF signal strength, loss profiles, and full spectral analyses for the intended operational RFID. Recommended practices prior to a survey include the following:

- Gather business requirements
- Interview managers and users
- Define information for security requirements
- Gather site-specific documentation
- Document existing network characteristics
- Gather permits, license (electrical) and zoning (fire marshal) requirements
- Note indoor or outdoor-specific information

Other operational considerations include

- Bill of material items comprising materials, components, subassembly, assembly, or product nature
- Packaged items—individual or mixed fundamental items in a single package
- Transport units comprising packages or other discernible items
- Unit load or palletized unit—to carry transport units or other discernible items
- Container units—for accommodating pallets or other discernible items
- Transportation vehicles—to carry container units or other discernible items
- Through-put speeds of pallets, cartons, or items on fork-lifts, pallet jacks, or conveyors

The site survey may identify the variables that would most affect your client's RFID project. Understanding the conflicts of these variables helps to ensure that the RFID technology will be optimized. Some of the variables that can affect positive RFID outcomes are

- RF-absorbing water/liquid content in food stuffs and manufactured goods
- RF-absorbing water content in corrugated cardboard packaging, with added variability due to changing relative humidity of indoor/outdoor climates
- RF-reflecting and/or RF-shielding metal content in both goods and packaging, even foils and metallic inks

12.A.1.1 Dock Doors Checklist

1. Dock door characterization
   a. Dimensions (height × width × depth)
   b. Note reflective surfaces
      i. Leveling ramp
      ii. Door
      iii. Side posts
   c. Are there protective posts/fences for the antennas?
   d. Does the door go straight up, parallel to the floor or roll?
   e. What is the distance between the dock doors?
   f. Are all docks the exact same configuration?
   g. Do they use a screen?
   h. Can the dock door or screen be partially open?
   i. Are there fire extinguishers present, which cannot block their view?
2. Study customers' work flow process and note exactly what each employee does
3. Note the kind of traffic, busy periods, quit period, volume
   a. Do they use fork lifts? What speed?
   b. Do they use pallet jacks? What speed?
   c. Do they manually load with totes? What speed?
4. Are wireless handhelds or forklifts used at dock? Note any communication devices including cellular phones
5. Are adjacent docks typically used simultaneously?
6. Do they stage pallets between docks? Do they park or store other items temporarily?
7. How do they load pallets in the trailer?
   a. Do they pivot pallets, e.g., one loaded from narrow side, one from the wide side of the pallet?
   b. What kind of product is loaded?
   c. Do they double stack (or more) pallets?
   d. Do they dual stack pallets, e.g., two adjacent?
   e. Do they quadruple load pallets, e.g., two adjacent, double stacked?
8. **Mounting**
   a. Best location for reader mounting
      i. Center of dock is not very serviceable
      ii. Are antenna cables sufficiently long?
      iii. What kind of brackets will be required?
      iv. How can you protect the antennas from fork lifts?
9. **Input sensors**
   a. Where can photoelectric sensors be mounted if required?
   b. Do you need a narrow detection beam or is a less expensive wide beam sensor okay?
   c. Can the sensors be inadvertently activated? Is that an issue?

   d. Where can magnetic door or screen switches be mounted if required?
   e. Is direction detection required (this may be complex—don't want to push for this)?
10. **Output sensors**
    a. Are visual indicators required?
       i. Does the site have restrictions on visual indicator colors?
11. Are audible outputs required?
12. Equipment requirements
    a. How many readers per dock?
    b. How many antennas per dock?
    c. What types of antennas per dock?
    d. Mounting brackets required for readers and antennas?
    e. How do you keep the antennas from getting misaligned?
    f. Are spare readers and antennas required?
    g. Will you need to attenuate power?
    h. Network design and connectivity to each reader
    i. LAN test panel
    j. Electrical power for readers
    k. Security issues for location
    l. Grounding availability

12.A.2 RFID Environment Evaluation

Many operations contain metal, carbon, and absorptive instruments in their physical environment. Operations such as maintenance warehouses, electronics manufacturers, and some distribution operations operate in these conditions. System design should be well thought out for RFID technologies to be successful. Dirty environment is defined in the text as operation in which RF transmissions may be difficult and provide an unreliable environment for RFID technologies to operate. Tagged materials that present difficulties for RF include lucent materials and opaque materials.

Lucent materials: Lucent materials or materials that allow RF energy to penetrate, such as paper, plastics, cloth, and cardboard, have the highest successful confirmed successful read in an RFID system. These items can be tagged arbitrarily and can be scanned successfully.

Opaque materials: Opaque materials present the most problems. First, conductive materials that block or reflect RF energy such as metals, pastes, carbon-impregnated plastic (black), conductive plastics, and foil-lined packaging scatter or block RF signals. Second, absorptive materials, which weaken RF energy, such as most liquids and moist fibers, green wood, moist wipes, and damp paper absorb RF energy, preventing the tag from reaching full capacitance.

The best practice is to containerize the opaque materials into some type of RF lucent material and scan those products.

12.A.2.1 Common Problems

1. Liquid items such as water do not scan
2. Metal items such as metal cans and toothpaste do not scan
3. Items located in the middle of pallets do not scan
4. Fast-moving pallets do not scan all tagged items

12.A.2.2 Best Practices

1. Re-containerize into scannable containers.
2. Use alternate scanning method.
3. Move tags to outside of items.

4. Speed through the portal is insignificant.
5. In a dense reader environment, change the reader from Talk first, then listen to Listen first, then talk. This solution will affect your read rate causing longer time to attain reads, but you may still attain the scan.
6. Limit speeds for mobile reader to allow time for the waves to manipulate the environment.
7. Tag pallets with difficult items such that the tags are facing outward toward the reader.
8. Tag materials such as water where there is conductive material such as the bottle cap.
9. Antennas may have to be adjusted in order to capture effective read area for your specific application.
10. The reader may need to be configured and adjusted similar to a portal configuration to account for dead areas within the scanning area. Antennas may also have to be adjusted such that the wrong product is not scanned at the wrong area.
11. Mount RFID tags to allow for the greatest possible surface area presentation to the reader.
12. Configure pallet to minimize shadowing, which is when tags are oriented on top of each other with a container in between. This shadowing causes the signal of one tag to mask the signal of all the tags situated behind it.
13. Consider the liquid line and air space within cases and pallets to enable better reads on absorptive materials.
14. Utilize alternative tag types for better results on absorptive products.
15. For opaque items, seek to add a scannable layer to the tag, or change the tag substrate to be reflective so the tags will scan.
16. Tag products on the outer side of the carton facing the antennas.
17. Test the pallet portal limits and dead zones.
18. Alter the multiplexing sequence of multiple antennas to give preferential read by the antenna better suited for capturing difficult tag geometry.
19. Ensure that the reader wake and sleep settings are optimized to capture scans.
20. Evaluate the link margin if the reader field appears limited. Evaluate settings by
    - Increasing gain in antenna
    - Increasing output power
    - Evaluating VSWR for reader power airborne
    - Evaluating connectors and coaxial cables
    - Evaluating tag to reader polarization
    - Enhancing tag with packaging on hard to scan packages, adding 1 in. of airspace with packaging or spacers such as foam, paper, or other more conducive materials
21. Include movements such as rotation to improve scanning. Utilize stretch wrap locations that rotate 360° multiple times for hard to scan products on mixed pallets.

12.A.3 Using a Spectrum Analyzer to Test EMI

Oftentimes, electronic devices operate by radiating excessive amounts of electromagnetic energy, and are susceptible to such energy from internal or external sources. Thus, the need for determining electromagnetic compatibility or EMC is necessary. EMI occurs when radiated or conducted energy adversely affects circuit performance and disrupts a device's EMC. Many types of electronic circuits radiate or are susceptible to EMI and may need to be shielded to ensure proper performance. Establishing basic EMC in any electronic device generally requires detail engineering. The first goal is to identify and reduce EMI generated from internal sources. Many manufacturers accomplish the reduction through engineering designs in which an electronic circuit is shielded in such a manner to generate less EMI. Residual EMI may then be suppressed or contained within the enclosure by appropriate filtering and shielding methods. Filtering cables at the point where they enter or leave the enclosure will reduce conducted emissions. A tool that is commonly utilized

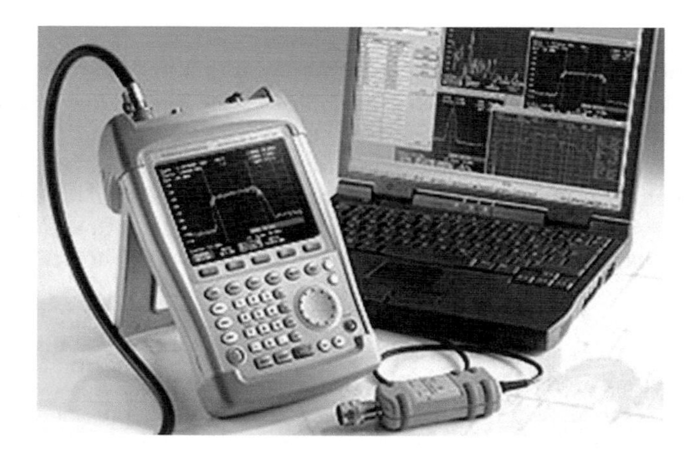

FIGURE 12.A.1 Sample spectrum analyzer.

to identify the EMI levels in a facility is known as a spectrum analyzer. A sample spectrum analyzer is shown in Figure 12.A.1.

12.A.3.1 General Directions for Using a Spectrum Analyzer

The analyzer displays the frequency spectrum from 100 kHz to 3 GHz. See Figure 12.A.2 for the sample start screen.

- At 100 MHz, the generator signal is displayed as a vertical line. Generator harmonics can also be seen as lines at frequencies that are multiples of 100 MHz.
- To analyze the generator signal at 100 MHz in more detail, reduce the frequency span. Set the R&S FSH's center frequency to 100 MHz and reduce the span to 10 MHz.

Centering the frequency to 915 MHz to capture RFID frequencies is necessary for identifying problems in the right frequencies.

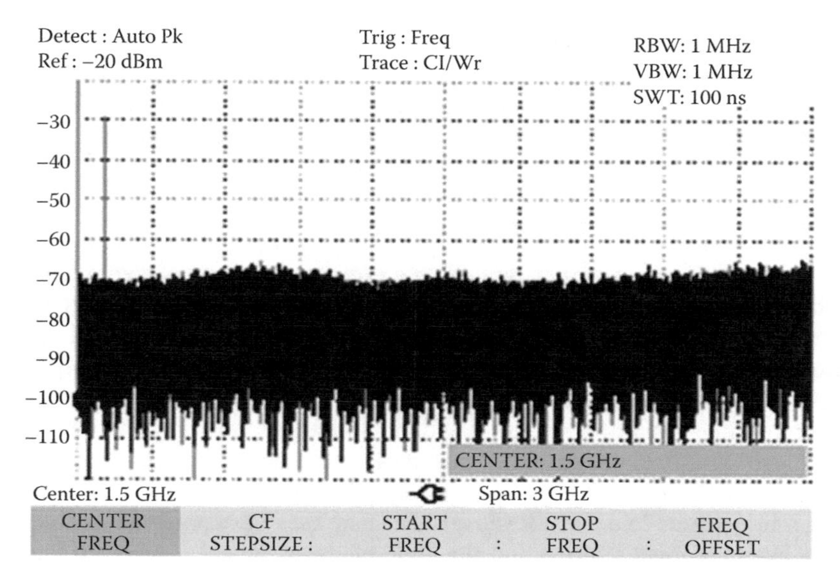

FIGURE 12.A.2 Sample start screen.

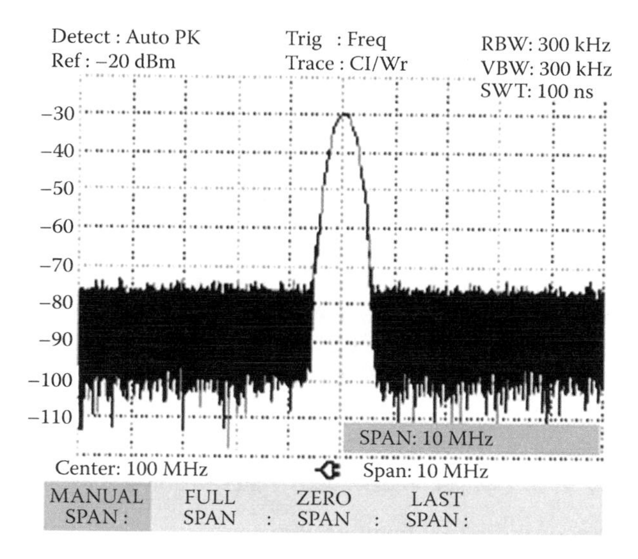

FIGURE 12.A.3 More details.

The general procedures are

- Press "Frequency" button
    - Enter 915 MHz
    - Set Span = 30 MHz
    - Press "Span" Button
    - Enter 30 MHz
- Press the Marker button
    - The marker automatically selects the trace maximum
    - Use the soft keys to set marker
    - Or rotate dial to move marker

See Figure 12.A.3 for more details.

12.A.3.2 General Guidelines and Notes

Setting the resolution and bandwidths (BWs)

- Resolution bandwidth controls the density of frequencies displayed. Lower resolution BWs will have a cleaner display but at the expense of accuracy
- BW settings are displayed on the top right of the screen
- A video BW setting rule of thumb is ResBW/10

General Method to set BW

1. Press the BW function key
2. Press the manual Res BW soft key
3. Use the up/down arrows or alphanumeric keypad to enter 100 kHz
4. Press the "manual video BW" soft key
5. Use the up/down arrows or the alphanumeric keypad to enter 10 kHz

- To facilitate our measurement readings, we will activate "Video Averaging"
  - Press the Trace function key
  - Press the Trace Mode soft key
  - Use the up/down arrow or the dial to select average
- Save settings

12.A.3.3 Other

- Monitor the relative interference amplitude using the spectrum analyzer
- Your measurements should correspond with the gridlines
- Throughout your measurements, maintain consistency with
  - Relative orientation with respect to the spectrum analyzer, e.g., antenna orientation
- Your body position relative to the spectrum analyzer antenna
- Observers should maintain a good distance away so as not to interfere with the measurements

12.A.4 Overview of Work Measurement and Times Studies

12.A.4.1 Brief History of Industrial Engineering

A good, detailed overview and introduction to industrial and systems engineering is provided by Turner, Mize, and Case (1987); we provide a brief overview. Work measurement and time studies is a common skill that is developed by industrial engineers. Though it is a foundational skill, most industrial engineers utilize this understanding to implement ERP, WMS, and other decision support systems in today's workforce. Operations research, statistical quality control (SQC), and logistics training provide future decisions support decision designers for complex computer systems, six sigma black belt's and champions, and logistics engineers, respectively. Before our overview of work measurement, we will provide an overview of today's industrial engineer.

The following formal definition of industrial engineering (IE) has been adopted by the institute for industrial engineers:

> Industrial Engineering is concerned with the design, improvement, and installation of integrated systems of people, materials, information, equipment, and energy. It draws upon specialized knowledge and skill in the mathematical, physical, and social sciences together with the principles and methods of engineering analysis and design to specify, predict, and evaluate the results to be obtained from such systems.

Although the term industrial is often associated with manufacturing organizations, here it is intended to apply to any organization. The basic principles of IE are being applied widely in agriculture, hospitals, banks, government organizations, and others.

There is considerable commonality among the different branches of engineering, each branch has distinguishing characteristics that are important to recognize. IE emerged as a profession as a result of the industrial revolution and the accompanying need for technically trained people who could plan, organize, and direct the operations of large complex systems. In 1880, industrial operations were conducted in more of specialty jobs shop manner and many first-line supervisors commonly abused the workers. Most operations did not provide training or procedures and so supervisors managed from personal perspective, commonly abusing good workers who they personally did not care for. Supervisors were expected to work as hard as he could. Any improved efficiency in work methods usually came from the worker himself in his effort to find an easier way to get his work done. There was virtually no attention given to overall coordination of a factory or process.

12.A.4.1.1 Emergence of Work Measurement

Frederick W. Taylor commonly referred to as the father of scientific management and IE is credited with recognizing the potential improvements to be gained from analyzing the work content of a job and redesigning the job for greater efficiency. Taylor's methods brought about significant and rapid increases in productivity. Later, developments stemming from Taylor's work led to improvements in the overall planning and scheduling of an entire production process.

Frank B. Gilbreth extended Taylor's work by his contribution work management, which involved the identification, analysis, and measurement of fundamental motions involved in performing work. Work motions were classified as "reach," "grasp," "transport," and so on, and by using motion pictures of workers performing their tasks, Gilbreth was able to measure the average time to perform each basic motion under varying conditions. This permitted, for the first time, jobs to be designed and the time required to perform the job known before the fact. This was a fundamental step in the development of IE as a profession based on "science" rather than "art."

Time studies, as it is practiced today at companies such as United Parcel Service (UPS), consist of understanding operations, following best practices, and allowing for a normal work pace. Work measurement and standardizing operations is the precursor to performing a time study. Workers who are timed should be seasoned workers who have been trained in prescribed work methods, and should be timed on a normal day. Consequently, time studies are mostly used to confirm a predetermined motion system.

Gilbreth's work derived into databases that have captured this information over many different environments into databases that can create predetermined work standards. These predetermined work measurement systems such as MODAPS, MTM, and others are computerized and consulting firms implement standards using these systems. Most consultants boast a 20% increase in worker productivity implementing these standards.

Another early pioneer in IE was Henry L. Gantt, who devised the so-called Gantt chart. The Gantt chart was a significant contribution in that it provided a systematic graphical procedure for preplanning and scheduling work activities, reviewing progress, and updating the schedule. Gantt charts are still in widespread use today.

W.A. Shewhart developed the fundamental principles of SQC in 1924. This was another important development in providing a scientific base to IE practice. Many other IE pioneers contributed to the early development of the profession. During the 1920s and 1930s, much fundamental work was done on economic aspects of managerial decisions, inventory problems, incentive plans, factory layout problems, material handling problems, and principles of organization.

12.A.4.1.2 More on Industrial Engineering

The period from 1900 to 1930 is generally referred to as scientific management. The next IE period begins in the late 1920s and is shown extending to the present time. This period is when operations research begins to influence IE practices. Next period of IE includes computer systems and distribution and logistics and continues to grow. Though not an exhaustive list of all IE teachings such as total quality management, Six Sigma, Lean, JIT, manufacturing engineering, and others are commonly taught in IE, the foundational periods provide foundations for the latter.

12.A.4.2 Industrial Engineering Organizations

Much can be learned about any profession by tracing the organizations that members of the profession form and/or join. The American Society of Mechanical Engineers provided the first forum for a discussion of the works of die early pioneers, particularly Taylor and his associates. Then, in 1912, the Society to Promote the Science of Management was formed. The name was changed in 1915 to the Taylor Society.

The Society of Industrial Engineers was formed prior to 1920. The American Management Association was formed in 1922, and many industrial engineers were active in this organization. In 1934, the Taylor Society and the Society of Industrial Engineers were combined to form the Society for die Advancement of Management.

The American Institute of Industrial Engineers (AIIE) was founded in 1948. The AIIE provided, for the first time, a professional organization devoted exclusively to the interests and development of the IE profession. Previously, IEs were associated with organizations whose main interests were in management or another branch of engineering.

The AIIE was an instant success in all respects. Within 1 year, student chapters were formed at 11 major universities. The official publication of the AIIE, The Journal of Industrial Engineering, also made its initial appearance the following year, the first issue being published in June, 1949. In 1969, the journal was divided into two publications: Industrial Engineering is published monthly and is devoted primarily to IE practice; UE Transactions is published quarterly and is devoted primarily to research and new developments within the profession.

In 1981, through a vote of its membership, the institute changed its corporate name from AIIE to IIE. By dropping the word "American," the institute officially recognized the international nature of its activities. The HE has members in more than 80 countries around the world.

Many practicing industrial engineers belong to other organizations that are related to the IE field. Some of these are

1. Operations Research Society of America
2. The Institute for Management Sciences
3. Association for Computing Machinery
4. American Society for Quality Control
5. Society for Decision Sciences
6. American Production and Inventory Control Society
7. Society of American Value Engineers
8. American Association of Cost Engineers
9. Society of Manufacturing Engineers
10. Robot Society of America

The IIE is the technical society for all industrial engineers, beginning with university students who are majoring in IE. A very important part of the overall organization is the university chapters. Students participating in these chapters receive the HE publications and are considered an integral part of the overall organization.

12.A.5 Cognitive Turnover Job Satisfaction Survey

12.A.5.1 Background

Because of the difficulty of measuring knowledge worker production, dissatisfied knowledge workers may take advantage of the situation. This mind-set of dissatisfaction may produce behavior in which personnel seek more financial satisfaction by giving themselves a stealth raise—cutting back the effective hours in which they perform knowledge work at the office. They may dedicate more mental effort to another activity that is not job related that brings them more satisfaction (Barber and Weinstein, 1999). Businesses lose $150 billion annually in health insurance and disability claims, lost productivity, and other expenses attributable to burnout, stress-related problems and mental illness (Bassman, 1992). Further quantification of the bottom-line impact of indirect cost is demonstrated by the high cost of absenteeism, which is estimated at approximately $40 billion per year in the United States (Gaudine and Saks, 2001).

Previous studies on turnover and burnout categorize costs into three groups: direct costs, indirect costs, and opportunity costs. Direct costs include disability claims; worker's compensation claims; increased medical costs; and litigation costs, (including wrongful discharge; hiring new personnel; training cost; advertisement for new personnel; and time spent interviewing new personnel). Indirect costs include costs associated with poor quality, high turnover, absenteeism, poor customer relationships, or even sabotage. Opportunity costs include costs associated with lowered employee commitment, lack of discretionary effort, commitments outside of the job, time spent talking about problems instead of working, and loss of creativity.

Cognitive turnover (CT) is a term copyrighted in this research to describe a mind-set that is created by a combination of turnover thoughts/cognitions brought about by burnout conditions. While everyone may manifest this mind-set periodically, excessive CT (eCT) may be detrimental to the individual and the organization they work for. Subtle acts such as absenteeism, poor quality, and lack of discretionary effort have been related to worker burnout and are common predecessors to quitting and becoming another turnover statistic. Noncommitment-type behavior may stem from employee stress and burnout created by management or organizational abuse; hence an eCT will have lowered productivity due to the lack of commitment.

Engineering managers may be able to avoid the negative consequences to the organization and employee by identifying the nonproductive knowledge worker that is experiencing eCT. However, it is probably more productive to seek aggregate or group information that will facilitate improvements in attitude, innovation, productivity of the organization, and may prevent ineffective events such as reduced employee productivity and sabotage.

CT is a combination of a turnover thought process and the results of burnout. Similar to pre-turnover thought processes, high degrees of burnout among major proportions of a group suggest low productivity. High burnout implies little slack in a person's coping capacities, and perhaps deficits in them. High measures of burnout are strong indicators of these phenomena, but the inverse, low burnout, does not necessarily indicate high productivity (Golembiewski, 1982). This research focused on the high measures of burnout in conjunction with pre-turnover indicators.

The researchers have developed a methodology that is being explored as a means to consistently measure knowledge worker's CT. SECtCS, or statistical evaluation of cognitive turnover control system, is a methodology that attempts to identify, measure, and document CT. This copyrighted methodology can only be used by the researchers in this study. The items produced from this study can be used by engineering managers and will be further described in the results and conclusions. The following are the six phases of the SECtCS Research Methodology for Knowledge Workers (note that this chapter focuses on the first two) is shown in Table 12.A.1.

12.A.5.1.1 Phase 1: Develop Test Instrument

The test instrument that was developed to test CT is based off of variables that organizations can actually do something about. Unlike other satisfaction questionnaires, the variables or constructs

TABLE 12.A.1
SECtCS Method Phases

1. *Phase 1: Develop test instrument*—Develop a customized test instrument (questionnaire) for the knowledge worker population, administer the questionnaire, and collect and record scores. Conduct reliability testing on the questionnaire. This testing continued until the questionnaire was reliable (SECtCS Questionnaire)
2. *Phase 2: Develop mathematical model*—Use the data collected in phase 1 and incorporate it into a mathematical model to give a valid CT index score (SECtCS Model)
3. *Phase 3: Statistical process control charts*—Use data from the model developed in phase 2 for the statistical measurement of individuals with respect to all respondents and identify at-risk CT index scores (SECtCS Evaluator-i). Establish a tracking mechanism for "at-risk," and "low-risk" respondents. The respondents are required to retake the questionnaire every 3 months in order to complete the SPC charts
4. *Phase 4: Intervention*—Educate, implement, and monitor the solution (SECtCS intervention)
5. *Phase 5: Intervention measurement*—Remeasure the respondents after they have been subjected to the intervention and compare to the results of phase 3 (SECtCS Evaluator-r)
6. *Phase 6: Evaluation of intervention*—Document the results and conclusions and add to solutions database

Intervention note: Any intervention, like organizational mentorship, has to be coordinated for effectiveness. Intervention contributors must be provided with guidelines so that there will be data consistency. These guidelines will also allow for efficient collection of feedback.

TABLE 12.A.2
General Definitions of Constructs

Cognitive Turnover Determinant	Construct	Construct Definitions
Burnout (B)	Depersonalization	Distancing oneself from others
Burnout	Personal accomplishment	Performing well on things that matter
Burnout	Emotional exhaustion	Ability to cope in high stress situations
Turnover (T)	Overall job satisfaction	Job satisfaction that determines turnover
Turnover	Goals	Feeling that goals are attainable and have meaning
Turnover	Comfort	The space and physical conditions of the job are adequate to perform at the job
Turnover	Challenge	Feeling that job is not boring and has reasonable challenges
Turnover	Financial rewards	Financial compensation is reasonable and fair
Turnover	Relationship with coworkers	Ability and willingness to work with others
Turnover	Resource adequacy	Organization provides adequate supplies and training to perform at job
Turnover	Promotions	Opportunity for fair chance at promotions

were chose from organizational variables that organizations know what to do when there are problems. For instance, if pay is a problem, then organization can raise pay, or if facilities are problems, then organization can choose to update facilities. The constructs are shown in Table 12.A.2. A questionnaire is given to measure these variables and then tested for reliability and reduced for the measured population or company.

Subjects completed three questionnaires. The first was comprised of 109 questions concerning the job satisfaction constructs and the burnout constructs. A second questionnaire asked for the person's more direct appraisal of their level of CT. Respondents were assured that their answers would remain anonymous. Respondents were given both a verbal and written description of the CT and the levels of CT. A description of each range is given in Table 12.A.3. The respondent then self-scored their level of CT given the range from 1 to 10. Subjects were asked to rate, on a scale of 1 (strongly agree) to 5 (strongly disagree), statements that indicated how they felt about their employer. An example is "My employer is concerned about giving everyone a chance to get ahead." They were also asked to rate specific job satisfaction questions on a scale of 1–5, representing 1 (very dissatisfied) to 5 (very satisfied). An example is, "On my present job, how do I feel about my pay and the amount of work I do." The mean values were calculated for each construct and for an overall value.

As was mentioned before, the initial version of the questionnaire and rating scale was pilot tested and critiqued by other researchers. After feedback from other researchers, ambiguous or confusing items were identified and eliminated. This was an effort to achieve face validity.

TABLE 12.A.3
Description of Each Range

Score	CT	Considering Leaving	Description
1–2	No	No	Not burned out
3–4	No	Occasionally	Light burnout
5–7	Yes	Open for other jobs	Medium to high
8–10	Yes	Strongly considering	High

Note: CT scores range from 1 to 10, with 1 representing low level of CT and 10 representing high levels of the CT.

12.A.5.1.2 Phase 2: Mathematical Model Results

For this research, a population of engineers across eight different companies was measured. The mean value for each construct of turnover and burnout was determined. The mean scores from the questionnaire constructs were calculated from the values attained from the responses to the questions. The mean value and standard deviations are listed in Table 12.A.4.

The result of the analysis of variance shows that only four variables had a significant effect on the CT. The p-values indicate that these variables have a significant effect on CT at an alpha level of 0.10.

The turnover variables, challenges and promotions, and the burnout variables, depersonalization and personal achievement, were significant for predicting engineer's CT in this study. Most of the job satisfaction constructs were not shown to be valid (goals, comfort, financial rewards, relationships with coworkers, and resource adequacy), and only one out of the three burnout constructs was not shown to be valid (emotional exhaustion). These constructs had a weak impact on CT for this group.

Based on the results, the mathematical model for predicting CT for engineers was given by the following equation:

$$F(x) = 1.199(\text{Challenges}) + 1.575(\text{Depersonalization}) - 1.712(\text{Personal achievement}) - 0.935(\text{Promotion}) + 5.122 \qquad (12.A.5.1)$$

Table 12.A.5 summarizes the four variables in the model that showed significant prediction for CT and describes the impact of each on CT.

The function $F(x)$ will be a number between 1 and 10. Scores that are approximately 1–4, represent low cognitions to leave and generally low burnout indications. Scores 5–8 represent moderate burnout and leaving cognitions. Scores 9 and above, represent eCT that may lead to detrimental burnout and possible sabotage if departure is not eminent (refer back to Table 12.A.3 for chart).

It is important to note that because only four variables were necessary to determine the CT level, it might be possible for the engineering manager to reduce the 59-question questionnaire to 19 questions. The danger in utilizing the 19-item questionnaire is that these results are based on a small sample size from exploratory research. The engineering manager should evaluate the limitations before using the 19-question questionnaire. The benefit of using the reduced question form is that it may be easier to

TABLE 12.A.4
Construct Means and Standard Deviations

Construct	Mean	Std. Deviation
Depersonalization (B)	2.55	0.68
Emotional exhaustion (B)	2.60	0.62
Personal achievement (B)	3.68	0.57
Goals (T)	3.62	0.67
Comfort (T)	3.34	0.83
Challenges (T)	3.46	0.84
Finances (T)	3.23	0.66
Relationships (T)	3.73	1.11
Resources (T)	3.69	0.82
Promotions (T)	2.94	0.83
Satisfaction (T)	3.60	0.72

Note: These scores are measured on a 5-point Likert scale with 1 = not very satisfied and 5 = very satisfied and 1 = strongly disagree and 5 = strongly agree.

TABLE 12.A.5
Model Translation Description

Construct	What It Measures	Type of Effect
Challenges	Feeling that the job is not boring and has reasonable challenges	Direct impact on CT. If you feel the job is too challenging, then you will have a higher CT index score
Depersonalization	Distancing oneself from others	Has largest direct effect on CT. If you feel that you are involved as part of the team, you will have a higher CT index score
Personal achievement	Performing well on things that matter	Has largest OPPOSITE effect on CT. This means if you believe you perform well, your CT index score will be lower
Promotion	Opportunity for fair chance of promotion	Has OPPOSITE effect on CT. If you believe you can be promoted. You will have a lower CT index score

implement. The researchers suggest using the current developed questionnaire and performing phase 2 for the engineering manager's specific knowledge worker group in order to attain the most effective questionnaire.

12.A.5.2 Study Limitations

Some limitations to this research were the sample size and questionnaire biases. This study used only 51 knowledge workers across organizations for the creation of the mathematical model. Currently, more populations are being targeted for further validation of the mathematical model. This should be taken into account when utilizing the model for possible sample bias when using the questionnaire and modeler. Also, questionnaire biases can occur when implementing the testing of the questionnaire. Respondents may not answer the questionnaire honestly if they feel threatened by what will happen if they score on the high end of the index. The researchers recommend utilizing tools such as a digital simulator or online questionnaire software to offset some of the fears of being identified and possible ramifications. Future research includes the development of a manager's checklist, which will allow managers to observe specific behaviors enabling the manager to score the employee for CT. If the 19-question questionnaire is utilized, it is important to note that this research is exploratory and caution should be used before the results are acted upon. Future research may focus on industry-specific models.

12.A.5.3 Lessons Learned and Recommendations

Our findings yielded several lessons learned and many recommendations. First, knowledge worker management is difficult, but crucial to companies' future growth and bottom line. Second, analyses of the empirical data on the CT indices presented here suggest that companies need to focus their current practice away from solely financial measures and toward providing challenging work, reduce isolated tasks that cause depersonalization and increase team activities, increase recognition of personal achievement, and provide realistic promotion opportunities. In this study, the high level of depersonalization may suggest that when knowledge workers perform isolated tasks, they could have higher levels of CT. The engineering manager may be able to improve this component with team-based tasks. Further, the high negative coefficient for personal achievement on CT indices suggests that recognition of knowledge workers can have a strong positive effect on CT. The other two significant variables were promotion and challenges, which may not be under the direct control of the engineering manager.

Finally, one opportunity for improvement that companies miss is giving real feedback to employees. Companies should address the problems with performance by brainstorming and communicating with employees about possible solutions. By using the first two phases of the SECtCS methodology, the engineering manager has a method for identifying some of the main components of the CT. The first two phases allow the organization to identify the most significant measures of eCT for the chosen group of knowledge workers. The complete methodology, which is not fully presented in this article, is designed to measure relevant components of eCT, remeasure implemented solutions effectiveness, document efforts, and provide feedback.

Feel free to contact the authors for a copy of the questionnaires.

13 RFID Supply Chain Planning Levels

13.1 INTRODUCTION

In the following section, we demonstrate the RFID logistics application framework. The squares represent the planning phases strategic, intermediary, and tactical. The left side of Figure 13.1 demonstrates the research and assessments performed at the different planning stages, which include supply chain network design decisions, transportation policies, and inventory control policies. The decision flow indicates that oftentimes decisions are made from top-down decision policies.

The right side of Figure 13.1 represents the validation and applications at these levels, which include facility location optimization decisions, transportation optimization, and inventory management. The information flow indicates that optimization takes place in a bottom-up manner given that actual or real world data has to be used as opposed to abstract scenarios, and at the base level, the real information has to flow from operations for executive management to make optimized decisions.

13.2 RFID SUPPORTS SUPPLY CHAIN PLANNING AND OPERATIONAL OPTIMIZATION

The problems encountered in the design and operations of complex logistics networks need to be solved at three levels: *tactical*, *intermediary*, and *strategic*. At a strategic level, a company must answer numerous design questions such as

1. How many facilities need to be built or leased
2. Where to build or lease them
3. The customers served by a facility
4. What segments of the transportation network to outsource to third-party logistics (TPL) companies
5. What segments to retain in-house
6. How many trucks to own, etc.

Because these factors are continuously changing, the company also must determine at an intermediate-term and short-term (tactical) level, the following:

1. How many and what transshipment points to operate
2. How many trucks to maintain
3. Driver staffing requirements, route selection, inventory positions and reorder points, etc.

Different operations may seek to minimize their cost, or maximize their profit, and pass on inefficiencies to the next operation. For example, in our framework, a warehouse manager making a tactical-level decision may seek to minimize receiving labor and create a large queue of inbound trailers (an intermediate-level decision) to wait for unloading. Also, a transportation manager (intermediate) may set trailer load plans that require each trailer to occupy a facility dock door, inevitably creating the need for a larger facility (strategic-level decision). The bullwhip effect on inventory is

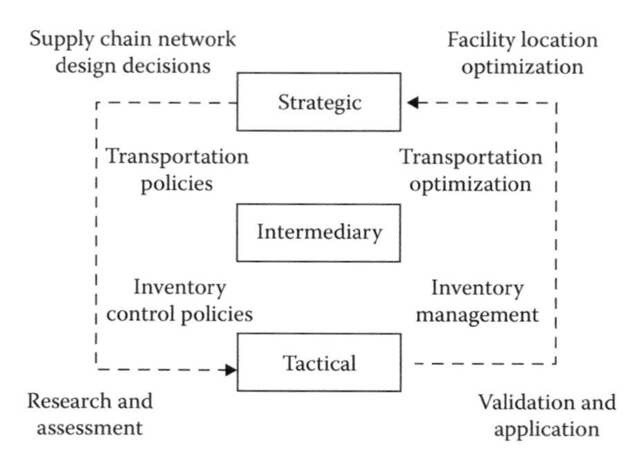

FIGURE 13.1 RFID logistics application framework.

demonstrated by excess inventory in warehouses (tactical) due to lack of confidence in forecasts given by corporate sales (strategic). We now review the levels in more detail.

13.2.1 Tactical Level

At the tactical level, using real-time data provided by RFID systems will mean on-demand availability of the most current information. Most pundits suggest that there will be a major impact on dynamic inventory replenishment for manufacturing and warehousing operations. The use of RFID as a closed-loop passive system has been discussed throughout in earlier sections. The ability for information to be used in an open system, which provides information throughout the different levels, provides unique opportunities.

As information captured is integrated into an open system, external and internal disturbances can be evaluated. Internal disturbances are caused by breakdowns in company-managed assets; external disturbances are caused by factors beyond the organization control such as hurricane, major snowstorm, tornado, earthquake, significant disruptions in fuel supply, and so on.

Due to the fact that current information is available on-demand, it is possible to develop intelligent agent-based, real-time decision support systems to dynamically realign supply chain planning such as adjusting the transportation network, or opening temporary facilities.

13.2.2 Intermediate-Level Problems

As we discuss intermediate level planning, we are commonly referring to transportation planning. The main effect is for RFID real-time capabilities to reorganize transportation operations. We introduce transportation in the supply chain in the next section to provide background.

13.2.2.1 Transportation Strategy

As a supply chain driver, transportation has a large impact on customer responsiveness and operational efficiency. Faster transportation allows a supply chain to be more responsive but reduces its efficiency. The type of transportation a company uses also affects the inventory and facility locations in the supply chain. The role of transportation in a company's competitive strategy is determined by the target customers. Customers who demand a high level of responsiveness, and is willing to pay for the responsiveness, allows a company to use transportation responsively. Conversely, if the customer base is price sensitive, then the company can use transportation to lower the cost of the product at the expense of responsiveness. Because a company may use transportation to increase responsiveness or efficiency, the optimal decision for the company means finding the right balance between the two.

13.2.2.2 Common Transportation Decisions

The transportation design is the collection of transportation modes, locations, and routes for shipping. Decisions are made on whether transportation from a supply source will be direct to the customer or will go through intermediate consolidation points. Design decisions also include whether multiple supply or demand points will be included in a single run or not. Also, companies must also decide on the set of transportation modes that will be used.

13.2.2.3 Transportation Mode

The mode of transportation describes how product is moved from one location in the supply chain network to another. Companies can chose between air, truck, rail, sea, and pipeline as modes of transport for products. Each mode has different characteristics with respect to the speed, size of shipments (parcels, cases, pallet, full trucks, railcar, and containers), cost of shipping, and flexibility that lead companies to choose one particular mode over the others. Typical measurement for transportation operations includes the following metrics:

- *Average inbound transportation cost* measures the cost of bringing product into a facility as a percentage of sales or cost of goods sold (COGS). Cost can be measured per unit brought in, but typically included in COGS. It is useful to separate this cost by suppler.
- *Average incoming shipment size* measures the average number of units or dollars in each incoming shipment at a facility.
- *Average inbound transportation cost per shipment* measures the average transportation cost of each incoming delivery. Along with the incoming shipment size, the metric identifies opportunities for greater economies of scale in inbound transportation.
- *Average outbound transportation cost* measures the cost of sending product out of a facility to the customer. Cost should be measured per unit shipped, oftentimes measured as a percentage of sales. It is useful to separate this metric by customer.
- *Average outbound shipment size* measures the average number of units or dollars on each outbound shipment at a facility.
- *Average outbound transportation cost per shipment* measures the average transportation cost of each outgoing delivery.
- *Fraction transported by mode* measures the fraction of transportation (in units or dollars) using each mode of transportation. This metric can be sued to estimate if certain modes are overused or underutilized.

13.2.2.4 Intermediate-Level Summary

The fundamental trade-off for transportation is between the cost of transporting a given product (efficiency) and the speed with which that product is transported (responsiveness). Using fast modes of transport raises responsiveness and transportation cost but lowers the inventory holding cost.

13.2.3 Strategic Level

RFID effects strategic level planning in the form of location of facilities. The impact of RFID technologies is envisioned to have large reduction of inventory and labor such that the number of facilities can be reduced. Also, models for dynamically erecting temporary buildings and portable facilities based on demand shifts. The previous concepts of open-loop nesting of automatic data capture (ADC) technologies with RFID and effective communication would support these types of operations. From a closed-loop perspective, this idea can be achieved on a smaller scale with the common passive RFID EPC protocols that are being standardized and accepted. The open-loop concept may be achieved in the future with this standardization and adoption of protocols. This type of mobile supply chain is the future, using RFID technologies.

We provide a brief overview of facilities planning and consideration next.

13.2.3.1 Facilities Strategy

Facilities strategies involves identifying the locations to or from which the inventory is transported. This includes within a facility where inventory can be transformed into another state during manufacturing, stored in warehousing, and selected for customers in order fulfillment operations. External facilities determine how to profitably fulfill customer orders using transportation and effective inventory location.

Strategies such as using economies of scale when a product is manufactured or stored in only one location; this centralization increases efficiency. The cost savings may reduce responsiveness, as many of a company's customers may be located far from the production facility. On the other hand, locating facilities close to customers increases the number of facilities needed and, consequently, reduces efficiency.

13.2.3.2 Facilities Decisions

Facilities decisions regarding facilities include facility function, location, and capacity.

13.2.3.2.1 Facility Function

Production facilities designs are based on whether they accommodate production operations that are dedicated, or a combination of the two. Flexible designs can accommodate many types of products but is often less efficient, whereas facilities designed for dedicated products are more efficient. Also, designs distinguish whether product will support a product focus or a functional focus. A product-focused facility considers functions such as fabrication and assembly when producing a single product.

Warehouses and distribution centers (DC) must design facilities to accommodate a cross-docking or storage strategy. Cross-docking facilities design accommodates for inbound trucks from suppliers to be unloaded, broken down into smaller lots, and reloaded onto outbound store-bound trucks. For storage facilities, design decisions about reserve storage, primary picking location and replenishment from reserve storage to primary pick location must be made.

13.2.3.2.2 Location

Facility location is commonly a trade-off here on whether to centralize in order to gain economies of scale or to decentralize to become more responsive by being closer to the customer. Economic factors including quality of workers, cost of workers, cost of facility, availability of infrastructure, proximity to customers, the location of that firm's other facilities, tax effects, and other strategic factors are important prior to final decisions.

13.2.3.2.3 Capacity

Facility capacity determines flexibility and the ability to respond to wide swings in the demand. Excess capacity will likely be less efficient per unit of product it produces than one with high utilization; however, it will have the ability to respond to demand fluctuations. Common metrics include

- *Capacity* is the maximum amount a facility can store or process.
- *Utilization* is the percent of capacity that is currently being used in the facility.
- *Production cycle time is* the time required to process a unit if there are no delays at any stage.
- *Actual average cycle time* is the average actual time taken for all units processed over a specified duration such as a week or a month.
- *Cycle time efficiency* is the ratio of the theoretical flow time to the actual average flow time.
- *Product variety* is the number of product processes in a facility.

- *Top 80/20 analysis* is the percent of total volume processed by a facility that comes from the top 20% SKUs or customers. An 80/20 outcome in which the top 20% contribute 80% volume indicates likely benefits from focusing on the facility where separate processes are used to process the top 20% and the remaining 80%. Generally referred to as Pareto analysis.
- *Process down time* is the percent of time that the facility was processing units, being set up to process units, unavailable because it was down, or idle because it had no units to process.
- *Average production batch size* is the average quantity produced in each production batch.
- *Production service level* is the percent of production orders completed on time and in full.

13.3 RFID BEST PRACTICES FOR SUCCESS

In this chapter, we have reviewed many insights and concepts that will allow the logistics, industrial engineer, and operations manager to understand the opportunities and challenges with using RFID technologies. Some of the best practices for implementing systems include

1. Understand the need for implementation
   a. Mandates and compliance with customer
   b. Strategic cost reductions
   c. Perform a SWOT analysis to review integration into company operations
   d. Strengths, weaknesses, opportunities, and threats
2. Identify the process and operation wherein RFID implementation will be most cost effective. Generally, the higher up the supply chain that it is implemented, the more cost effective and the greater the complexity of the implementation
3. Create a prototype implementation in the identified operation
4. Test and evaluate the prototype
5. Improve the prototype
6. Retest the prototype
7. Roll out the RFID system to operations

In the next chapter, we suggest implementing steps 2 through 7 using a design for Six Sigma research approach.

13.4 SUMMARY

In summary, we describe how RFID supports information in the supply chain by enabling visibility. This visibility enhances supply partners' ability to optimize inventory, orders, raw materials, and delivery points. Automatic identification or auto-id technologies such as RFID into a common RFID nomenclature created by standardized technology protocols will provide large supply chain savings.

We introduce a planning structure that provides opportunities at different levels to reduce inventory costs with more effective labor policies, more effective scheduling, and the reduction of expensive assets such as facilities transportation containers. The ability of RFID to provide timely information and visibility into the supply chain is based on three aspects of RFID technologies that include ADC, real-time information, and real-time location status. This information can be used by the industrial engineer to provide successful RFID initiatives.

14 Implementing RFID Systems

14.1 INTRODUCTION

In this chapter, we discuss how radio frequency identification (RFID)

- Needs to be tested in prototype environment
- Introduction of Design for Six Sigma Research (DFSSR) for RFID testing
- How to execute DFSSR to test and evaluate prototypes
- Other methods for testing RFID environments

Recommended steps for a proactive implementation of RFID systems include

- Make the return-on-investment (ROI) case for RFID.
- Choose the right RFID technology.
- Anticipate RFID technical problems.
- Manage the IT infrastructure issues.
    - Data management concerns
    - Integration with back-end applications
- Leverage pilot project learning experiences.

14.1.1 Make the ROI Case for RFID

This first step appears to be a logical step for implementing any technology including RFID. The ROI analysis is commonly utilized to justify capital investments such as costly systems. Due to the fact that mandates are driving suppliers to use passive RFID technologies, many cost justifications are not performed prior to implementation. The mandates have moved RFID implementation as a cost of doing business as opposed to a justified business expense. At a minimum, cost justifications should be made to evaluate the lost value of implementing an RFID solution or when the investment may eventually pay for itself.

14.1.2 Choose the Right RFID Technology

RFID has been put into practice within industry for a number of years. To better assess its potential applications and implementation, this chapter presents an approach called Design for Six Sigma (DFSS). DFSS has many applications, including creating real-world warehouse testing environments, measuring and analyzing the implementation of RFID technology, and developing and optimizing the conditions for the most efficient readability. Finally, DFSS can be used to validate RFID technological principles and draw conclusions under this real circumstance.

RFID has been called the "bar code of the next generation." It is currently in use with applications ranging from libraries to toll booth e-passes. The greatest advantage in RFID systems, which are composed of an interrogator (reader) and transponders (tags), is that they do not require a line-of-sight (LOS) between the reader and the tags. Currently, thousands of items are tracked by manually updated databases or LOS bar code scans, which scan one item at a time. With an RFID system in place, entire bags of items affixed with RFID tags can be audited in seconds without ever having to open the bag.

14.1.3 RFID System Details

RFID systems consist of an interrogator (also referred to as a reader) and transponders, or tags. In other words, the tag can be hidden or imbedded within the item and the item will still be identified. In an RFID system, the antenna of the reader emits radio signals. The signals are received by the tag's antenna, which can be powered via a battery or by the radio frequency (RF) energy from the reader's pulse. The tags respond with a unique code, which is preprogrammed in the tag's microchip. After the reader antenna receives and decodes the signal, the antenna sends the information to a computer via a standard interface; this information is accessed through a database. Since RFID systems do not require an LOS between the tag and the reader, a space station crew-member could potentially initiate a reader in the general vicinity of a cargo transfer lab, that is full of tagged items, and record an accurate count of all items within the bag in seconds. In this chapter, we will use a simulated warehouse as an environment for DFSSR techniques.

14.1.4 Six Sigma Methodology

With the advantages illustrated above, a series of experiments and tests will be conducted by RfSCL lab at the University of Nebraska-Lincoln. A blend of methodologies will help the RfSCL lab use information to better develop a solution that best meets the needs of customers. The methodology used in this chapter is the integrated DFSS (Breyfogle, 2003). DMADO (De-may-doh) is the acronym for design, measure, analyze, develop, and optimize (Breyfogle, 2003). The planning function is categorized as defining the objectives and determining the correct measurement parameters. The predictive functions are analyzing viable options, designing experiments that lead to the development, identifying gaps, reanalyzing options based on performance, and creating a final design that meets the stated objectives. The performing functions are optimizing the performance of the design and verifying that the prototype is operational. This new approach to research will help bridge the gap between academic organizations and industry. The experiments and analysis will be conducted following the order of scientific methodology.

14.2 3P's THEORETICAL MODEL

In this chapter, we introduce a research framework, DFSSR, that is based on a common operational prototype theme that requires development teams to plan, predict, and perform. This 3P's methodology is utilized to encapsulate our DFSSR framework. For RFID technology, to better serve industrial applications, we need to conduct a series of experiments to validate the principle of facility layout of RFID into real case scenarios. The Six Sigma methodology helps us build a scientific procedure and makes sure it is the optimum layout for real warehouse scenarios. We use this framework so that operations can identify the status of projects and investigate the detailed processes within the framework. The DFSSR process steps are organized within the 3P framework as shown in Figure 14.1. The results or lessons learned can be used to effectively implement the technology in this environment. Further, the compiled lessons learned can be used to determine the best practices for implementations in the future. The methodology allows for the defining of the correct prototype environment, RFID subsystem testing, and integrated system testing for the prototype environment.

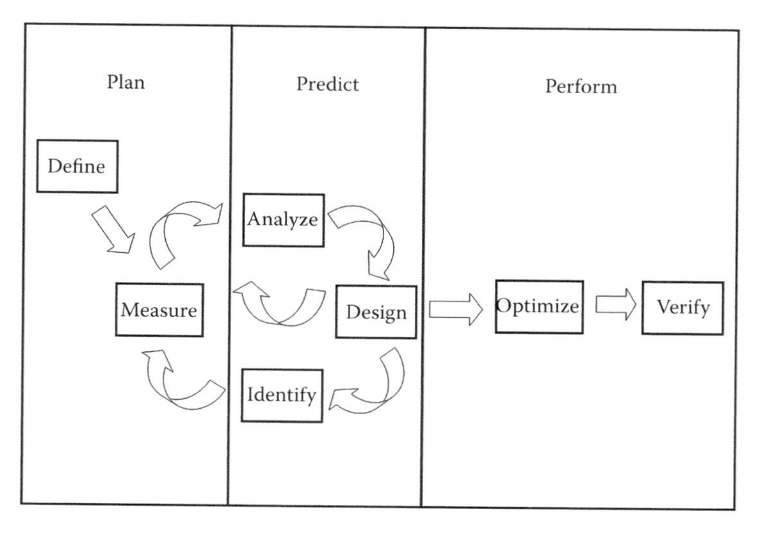

FIGURE 14.1 DFSSR 3P's methodology framework.

14.3 PLAN

In this phase, we need to identify the critical path for both information and material flow. As a beginning of plan phase, the first thing is to define the problem in real case, for example, what type of product do they use for inventory? What is the frequency of transportation they have everyday? That is all related to our test design for warehouse.

14.3.1 Define

In the define phase, it is necessary to compile the real environmental requirements into the test parameter. This makes it necessary to show the theoretical model in the design and analyze phases as an explanation and foundation for our future experiments. In this step, we describe facility layout process is based on input data, an understanding of the roles and relationships between activities, a material flow analysis, and an activity relationship analysis. The defining step is shown in the first phase—Plan; the clear material flow should be identified in this step as a basis to predict and perform. Figure 14.2 shows us a clear view of the thought process.

14.3.2 Measure

Multi-objective radio frequency (RF) warehouse architecture is the overall RFID warehouse implementing system. It includes three main parts: an RFID edge layer, an RFID physical layer, and an

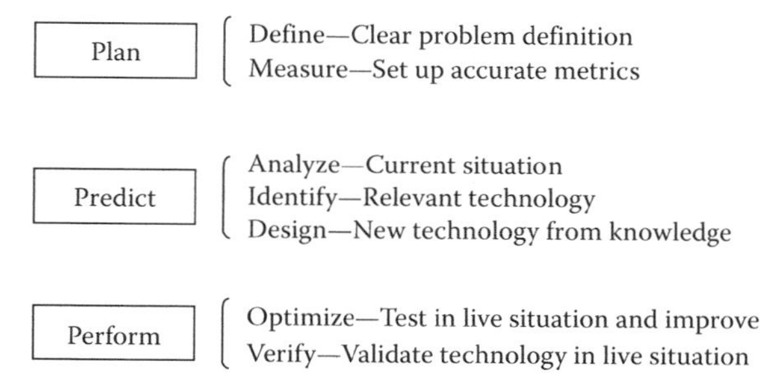

FIGURE 14.2 3P's thought process.

enterprise integration network. The RFID physical layer is the connection of the other two layers. The RFID system is designed to process streams of tags, or sensor data, coming from one or more readers. The edge layer has the capability to filter and aggregate data prior to sending it to a requesting application. For example, an action (tag read) is triggered when the object moves or a new object comes into the reader's view. The RFID edge servers filter and collect the tag data at each individual site and send it over the Internet to the third layer—enterprise integration layer. The localized data is identified by moving actions and stationary actions separately. This difference divides the usage of RFID antenna into two types of equipments, one is a fixed reader for warehouse portal door, and another is a mobile reader for tracking inventory. The fundamental tenet of the warehouse portal distribution system is that they must be able to accommodate changes that may occur on a network. The portal devices provide real-time, positioning access capabilities to user communities, and delivering and searching personal data. It allows external customers and partners access to data secure access.

We can now divide the RFID warehouse system into three parts as we discussed before: the physical layer, the logic layer, and the system integration layer. Each layer has different components depending on what functions the RFID system needs. By understanding the flow in the warehouse, we can determine the types of tags and antennas needed.

In short, RFID implementation in any process has two or three layers. The physical layer produces log events for RF sensors during process executions. The logic layer records the log events-related data including filter and integrate functions. The analysis of the physical layer activity has been discussed in facility layout research. Previous research with RFID facility layout does not include data flow as a factor that influences the RFID warehouse efficiency and performance.

14.4 PREDICT

In this phase, the major issue is to analyze the outcome and process of our RFID operation. As we know, the critical results of experiment or test are very important for the company that wants to implement RFID; the situation may vary from each company. Combine the real environment and site requirements, the test should be conducted in appropriate and cost-effective ways.

14.4.1 Analyze

Data environment analysis: One of the design components of the RFID warehouse layout is the data flow through the distribution process. This is included in the experimental design phase. The goal of such an activity is to define the input and output data in order to confirm the efficiency of data flow and its physical flow. Data standards can be smoothly exchanged within the supply chain because the data is already formatted and organized. The work flow and data flow are both generated by the production flow from the physical layer to the logical layer. All the data chosen through the distributing process are generated by RFID equipment including the tags on each pallet or the antenna on the portal. Therefore, the location of the RFID equipment is influential on the accuracy of the distribution process, which forms the individual data flow according to the workflow. The location of the RFID antenna, also called a sensor, will be discussed later on.

14.4.2 Design

According to our analysis of site environment and data types, we choose to use passive tag as our technology in the warehouse, which means low cost but high volume information flow.

First, the "sensor" is used to refer to a device that is connected via network or RF communication medium to other sensor devices in the network. The location of the sensors in the warehouse relate to either its environment or its data traffic flow, which is detected by a fixed antenna on the portal door. Similarly, the data flow will be employed by sensors specifically in the picking entrance portal and the distributing portal. Therefore, the data traffic through the two portal doors and its layout will be considered in this chapter. The other communication between the nodes in the warehouse will be discussed in future work. In order to measure the accuracy and efficiency of RFID performance in the warehouse, we are using a ratio to evaluate the relationship of performance and efficiency of RFID readability. This is σ that equals the simple relationship between input and output data, which will be related to a regression analysis to show fair performance:

$$\sigma_r = \frac{\alpha_I}{\beta_o} \tag{14.1}$$

where
σ_r is the ratio
α_I is the input
β_o is the output

However, this ratio only gives an average performance for RFID readability. The components of the input require the precise data to evaluate the environment and performance. However, we measured the benchmark of the performance used to compare the different input data and data flow. For example, the different amounts of workflow reflect the different data flow in the warehouse, but the benchmark gives us reliable data to measure different warehouse environments and workflows. In the next stage, the optimize step, we will use a case study to illustrate our theory explained in the design and analyze phases.

14.5 PERFORM

The last phase of our study will be Perform, after previous experiments and design applications. The next stage is to prove the feasibility of our design by using design of experiment (DOE) and optimize the system performance according to the current configurations.

14.5.1 Optimize

Based on our theoretical model, we will conduct experiments following the order of the optimum facility layout and cost-effective equipment. As a major part of statistical analysis in Six Sigma methodology, we use DOE as a possible improvement through process. In DOE, the effects of several independent factors (variables) can be considered simultaneously in one experiment without evaluating all possible combination of factor levels. For experiment 1, two independent variables (factors) and one dependent variable were used. The two independent variables include the tag replacement and the number of antennas. The dependent variable was the readability of the tags. The observed results of the experiment had an effect on the independent variable:

$$\text{Readability} = f(\text{TP}, \text{AN}) \tag{14.2}$$

where
TP is the tag placement
AN is the antennae number

For experiment 2, we needed to regulate some of the variables until the results achieved full-read efficiency. This was done to satisfy the customer's requirement of 100% readability. The model of the experiment is

$$\text{Readability} = f(\text{AP}, \text{PP}) \tag{14.3}$$

where
AP is the antenna's position
PP is the portal's position

14.5.1.1 Factors and Levels

When executing a full factorial experiment, a response is obtained for all possible combinations of the experiment. Because of the large number of possible combinations in full factorial experiments, two-level factorial experiments are frequently utilized. Experiment 1 was a two-factor, two-level experiment. As described in Table 14.1, a total of four trials were needed to address all assigned combinations of the factor levels.

TABLE 14.1
Factorial Design

Factors: 2	Replicates: 10
Base runs: 4	Total runs: 40
Base blocks: 1	Total blocks:1
Number of levels: 2, 2	

The specific situations to which a DOE is applied will affect how factors and levels are chosen. Factor levels also can take different forms. In this experiment, levels are quantitative. Experiment 1 should allow for a systematic observation of a particular behavior under controlled circumstances.

Therefore, the two factors utilized in the experiment were as follows:

- *Tag placement*: Top, side
- *Number of antennae*: One antenna and two antennas (on each side of portal)*

For experiment 2, in order to test the readability of the tags and metrics performance, the same trials were performed as experiment 1, with the following variables:

- *Position of antennas*: We installed two antennas at the same height on each side and at two different heights (3, 5 ft).
- The distance between each side of the portal: 5 ft; 7 ft.

The standardized time scale was 30 s in consideration of limited real-world data acquisition times. All of the specifications were conducted 10 times with three different replacements of tags and 10 items in each trial. The experiment factors and levels are summarized in Table 14.2.

14.5.2 Verify

We have several ways to optimize the current layout design to improve the reading accuracy: For experiment 1, the placement of the tags on the pallet or item can be classified in three ways: top, front, and side. The performance of each classification is different. To sum up, the best position of a tag on an item is on the side, (face to the antennae) because of the polarization and magnetic field (polarization test report). The performance of the antennae and tags is totally different with these three classifications. The third classification, tag on the side, had the best performance. Compared with the other two classifications results, the readability of the tags, when they are on the side, can be up to 60% of full satisfaction. This classification is still not very satisfying. We also determined that a significant change occurred when the number of antennas was varied. We ran the experiment

TABLE 14.2
Experiment Factors and Level

Factors and Designations		Levels (1)	Levels (2)
Experiment 1	A1:Tag placement (TA)	Top	Side
	B1:Antenna number (AN)	1	2
Experiment 2	A2:Antenna's position (AP)	Horizontal	Non-horizontal
	B2: Portal's distance (PD)	5 ft	7 ft

* Two antennas were only used in one trial within the experiment.

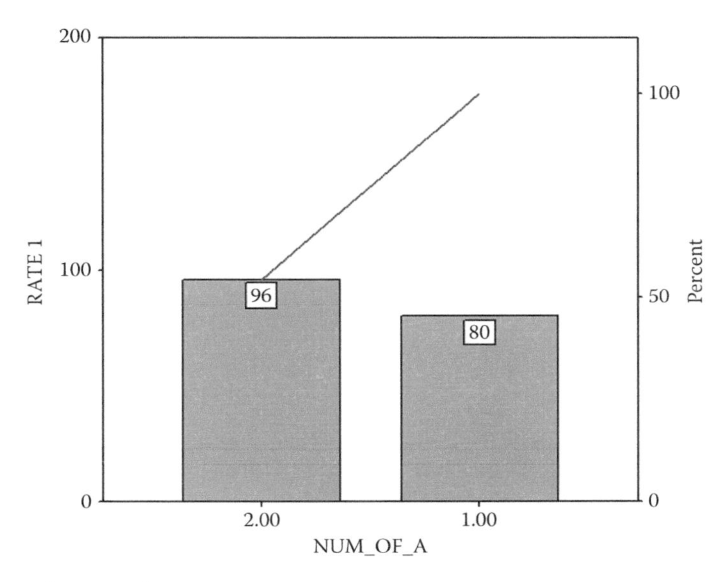

FIGURE 14.3 Read rate graph.

using one antenna with 10 items first. The readability of the tags was only 80%, compared with almost 100% when using two antennas (AVE: 96%). The read rate graph is shown in Figure 14.3.

For experiment 2, the influence of the variable tag placement in the model is the same as in experiment 1. The antenna position had different effects on the two experiments. The experiment hypothesized that the antenna, when placed in different horizontal planes, would have a positive influence on readability. The non-horizontal antenna showed better results when distance between antennas increased. For example, the experiment conducted from 5 to 7 ft demonstrated better results at 7 ft than at 5 ft. Portal dock's position is an important factor that has an influence on reading accuracy, especially in experiment 2. The hypothesis was that the shorter the distance is, the better read efficiency would be. The experiment simulates real-world circumstances and is designed for two hypothesized cases. The requirements are the distance between each side of the portal must be 5 ft with the antenna on the same horizontal line and the distance between readers must be 7 ft with different heights. Therefore, the objective of the experiment was to verify the hypothesis about whether the performance of readability will be better with "non-horizontal line" orientation. If the hypothesis was proved to be true, the improvement on the antenna reading efficiency would increase by varying the height of the two antennas when other factors are fixed.

Finally, the results of the experiment supported the above hypothesis. We identified the normal distance between each side as 7 ft, but the optimum and effective distance is approximately 5 ft. The factors of distance between each side and the non-horizontal antenna both have an influence on the effectiveness of readability.

14.6 CONCLUSION

The reaction time of the antenna on the tags was almost the same in these two cases. It can be determined that readability can achieve a full-read expectation when performed under the following specifications:

- The full-read range is 3–5 ft when antennas are fixed at the same horizontal line on each side of portal.
- The full-read range is 6–8 ft when the antennas are fixed with different heights on each side of portal.
- The full-read requirement needs to have the tags on the sides of items or facing toward the antennas.

Frequency, distances and angles, type of tag, location and replacement, influences of moisture and metals, and pallet patterns all played part in the readability of the tags. The effective reading distance was analyzed in MATLAB® (6.0, Release 13), for visualizing the results and provided documents for future research. The data points on the graph showed random variation, but the visualization graph gives a clue that the most effective scale for the antenna is between 2 and 3 ft around the middle line. The color bar on the right side indicates the read rate, which is based on our experiment specification of 10 tags per pallet. The reading rate can be reached at 90% or better under this specification (Figures 14.4 and 14.5).

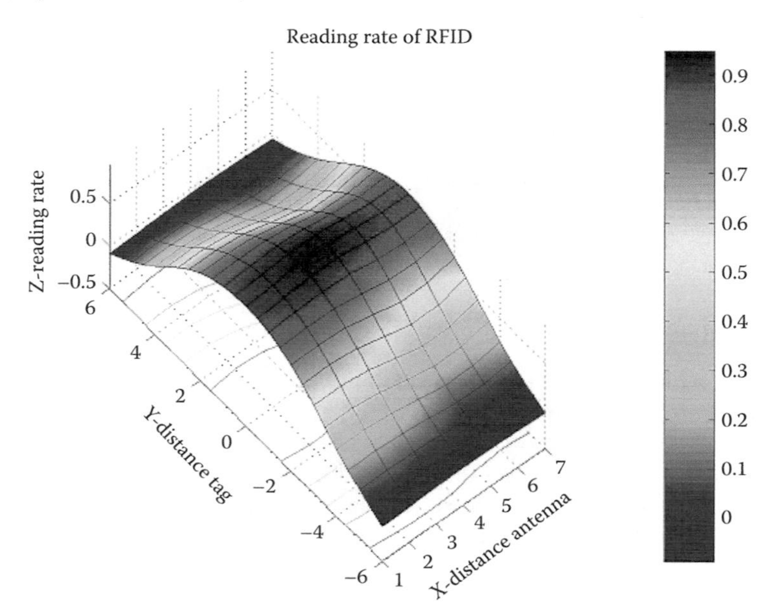

FIGURE 14.4 3D graph for effective reading rate.

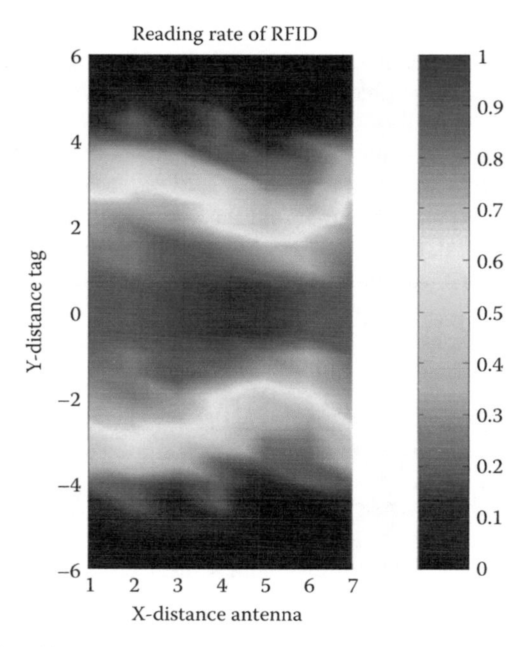

FIGURE 14.5 2D graph for effective reading scale.

To sum up, the total conditions for receiving more than 90% readability includes several considerations as follows:

- Placement of tags
- Distance between antennas (if there is more than one antenna)
- Appropriate stop-by time when going through the portal (at least 3 s)
- Change of the position of the antenna when other limitations are fixed

Using Six Sigma methodology increases the efficiency of test procedures and validates the influence of real warehouse case layout scenarios.

15 Obsolete Inventory Reduction with Modified Carrying Cost Ratio

15.1 INTRODUCTION

This chapter considers a medical supplies and pharmaceutical inventory management problem faced by a city health and human services department and shows how inefficient warehousing can become if inventory is not periodically checked for obsolescence. The department purchases and distributes medical supplies used by service centers spread throughout the city. To support the service process, the department operates several warehouses in the city. Furthermore, the warehouses order large quantities of goods at a negotiated price and store the items until a service center places an order for a relatively smaller quantity. Then the warehouse fills the orders. If the order quantity is on hand, the items are picked by warehouse employees and delivered to the offices. If the warehouses do not have an item in stock, the item is placed on backorder and is delivered when it becomes available. The department estimates that the out-of-stock inventory is very costly.

This chapter is organized in the following manner. The next section will discuss the carrying costs associated with two-echelon supply chain inventory model. Then a one-echelon supply chain inventory model that utilizes a just-in-time (JIT) procurement will be discussed. Next, we will introduce the modified carrying cost ratio model as a decision tool to evaluate which system to use in practice, followed by a case study that demonstrates the practical use of this ratio in making a decision in the aforementioned governmental operation.

15.2 TWO-ECHELON MODEL

In 2003, Caglar presented a two-echelon supply chain model that we consider very useful in making cost-effective decision about warehouse inventory levels. We utilize this model to demonstrate the current two-echelon supply chain in practice by the city department. First, we will consider a two-echelon multi-consumable goods inventory system consisting of a central distribution center and multiple customers that require service illustrated in Figure 15.1.

Each service center office acts as a smaller warehouse. This is because they each supply many customers and maintain a stock level SiM for each item. Therefore, each office consists of a set I of n items that are used at a mean rate. When an item is used by a customer, the customer replenishes itself by taking item i from office M's supply stock if the item is in stock. If the item is not in stock, the item is back ordered and the customer has to wait for the item to become available at the office.

The goal of this chapter is to make a decision of supply chain type based on basic purchasing and holding cost information, while maintaining an average response time that will not negatively affect the customers. This may include the elimination of the central warehouse.

Using the notation in Table 15.1, a model of the cost of operating a warehouse and implementing a JIT system was derived. This information can then be used to determine if the organization benefits from operating the warehouse.

There are many operating costs associated with warehouse management. These operating costs include fixed costs such as racking, utilities, labor, vehicle fleet maintenance, property maintenance,

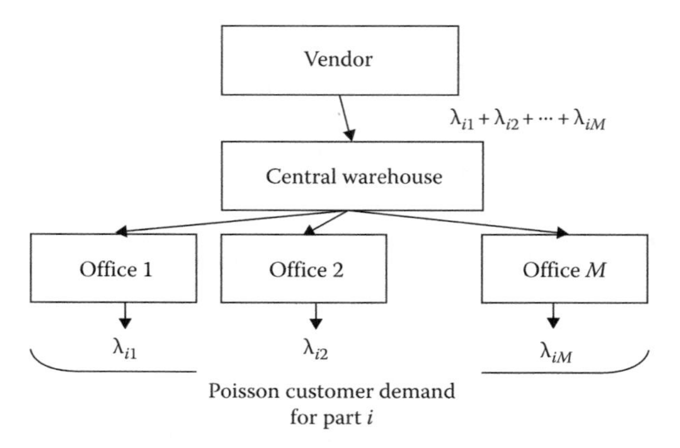

FIGURE 15.1 Two-echelon inventory chain.

TABLE 15.1
Notation for Models

Notation	Meaning
A_w	Annual fixed cost of warehouse operation
C_I	Total cost of holding inventory
C_{Lj}	Labor cost at warehouse j
C_V	Cost of vehicles and maintenance at office j
C_{Uj}	Cost of utilities at office j
C_W	Lease price or depreciation and cost of capitol of warehouse
C_{Mj}	Annual property maintenance for warehouse j
$J = \{1, 2, \ldots, M\}$	Set of offices
K_j	Customer at office j
l_i	Demand rate of item i
L_{JITij}	JIT lead time for an expedited order of item i at office j
$\lambda_{ij} = K_j l_i$	Demand rate for item i at office j
θ_c	Organizations cost of capitol
θ_{Oij}	Obsolescence rate for item i at office j
θ_S	Shrinkage rate based on total inventory in system
P_{Wi}	Purchase price using warehouse system of item i
P_{JITi}	Negotiated JIT purchase price for item i
S_{ij}	Base stock level for item i at office j
SS_{ij}	Safety stock of item i at office j
V_{Wj}	Value of warehouse j
W_{ij}	Waiting time for a customer ordering item i from office j
W_j	Waiting time for a customer ordering from office j

property depreciation, and a lease or any other tied-up capital. Let A_w be all periodic fixed costs that the savings of purchasing in large quantities have to justify in order to minimize the total cost of the operation. For this model, we will use annual costs:

$$A_w = \sum_{j \in J} C_{Wj} + C_{Uj} + C_{Lj} + C_{Vj} + C_{Mj} + \theta_c * V_{Wj}$$

These fixed costs, in addition to item-associated costs, make up the total cost of having a warehouse in operation. Many of these costs are hidden and are frequently overlooked when procurement managers decide the level of quantities to purchase. Shrinkage in the form of lost items, stolen items, or damaged items, obsolescence, and the cost of capitol on the inventory is typically among these hidden costs. These costs can be modeled as a percentage of the total inventory on hand.

15.3 ONE-ECHELON MODEL

The second model used for reference is the common one-echelon JIT system. JIT requires better planning of demand from customers and can sometimes make management feel uneasy about the extra procurement cost of items on a per unit basis. But there are many cases where the elimination or significant downsizing of a warehouse operation can save money without sacrificing service to the customer.

In the JIT system depicted in this model, ordered items go directly from the vendor to the office, where a smaller stock level is used versus the warehouse. One-echelon systems will differ in that there is no intermediary between the vendors and the offices (Caglar et al., 2003; Lee, 2003; Wang et al., 2000). This system is shown based on a simplification of the Caglar model in Figure 15.2.

The JIT contracts that will need to be made with the vendors be established based upon demand rate λ_{ij}. We determine the expected time of backorders of item i in office j by the following:

$$W_{ij} = E\left[L\left(S_{ij}\right)\right] = \sum_{j \in J} \sum_{i \in I} \left(L_{\text{JIT}ij} * \left(1 - \sum_{i=0}^{SS_{ij}} \left(\frac{\lambda_{ij} L_{\text{JIT}ij}}{n!} \right)^{n} \exp\left(\lambda_{ij} L_{\text{JIT}ij}\right) \right) \right)$$

In this case, items are delivered to the offices at the same rate the items are being used. The symbol t_{ij} represents time between deliveries for item i at office j. Therefore, by substitution, $\lambda_{ij} t_{ij}$ is also the order quantity:

$$S_{ij} = \lambda_{ij} t_{ij} + SS_{ij}$$

Keeping the expected wait time for the customer for each system the same will allow for a comparison of costs without changing the response time to the customer.

Costs associated with the JIT system will contain all of the fixed costs of the system as well as any additional costs of requiring more service from vendors. In some instances, the unit price can remain constant by ordering a couple of large-quantity orders or several small-quantity orders. However, shipping rates for the smaller orders may increase. Due to this, it may be important to select vendors that are close to the offices.

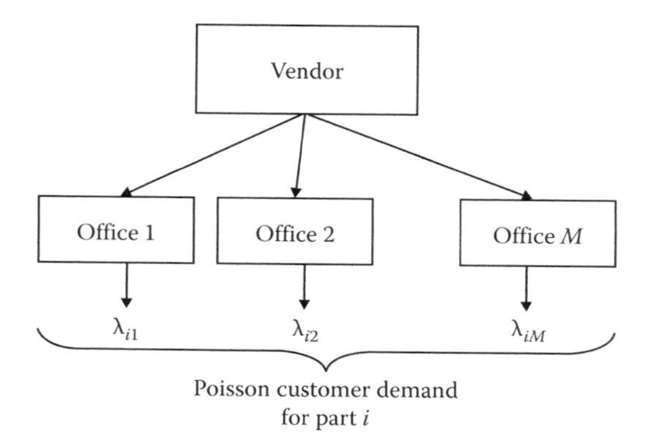

FIGURE 15.2 One-echelon inventory chain.

After factoring in a possible increase in purchase and shipping prices, we suggest that the total cost for the JIT system will be as follows:

$$C_{\text{JIT}} = \sum_{i \in I} \sum_{j \in J} P_{\text{JIT}i} \lambda_{ij} + C_I$$

when

$$C_I = \sum_{i \in I} \sum_{j \in J} \left(I_{ij} * \left(\theta_c + \theta_S + \theta_{Oij} \right) \right)$$

Once again, in many situations, the data needed to use this optimization may not be available in the time allotted to the project. This is where our simplified carrying cost ratio model simplifies the decision to move to a two-echelon system.

15.4 MODIFIED CARRYING COST RATIO MODEL

The model focuses on comparing the two systems and selecting the best choice operational model. As long as the total cost for purchasing, storing, and delivering items to the customer can be derived, we can determine which system is a better economic choice with our decision model.

The ratio simply compares the total cost of the purchased inventory to the amount of money spent holding and delivering it to the offices. This cost ratio has been developed to evaluate and analyze supply chain costs for operations relying on inventory delivery from a supplier. The purpose is to provide a methodology for determining cost incurred over the supply chain process from the time an inventory item is loaded on a truck from the original vendor to the time an operation buys or requisitions the item for use in their business. The merits of understanding these incurred costs include (1) an understanding of the cost of each item, (2) knowledge of the cost the operation would be required to overcome, and (3) guidelines for which actions an operation can take to decrease the cost/dollar spent ratio.

We hypothesize that the cost of inventory plus the fixed costs comprises the total cost of the warehouse operation, given by the following equation:

$$\text{Total warehouse cost} = A_w + C_I$$

We suggest that after identifying the stock levels using the aforementioned formulas or current accounting information, the next step would be to use our ratio to determine which system is better for the operation. We present the ratio as a calculation that can be used in operations. Put simply, it is the ratio of the total cost of maintaining the inventory and the total inventory purchase price:

$$\mu_W = \frac{A_W + C_I}{\sum_{i \in I} C_{Wi}}$$

when

All costs are annual.

$$\sum_{i \in I} C_{Wi} = \text{total dollars purchased}$$

The above relationship defines the total cost determined over the course of the supply chain. It combines the cost of delivering an item with the cost incurred during the process of holding that item in inventory. This equation is the ratio of warehouse cost per item to purchase price per item. This effectively demonstrates the ratio of money a supplier spends storing and shipping an item to the actual monetary investment put into each inventory item, represented by the ratio (CSystem)/CP.

TABLE 15.2
General Handling Cost

Cost Source	Purchase Price (%)
Insurance	0.25
Storage facilities	0.25
Taxes	0.50
Transportation	0.50
Handling	2.50
Depreciation	5.00
Interest	6.00
Obsolescence	10.00
	25.00

This ratio, when combined with holding cost, can be extremely effective in determining the efficiency of a supply chain as well as providing an indicator of the inventory turn rate for the entire system. We will later show the results of our case study using data to perform the calculation of this ratio.

Another benefit of this formula is that it can be used as a baseline for the financial efficiency of the operation. This unitless number is a ratio of total dollars spent maintaining inventory versus the total purchase price of all the items in the inventory.

Most JIT contracts will increase purchase prices between 15% and 25%. Thus, if an organization's modified carrying cost ratio is above this target, JIT one-echelon options should be considered. Table 15.2 shows a widely cited breakdown of holding costs associated with warehousing merchandise (Johnson, 1999). If the percentages are above these baselines for any specific holding cost area, focus can be turned to that area. Some pundits suggest that in the event that the total cost for storage facilities is above the 25% baseline, lowering facilities costs through the elimination of facilities in conjunction with a JIT system should be recommended. We suggest the use of the carrying cost ratio as an alternative before making this type of decision.

15.5 RESULTS: CASE STUDY

This system was used in the analysis of the "City of X" health and human services (CoXHHS) department that had its own distribution network to service 30 offices. An analysis was performed to determine inefficiencies in the supply chain (slow inventory turn items, inefficient racking, etc.). This data was then used to create a cash flow analysis to determine which actions would be useful in reducing operational costs. We suggest these methods can be beneficial in determining which actions will yield the most positive results in reducing costs and increasing net profits for an organization.

15.6 FACILITIES COSTS

The second element of the holding cost calculation involved compiling the total facilities cost for each of the warehouses involved in the operation's supply chain. This data is included in Table 15.3. Additionally, CoXHHS was only leasing WH 2 at a price of $78,000 a year. This incremental price was another possibility for removal, as all the other warehouse facilities were owned by the city. These problems would be an important factor in determining which actions to take in an options analysis.

TABLE 15.3
Facilities Cost

WH *j*	Labor Cost	Utilities and Supplies	Lease Cost	Facility Total Cost
WH 1	123	356	0	480
WH 2	30	50	78	158
WH 3	26	74	0	100
WH 4	26	62	0	89
WH 5	12	28	0	40
Total	217	570	78	867

15.7 PURCHASING COSTS

With facilities costs and inventory turn rates by item calculated, it was possible to proceed to a more in-depth analysis of the data. Inventory turns refers to the number of times per time period an item is purchased and sold. For example, a turn rate of 1.0 indicates that all inventory that was purchased during the year was ordered. The first step was to calculate an average turn rate for all items for each facility in the CoXHHS supply chain. The desired result is that each facility would have at least a turn rate of 1.0, indicating that the inventory in each warehouse was turned once a year. The results are summarized in Table 15.4.

Table 15.4 shows the only facility that demonstrated the desired average turn rate was WH 2. The other buildings, especially the WH 3, featured extremely low turn rates. The most likely cause of this was the presence of vast amounts of obsolete inventory in each facility. Schnetzler et al. (2007) notes that in trying to achieve lower inventories and shorter lead times, operational costs are affected. These effects can be counteracted by reducing the amount of waste and obsolete inventory present in the system. The low receipts for the WH 5 shows that they were not ordering any items, a fact which is consistent with its role as an intermediary building in the supply chain. Thus, their low turn rate is acceptable given the building's role. However, the WH 4 and WH 1 each sent out a large number of orders but contained an unacceptably low turn rate.

15.8 MODIFIED CARRYING COST RATIO

The total cost incurred per item was calculated for the entire CoXHHS supply chain and compared back to the total purchase cost, resulting in the warehouse cost per dollar spent. This calculated value was also exceptionally high, netting an average of $0.97 per dollar purchased being spent to store and transport each inventory item. Lowering this ratio could be accomplished through a variety of methods, including consolidating inventory, increasing efficiency by standardizing procedures and optimizing storage use, and, most importantly, through elimination of obsolete inventory items from each facility. Table 15.5 shows the calculations for the CoXHHS modified carrying cost ratio.

TABLE 15.4
Purchasing Cost

Warehouse #	Turns/Year	Total Receipts
WH 1	0.36	$48,065.62
WH 2	2.18	$501,062.43
WH 3	0.07	$34,541.00
WH 4	0.49	$531,931.75
WH 5	0.15	$25,475.21
Total purchases		$1,141,076.0

TABLE 15.5
Carrying Cost Ratio of CoXHHS

Costs	Facilities	Shrinkage	Fleet	Sum
Annual	867	127	87	1081
Purchases	1115			1115
			$\mu =$	0.97

15.9 INVENTORY TURN ANALYSIS

The ratio showed that the facilities cost of the system was well above 25% of the total purchase price. So, in order to eliminate facilities and implement JIT, inventory turns data was needed. Inventory turns are defined as the average number of items kept in stock divided by the annual usage of the item:

$$T = \frac{S_{ij} + S_{i0}}{\lambda_{ij}}$$

TABLE 15.6
ABCD Analysis of Inventory

Category	Number of Items
A	104
B	150
C	476
D	2262
Total	2992

The ABC analysis compares all the items ordered and prioritizes them according to use. The results of the analysis are presented in Table 15.6. An ABC analysis evaluates the turn rates of inventory. An A mover is the fastest mover with the highest turn rate, a B mover is next, and so on. In most ABC analyses, the slower moving inventory is referred to as C and D movers. In our study, the order policy for each type of movers was set by movement category. Items that are deemed as "A" movers were placed on continual review for reordering. "B" movers have a review quarterly. "C" movers are reviewed annually. Items that had an inventory turn period greater than 1 year, we classified as "D" movers or obsolete inventory.

15.10 DECISION

After determining that the current cost ratio for the CoXHHS was above the expected 15%–25% procurement cost increase, a decision was made to switch from a two-echelon system to a one-echelon system. The switch had an earning before interest and taxes of $250,000 with a return on investment of just over 1 year. The cost ratio was reduced from 97% to 30%. Ordering policies were simplified and managed by each office, eliminating the need for a centralized logistics system. However, much of the savings was due to lowering the total volume of obsolete inventory in the warehouses. This reduction in obsolete inventory produced a 75% reduction in racking requirements.

15.11 CONCLUSION

Many organizations operate warehouses in order to reduce costs. Oftentimes in governmental operations, if not carefully managed, these warehouse operations become bloated with inventory that is no longer needed or is needed at a much lower demand. Unless managers periodically analyze the contents of their warehouse, the carrying cost of all items purchased can outweigh savings from procurement when purchasing in bulk.

In today's fast-paced business world, the time allotted to evaluate business operations is not available and quick decisions need to be made. This modified carrying cost ratio, based on easily found data, shows when warehouse operations are not cost effective and inefficient. This model speeds up the process and thereby speeds up change and cost savings in a company.

However, there are some limitations to this model. One limitation would be very large systems where JIT contracts would be too complicated. Organizations with a large service range such as a regional or larger retailer may not benefit from this ratio as is. However, for a smaller company or a city, this model can be very effective at recognizing overcapacity or inefficiencies in a supply chain.

Part V

Overview of Logistics Planning and Inventory Control

16 Engineering Economics of RFID

16.1 INTRODUCTION

In order to ensure cost-effective production of specialized products, many companies use calibrated tools. Failure to use the proper calibrated tool can result in scrap and rework. This inefficiency results in events such as audit costs, higher labor costs, and loss of customer trust that translates into lost sales. Further, customers often sue companies that produce defective parts that cause injuries. These events not only produce negative publicity for the companies but may result in costly lawsuit settlements.

Calibrated tools are essential to producing quality products and should be efficiently tracked for optimal labor productivity. However, this practice is not always demonstrated. In this study, the observed company's management attributes financial losses each year to lost or stolen calibrated tools. A stolen tool is considered a lost tool for this analysis. Defective parts due to using non-calibrated tools can trigger facility audits by customers as specified in contractual arrangements. Consequently, failed audits may result in a contract fine if an operator is found to be using a non-calibrated tool.

In addition to the cost of an operator using a non-calibrated tool in production, there are also the costs of losing calibrated tools. First, there is the cost of the tool itself if the tool is never found. Second, there is the cost of the lost labor time spent searching for the tool. And finally, the lost production time if the tool is needed immediately in production. We suggest that these latter costs could be alleviated by the use of radio frequency identification (RFID) which can provide real-time tracking of calibrated tools.

RFID can be used in an asset management system, such as those supported by enterprise resource planning (ERP) systems. Asset management systems should locate assets individually, allow for locating the correct assets at the correct time, and provide information about each individual asset and the physical status of the asset.

These uses are the foundation of RFID systems. A tag is placed on each asset that carries the information of the individual item. These tags contain an antenna, which allows information to be transmitted at the frequency identified by the reader. Readers are located throughout the production facility in coordination with their reading distance abilities. Software is used to capture the information transmitted to the reader. The reader sends the data to the inventory management system, which allows for parts to be tracked and located throughout the production facility.

16.2 PROBLEM STATEMENT

The fundamental question for this study is will implementing an RFID-based system reduce some of the cost of poor quality incurred due to lost calibrated tools? This analysis focuses on the labor costs, audit costs, and management time needed for implementing an RFID system. This chapter does not address lost quality due to other factors that management felt were not related problems that RFID could solve. These include training, wrong tool usage, and gauge precision. These are valid reasons for loss of quality, but not explicitly related to the tracking of lost tools.

This chapter formulates a cost analysis of implementing an RFID system to track calibrated tools throughout a production facility. By comparing two different scenarios, the best plan of action is defined (Evans et al., 2004). The net present value is used to evaluate an RFID system implementation. The goal of a new system is to save the company money with increased traceability. The RFID system put in place must provide savings greater than the cost to implement and be innovative in nature so as to put the company ahead in the industry.

16.3 BACKGROUND

An armament and technical products manufacturing facility, which has no current tracking system in place, is analyzed. Increased costs due to audits, rework, and customer dissatisfaction have been identified by management as costs incurred by using the wrong tools, or a non-calibrated tool. In order to place more control over the practice of wrong tool usage due to the loss of calibrated tools, the company evaluated an RFID tracking system. The facility has approximately 132,000 ft^2 and 200 total employees. Management personnel and production personnel costs are $40/h and $20/h, including indirect costs, respectively. Approximately 2000 calibrated tools are utilized in production.

A dedicated staff has the responsibility of calibrating and supplying the tools to the production workers. The communication between the production floor and calibration staff on the location of tools was identified as a problem; there was no effective tool tracking. Therefore, calibrated tools were difficult to find when needed and little feedback was available to production supervisors when production operators were unable to find a tool. This lack of feedback led to operators either looking for the tools or using the incorrect tool. The costs associated with using the incorrect tools including labor, scrap, rework, and failed audits are evaluated in this chapter.

16.4 COST JUSTIFICATION

This study presents two scenarios. The first scenario is the doing nothing option or the company remains status quo. This scenario describes the baseline costs for the study. Scenario 2 demonstrates the costs of implementing the RFID system over a 5 year period.

16.4.1 Scenario 1: Baseline

The first option the company has in regards to tracking its calibrated tools is to remain status quo, which we consider the baseline. This suggests that the costs the company is incurring will remain unchanged over the 5 year period considered in this chapter. In order to show the total cost, each cost is considered separately. These costs include audit costs, rework costs, scrap costs, management costs, and customer service costs.

16.5 AUDIT COSTS

External auditors periodically review processes and procedures at the company. These audits include governmental audits and environmental audits. This study focuses on the audits that review the products created from calibrated tools. For each of these auditors that visit the plant, there is a cost to the company. The initial audit is always obligatory, therefore, the cost of the first audit is not considered. However, problems identified during the initial audit can lead to an additional audit for production areas that do not pass inspection. These secondary audit costs create additional company efforts such as management and operator time for communications about failed audits, delayed contracts, and possibly layoffs due to lost contracts.

There may be one or two more secondary audits throughout the year, which would not have been needed previously. In order to estimate the cost of an additional audit, conditional probability trees are utilized to assess the probability of the company needing a second, third, or fourth audit. The probabilities utilized were established through management interviews. Due to company privacy

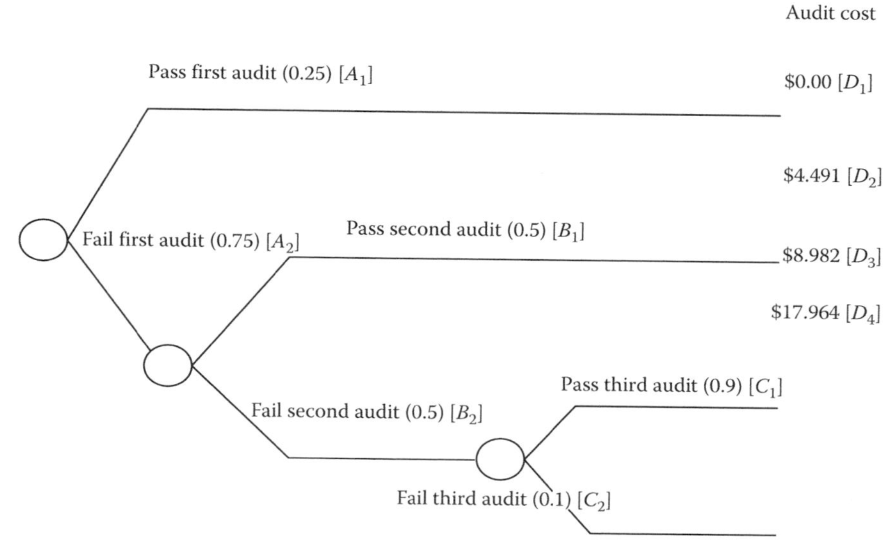

FIGURE 16.1 Audit conditional probability tree.

issues, this study does not display the histograms or show the distributions of the number of occurrences of error. Instead, only the distribution probabilities are shown in Figure 16.1 as conditional probability trees.

The expected value of the audit cost $E(X_i)$ is given as a function derived from the decision tree. The following equation will be in this study for the audit cost calculations:

$$E(X_i) = \mathrm{A1D1} + \mathrm{A2B1D2} + \mathrm{A2B2C1D3} + \mathrm{A2B2C2D4} \tag{16.1}$$

X_i is the audit cost of contract i. For the audit cost calculation, an additional penalty is given if the third audit is not passed, represented as D4 in Equation 16.1. The given cost is twice the cost of passing the third audit. This cost was included due to the additional cost of special efforts made by the company when an audit fails three times.

The costs include auditor labor and travel expenses. Each auditor is conservatively estimated to earn $25/h with 33% benefits or a total cost of approximately $33/h. Each audit takes approximately 3 days, and the auditors work 8h daily. Thus, the average number of hours worked per audit is 24h. Overall, the personnel cost for an audit team with three members is $2376 per audit.

In addition to the costs for the auditors, travel expenses were included. Travel expenses include air travel, lodging, and food. Given an audit team of three auditors, the total cost for airline tickets is $900 ($300 per airline ticket). Next, the audit team lodging and food costs were $100 per night for a hotel room and $35 a day for food over 3 days, such that the cost for three auditors would be $1215. These costs are summarized in Table 16.1.

TABLE 16.1
Audit Costs

Audit Cost Elements per Audit	Scenario 1
Auditor	$2376
Travel	$900
Lodging and food	$1215
Total cost	$4491

The total cost per audit is approximately $4491. The expected value for an audit with the given probability of failure is estimated to be $5389.20:

$$\begin{aligned} E(X1) &= (0.25)(\$0) + (0.75)(0.5)(\$4,491) \\ &\quad + (0.9)(0.5)(0.75)(\$8,982) + (0.1)(0.5)(0.75)(\$17,964) \\ &= \$5,389.20 \end{aligned} \tag{16.2}$$

The current company contracts that were audited the previous year were 27. The possible savings that can be achieved from audit reduction is the product of the expected audit cost and the number of audits. We use a conservative estimate of reducing the cost of 20 audits for a total cost savings for audit of $107,820.

16.6 REWORK COSTS

Another cost to consider is rework cost. Through this study, the company analyzed did not have data available to individually measure the different causes of rework, specifically due to a missing calibrated tool. Management estimates that 90% of defects were reworked and the other 10% become scrap. The previous year's total defects were 3800, including defects in the final product and in subassemblies. Given a fraction defective of 0.90, we conservatively estimate 3420 defects were reworked. Rework included such things as retooling in order to fix the defect, complete rework of a part, or fixing a broken piece. Management interviews support an average of 8 h per defect reworked. A production worker makes approximately $15/h plus 33% benefits, which equates to a fully total cost of $20/h for a production worker. Therefore, with 3420 defects per year reworked, 8 h lost per rework, and $20/h per operator, the estimate average total rework cost per year is approximately $547,200 (Table 16.2).

16.7 SCRAP COSTS

The estimated scrap cost was calculated by taking the median price of the final product and multiplying by a factor of 0.72. This factor was used because the profit and transportation costs were estimated at 28% of final product price. Other costs such as management, warehousing, and labor were inclusive in the cost factor. Parts in this company range from a price of $400 to $200,000 per part. There are approximately 2346 parts manufactured each year. These are the final products produced to be sold in the market. The median price was estimated at $50,300 and the cost is estimated at $36,216.

Currently, 380 defects are scrapped each year. We conservatively estimate 10% of these defects are related to calibrated tools. We use this conservative estimate due to the fact that though the defects may result in a scrapped assembly, it may not result in a scrapped product. It is estimated that 38 products are scrapped per year. Therefore, the final estimated cost of scrap is $1,376,208 (Table 16.3).

TABLE 16.2
Annual Rework Costs

Rework Cost Elements	Scenario 1
Defects to rework	3,420
Rework hours	27,360
Operator cost	$20/h
Total cost	$547,200

TABLE 16.3
Annual Scrap Costs

Scrap Cost Elements	Scenario 1
Cost per product	$36,216
Scrapped products	38 products
Total cost	$1,376,208

16.8 MANAGEMENT COSTS

Management costs were estimated as a percentage of total management time. Currently, there are five managers directly involved with the results of audits and other problems that arise from the production floor. Given that there are 2080 h of potential work time in a year for each manager or a total time for the five managers of 10,400 h per year, the percentage of total time utilized to address audit concerns and production floor problems was 5%. Of this 5%, we assume that a minimum of 50% of this time or 2.5% of total time is dedicated to working with problems of rework and scrap. This percentage is a conservative estimate. The total time per year for five managers is 520 h. Finally, the hourly cost per manager was determined to be $30/h with 33% benefits. Benefits increase the cost per manager to $40/h. The cost of management per year in regards to audits and production performance was found to be $20,748 per year (Table 16.4).

16.9 CUSTOMER SERVICE COSTS

The customer service cost considers the risk of losing the customer or contract. Conservatively, the chapter only considers the potential loss of a portion of a contract, not the loss of the complete contract. This cost would be occurred if defective parts were distributed to the customer or the customer was unsatisfied with the service of the company. The customer may either cancel or reduce the contract at the end of the contract period. Due to this risk, the range of values was considered. Contracts could potentially range from $1000 to $11.8 million. The conditional probability trees consider the reduction of a contract due to a defect reaching the customer and after two occurrences, the contract would be canceled. All these probabilities were considered in the conditional probability trees shown in Figure 16.2.

TABLE 16.4
Annual Management Costs

Management Cost Elements	Scenario 1
Hours per year	2,080 h
# Managers	Five managers
5% Total work time	520 h
Hourly cost	$39.90/h
Total cost	$20,748

The second conditional probability tree utilized in this chapter was for deriving the function for the expected customer service cost $E(Y_j)$:

$$E(Y_j) = \text{A1D1} + \text{A2B1C1D2} + \text{A2B2C2D3} + \text{A2B2C3D4} + \text{A2B2C4D5} \quad (16.3)$$

Equation 16.2 describes the expected customer service cost calculations. Y_j represents customer service cost for contract j. The conditional probability tree was based on the following assumptions from estimates of past occurrences. First, it was estimated that a minor defect on a first occurrence would cause the customer to reduce their contract by 10%. However, a major defect would cause the customer to reduce their contract by 25%. If a minor defect was found by a customer for a second time or more, the customer was estimated to reduce their contract by 75%, but if the defect was major, the customer was likely to seek another company to do business with. Therefore, the company could potentially lose the entire contract on a second major defect. This penalty cost is given as D5 in Equation 16.3. The penalty cost is quantified as the cost of completely losing value of an expected contract.

In order to complete the conditional probability tree, an expected value of the contract price needed to be found. The smallest contract was considered to be 90% of the total contracts, with 10% being the contract of $11.8 million. This gave a weighted average contract of $1,180,900 as calculated in the following:

$$(\$1{,}000)(0.9) + (\$11{,}800{,}000)(0.1) = \$1{,}180{,}900 \quad (16.4)$$

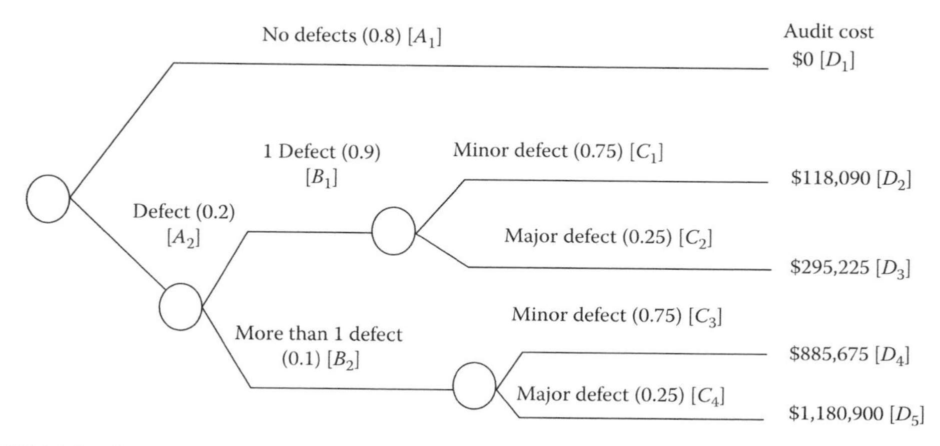

FIGURE 16.2 Customer service conditional probability tree.

These conditional probability tree calculations are demonstrated in Figure 16.2. The conditional probability tree was collapsed in order to find an expected cost for customer service, \$48,417:

$$\begin{aligned} E(Y2) &= (0.8)(\$0) + (0.75)(0.9)(0.2)(\$118{,}090) + (0.25)(0.9)(0.2)(\$295{,}225) \\ &\quad + (0.75)(0.1)(0.2)(\$885{,}675) + (0.25)(0.1)(0.2)(\$1{,}180{,}900) \\ &= \$48{,}417 \end{aligned} \tag{16.5}$$

Many assumptions were made with the inclusion of the customer service cost. However, it is a risk that the company takes every time they ship a product. Therefore, this cost needed to be included. This estimation has some limitations. The main one to note would be the lack of numerous occurrences. This cost was estimated on the assumption of one occurrence a year. The company is a government contractor and has lost no more than five contracts over the last 5 years to other plants, though audit costs have risen. This may be a low or high estimate and must be considered by other companies looking into this model. Overall, this expected value fits this company well.

16.10 TOTAL ANNUAL COST

The total costs for scenario 1 is outlined in Table 16.5. The total cost is approximately \$2.1 million dollars per year. The implementation cost of scenario 1 was zero due to the fact it is the baseline system that was not changed in the current system.

16.10.1 Scenario 2: RFID Implementation

The second scenario evaluates the cost of implementing an RFID system to track calibrated tools. This implementation would be the greatest cost to the company for this scenario, as will be seen. In addition to the implementation cost, there will be a maintenance cost each year to keep the system in good working conditions. Also taken into consideration with scenario 2 is the cost reduction linked to the implementation of the RFID system.

TABLE 16.5
Total Annual Cost Scenario 1

Cost Elements	Scenario 1
Audit	\$107,820
Rework	\$547,200
Scrap	\$1,376,208
Management	\$20,748
Customer service (contracts lost)	\$48,417
Implementation	\$0
Total cost	\$2,100,393

16.11 TAG COSTS

Tags are the transponders that carry the information in an RFID system. Passive RFID tag costs can range from \$0.40 to \$0.80 (United States Department of Defense, 2004). Through this study, passive class 1 labels were used. These tags cost approximately \$0.75 each. In order to implement the system, all 2000 calibrated tools would be tagged. The tags would be used to store and pass data between systems devices, in this case the tag and the reader. The estimated total tag cost is \$1500 (Table 16.6).

The cost of the tags will also include a yearly replacement of the same amount of \$1500 due to the annual replacement cost of tags. This cost will be discounted to the present value using net present value analysis shown later in this chapter. The tags will have the ability to be programmed with such information as the unique tool number, type of tool, planned location, current location, and next scheduled tool calibration.

TABLE 16.6
Annual Tag Costs

Tag Cost Elements	Scenario 2
Cost per passive tag	\$0.75
Number of tags	2000
Total cost	\$1500

16.12 READER COSTS

The reader is used to communicate with the RFID tags through electromagnetic waves. The information is relayed from the tag to the reader. There are both portable readers and fixed readers in the market today (Nobel, 2004). In this case, the company would be using fixed readers placed at strategic locations throughout the plant. Considering there are numerous rooms in the plant, 15 readers would be needed. All exits to the outside would be covered with a reader as well as other locations within the plant. Each reader would cost the company approximately $5000. Therefore, the initial cost of the readers would be approximately $75,000.

TABLE 16.7
Annual Reader Costs

Reader Cost Elements	Scenario 2
Readers	$75,000
Antennae	$7,500
Total cost	$82,500

In addition to the readers, 15 antennae would be purchased to boost the signal sent from the readers to the tags at various locations within the plant. A spectrum analyzer was used to conduct a site survey in order to determine the number antennae that would be used in this environment. The antennas would be used at ingress and egress locations to provide optimal data capture of RFID tags.

These antennae would be used in areas where transmission through the material may be an issue. By adding antennas in these areas, the entire plant could be covered by the RFID network in order for all tags to be read when needed. An inventory search can be initiated at any time by activating the readers and antennae. The search would identify tools that are missing from the plant. There would be no readers placed within the offices. Therefore, engineering or tooling staff would need to confirm that they were moving a calibrated tool to an office in order to prevent alarming the system. Each antenna costs approximately $500 dollars; a total cost of $7,500. Therefore, the total cost of the readers and antennae is $82,500. The installation cost is included in the implementation cost for the RFID system analyzed later in this chapter (Table 16.7).

16.13 SOFTWARE COSTS

The final component of the RFID system is the middleware software. This software serves as a traffic cop to send the correct data to the enterprise resource planning (ERP) system currently in place at the company. This software typically costs approximately $25,000 and includes the cost of upgrades for up to 3–5 years. Currently, the company purchases the tags and readers from a supplier, which will include the supplier's software, commonly termed edgewear. The integration between the reader's edgewear, middleware software, and ERP is the typical software cost. Another consideration when purchasing the middleware is that the software platforms must be compatible with the current ERP system. Possible costs include interface programs in such languages as C, C-sharp, and C++, which can be as high as $200/h and the costs can exceed $50,000 for this type of software integration. For this study, conservatively, we estimate $25,000.

16.14 IMPLEMENTATION COSTS

We estimate implementation cost by assuming two technical personnel are assigned the task to implement the RFID system over 6 months. These technical personnel cost $33/h fully burdened. Over 6 months (assume 20 work days per month) and assuming an 8 h work day, these two employees would work a total of 960 h. This cost is $31,680 per technical personnel or $63,360 for implementation of an RFID system over a facility of 132,000 ft^2.

16.14.1 Investment for Scenario 2

The total cost for year 1 of the second scenario is shown in Table 16.8. There is estimated to be a cost of approximately $172,360 for implementation of the RFID system.

In addition to the implementation costs of year 1, there will also be a yearly maintenance cost. This includes replacing the tags and the labor needed to perform this task. The cost to replace the tags will be considered to remain the same over the 5 years this study analyzes. Therefore, the cost of replacing 2000 tags will be $1500. The time of this project is estimated to take approximately 2 min per tool to both find the tool and replace the tag. With more than 2000 tools, it will take 4000 min, or 67 h. Employees performing this task will be making $15/h. After 33% benefits were added to this labor cost, the personnel cost rose to $20/h. This brings the labor cost of maintenance to $1334 per year and total costs to $2834, as shown in Table 16.9.

Next, we consider the change in costs per year that were incurred in scenario 1. This study estimated a reduction of 20% of total cost based on time studies of labor, audits, and rework (see Table 16.10). Therefore, the company would still be incurring $1,690,582 per year. This translates into a savings of $409,811.

TABLE 16.8
Initial Investment Scenario 2

Cost Elements for Year 1	Scenario 2
Tags	$1,500
Readers	$82,500
Software	$25,000
Implementation	$63,360
Total cost	$172,360

TABLE 16.9
Annual Maintenance Cost

Maintenance Cost Elements	Scenario 2
Tag replacement	$1500
Labor	$1334
Total cost	$2834

16.15 NET PRESENT VALUE COMPARISON

We use a net present analysis to compare the two alternatives. The assumptions applied here are that the cash flows are deterministic and they occur at the end of each period of time analyzed.

The discount rate used on this study is considered as being the current rate of inflation of 4%. This value makes the analysis more conservative and the conclusions more clear. The time period used is a 5 year period, which is the minimum considered by the company to compare the two projects. On this comparison, there is no salvage value. A further study analyzing the variability of the elements that compose the costs will be presented on the sensitivity analysis part. Cash flow diagrams for each scenario are shown in Figure 16.3.

The above cash flow diagram (Figure 16.3) illustrates the costs and returns of scenario 1. When brought to present value, the total net present worth is approximately $9.3 million demonstrated in the following equation using Equation 16.6:

$$\begin{aligned} \text{PW}(4\%) &= \$0 + \$2{,}100{,}393(\text{P/A}, 4\%, 5) \\ &= \$2{,}100{,}393(4.452) \\ &= \$9{,}350{,}949 \sim \$9.3\,\text{million} \end{aligned} \tag{16.6}$$

TABLE 16.10
Total Annual Cost Comparison

Cost Elements	Scenario 1	Scenario 2	Cost Reduction (%)
Audit	$107,820	$87,334	19
Rework	$547,200	$432,288	21
Scrap	$1,376,208	$1,114,728	19
Management	$20,748	$17,013	18
Customer service (contracts lost)	$48,417	$39,218	19
Total cost/year	$2,100,393	$1,690,582	20

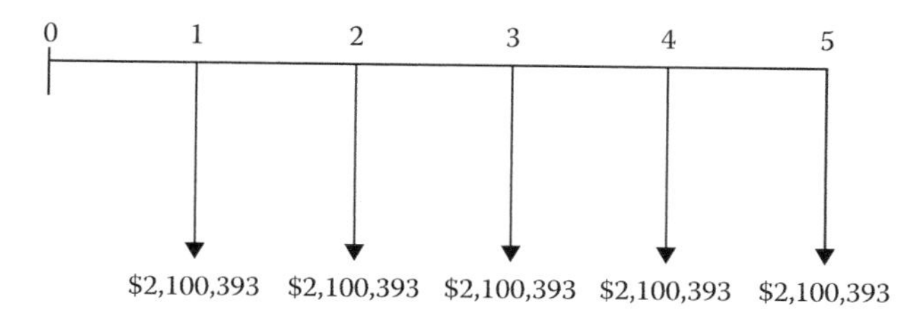

FIGURE 16.3 Scenario 1 cash flow diagram.

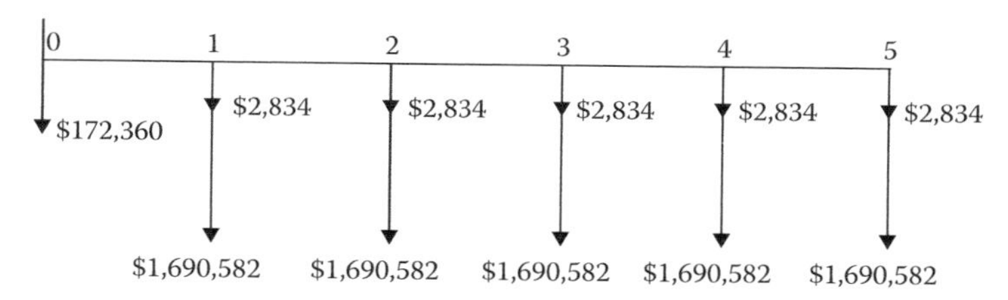

FIGURE 16.4 Scenario 2 cash flow diagram.

The above cash flow diagram (Figure 16.4) illustrates the costs of scenario 2. In this scenario, the net present value was approximately \$7.7 million as seen in the following equation using Equation 16.7:

$$\begin{aligned} PW(4\%) &= \$172,360 + (\$1,262,077 + \$2,834)(P/A, 4\%, 5) \\ &= \$172,360 + \$1,690,582(4.452) \\ &= \$7,698,831 \sim \$7.7 \text{ million} \end{aligned} \tag{16.7}$$

16.16 COMPARISON

From the results of the net present value analysis above, RFID will provide an economic benefit to the company. The difference in the two net present values is approximately \$1.6 million. Due to this difference in the net present values as well as the initial savings, RFID would be a good choice for the company to pursue.

16.17 SENSITIVITY ANALYSIS

In order to justify the results of this study further, a sensitivity analysis was developed. This was done in order to justify the study and compensate the large amounts of variability in the variables.

The term sensitivity analysis examines how uncertainty of the variables that compose the cash flows can influence the recommended decisions. The sources of uncertainty can be due to measurement error, unclear specification, or volatility of the future. The techniques used in this model are the spiderplot and the tornado diagram. The spiderplot has the advantage of showing the impact of each cash flow's uncertainty on the present worth of the project and also makes comparison between the individual variables influence easier. A spiderplot should contain the limits of uncertainty for each cash flow element; the impact of each element on the PW, and the identification of each element might change the recommendation (Eschenbach, 2003).

The tornado diagram, on the other hand, is to be used when a summary of the economic performance of the variables on the present worth is needed. This chart shows the variables that have the most influence on the present worth in descending order. This makes the diagram have a tornado shape. Normally, this analysis assumes that the variables are statistically independent (Eschenbach, 2006), which is the case.

TABLE 16.11
Limits of Variables

Variable	Scenario 1	Scenario 2	Lower Limit (%)	Upper Limit (%)
Investment	—	$172,360	80	115
Yearly cost	$2,100,393.00	$1,690,582	60	120
Maintenance cost	—	$2,834	75	120
i	4%	4%	95	115
N	5	5	75	150

TABLE 16.12
Values of Limits

%	Investment	Yearly Cost	Maintenance Cost	*i*	*N*
60		$909,667			
75		$1,183,328	$1,642,584		$1,219,228
80	$1,673,902	$1,274,548	$1,641,953		$1,304,924
95	$1,648,048	$1,548,209	$1,640,060	$1,649,654	$1,557,029
100	$1,639,430	$1,639,430	$1,639,430	$1,639,430	$1,639,430
115	$1,613,576	$1,913,091	$1,637,537	$1,609,292	$1,881,838
120		$2,004,311	$1,636,906		$1,961,070
150					$2,420,484

The lower and upper limits used in this case were estimated based on rules of thumb as taken by Eschenbach (2003); they are demonstrated in Table 16.11. As we are dealing with two mutually exclusive scenarios, the sensitivity analysis was created using the differences between the present worth of the two scenarios.

All calculations used to construct the spiderplot used the present value used in Equation 16.3. Table 16.12 summarizes those results and detail calculations are shown in Appendix 16.A.

We utilized this data to construct a spiderplot that demonstrates the effect on each variable when the other variables remain constant.

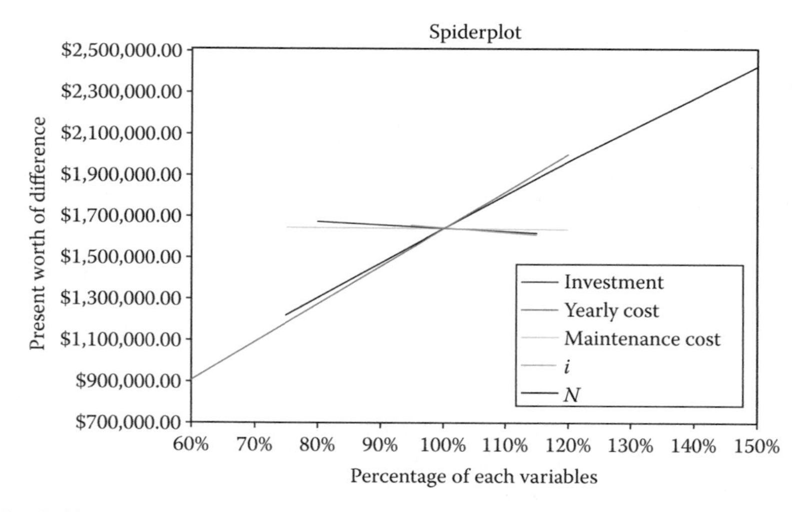

FIGURE 16.5 Spiderplot of PW of difference of the two scenarios.

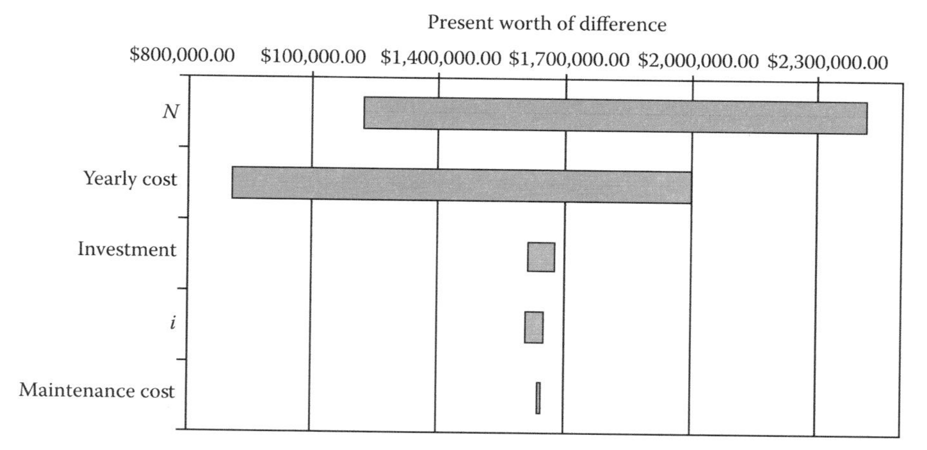

FIGURE 16.6 Tornado diagram.

From Figure 16.5, it can be seen that the factors that account for most of the variability between the two scenarios are yearly costs. The minimum difference of PW between the two scenarios is obtained when the number of periods is set to its lower limit.

To summarize the relative sensitivities of each variable, a tornado diagram (Figure 16.6) was then created with this propose. This diagram shows the variability accounted for each factor in descending order. From this chart can be seen the influence of each variable on the present worth.

From the above sensitivity analysis, this study shows that even with variability added into the model, there remains a substantial benefit in implementing an RFID system. This can be seen in reviewing both sets of calculations. If variability were to place the yearly cost for the scenario 1%–60% of its value, there would still be a benefit of implementing an RFID system of approximately $700,000. Even these savings would be very cost effective to a company. Therefore, the study further justifies on a more technical level, taking into account the reality of variability in the study.

16.18 LIMITATIONS

In this study, there were several limitations. The cost of scrap associated with not using a calibrated tool. Scrap costs can be further associated with inadequate training, improper workstation design, and other non-productive practices. In order to limit this study to the impact of lost tools, we made the simplifying assumption that 10% of scrap was associated with using the wrong tool due to lack of availability of the correct calibrated tool. The cost to upgrade the RFID systems is not included in this analysis due to the fact this will probably happen after 5 years which exceeds our cost time horizon.

Finally, the high-level time study that yielded a projected 20% labor reduction due to RFID implementations has a significant effect on the study. Though managers were comfortable with these saving estimates, a more detailed time study was recommended by the researchers.

16.19 CONCLUSIONS

This approach has outlined the current costs of using non-calibrated tools to the company. These costs have the potential to be reduced by implementing an RFID system to track the calibrated tools. With this technological advance, the company has the potential to use RFID in other areas besides calibrated tools. This could lead to other savings. This study outlined the cost of implementing an RFID system as well as the cost to maintain this type of system. We utilized the net present value analysis in order to evaluate the cost and benefits of RFID. The net present value analysis showed the difference between a do-nothing scenario and the scenario in which we implement RFID. The results demonstrated a savings of approximately $1.3 million if the RFID system is implemented. The savings, therefore, would be significant to the company and a good investment for the future.

APPENDIX 16.A

16.A.1 Investment

$$\begin{aligned} \text{PWFC, } 80\%(4\%) &= \$2{,}100{,}393(\text{P/A, } 4\%, 5) - (80\%)(\$172{,}360) \\ &\quad + (\$1{,}690{,}582 + \$2{,}834)(\text{P/A, } 4\%, 5) \\ &= \$1{,}673{,}902 \end{aligned}$$

$$\begin{aligned} \text{PWFC, } 95\%(4\%) &= \$2{,}100{,}393(\text{P/A, } 4\%, 5) - (95\%)(\$172{,}360) \\ &\quad + (\$1{,}690{,}582 + \$2{,}834)(\text{P/A, } 4\%, 5) \\ &= \$1{,}648{,}048 \end{aligned}$$

$$\begin{aligned} \text{PWFC, } 100\%(4\%) &= \$2{,}100{,}393(\text{P/A, } 4\%, 5) - (100\%)(\$172{,}360) \\ &\quad + (\$1{,}690{,}582 + \$2{,}834)(\text{P/A, } 4\%, 5) \\ &= \$1{,}639{,}430 \end{aligned}$$

$$\begin{aligned} \text{PWFC, } 115\%(4\%) &= \$2{,}100{,}393(\text{P/A, } 4\%, 5) - (115\%)(\$172{,}360) \\ &\quad + (\$1{,}690{,}582 + \$2{,}834)(\text{P/A, } 4\%, 5) \\ &= \$1{,}613{,}576 \end{aligned}$$

16.A.2 Yearly Cost

$$\begin{aligned} \text{PWYC, } 60\%(4\%) &= (60\%)(\$2{,}100{,}393)(\text{P/A, } 4\%, 5) - (\$172{,}360) \\ &\quad + (60\%)(\$1{,}690{,}582 + \$2{,}834)(\text{P/A, } 4\%, 5) \\ &= \$909{,}668 \end{aligned}$$

$$\begin{aligned} \text{PWYC, } 75\%(4\%) &= (75\%)(\$2{,}100{,}393)(\text{P/A, } 4\%, 5) - (\$172{,}360) \\ &\quad + (75\%)(\$1{,}690{,}582 + \$2{,}834)(\text{P/A, } 4\%, 5) \\ &= \$1{,}183{,}328 \end{aligned}$$

$$\begin{aligned} \text{PWYC, } 80\%(4\%) &= (80\%)(\$2{,}100{,}393)(\text{P/A, } 4\%, 5) - (\$172{,}360) \\ &\quad + (80\%)(\$1{,}690{,}582 + \$2{,}834)(\text{P/A, } 4\%, 5) \\ &= \$1{,}274{,}549 \end{aligned}$$

$$\begin{aligned} \text{PWYC, } 95\%(4\%) &= (95\%)(\$2{,}100{,}393)(P/A,\ 4\%,\ 5) - (\$172{,}360) \\ &\quad + (95\%)(\$1{,}690{,}582 + \$2{,}834)(P/A,\ 4\%,\ 5) \\ &= \$1{,}548{,}210 \end{aligned}$$

$$\begin{aligned} \text{PWYC, } 100\%(4\%) &= (100\%)(\$2{,}100{,}393)(P/A,\ 4\%,\ 5) - (\$172{,}360) \\ &\quad + (100\%)(\$1{,}690{,}582 + \$2{,}834)(P/A,\ 4\%,\ 5) \\ &= \$1{,}639{,}430 \end{aligned}$$

$$\begin{aligned} \text{PWYC, } 115\%(4\%) &= (115\%)(\$2{,}100{,}393)(P/A,\ 4\%,\ 5) - (\$172{,}360) \\ &\quad + (115\%)(\$1{,}690{,}582 + \$2{,}834)(P/A,\ 4\%,\ 5) \\ &= \$1{,}913{,}091 \end{aligned}$$

$$\begin{aligned} \text{PWYC, } 120\%(4\%) &= (120\%)(\$2{,}100{,}393)(P/A,\ 4\%,\ 5) - (\$172{,}360) \\ &\quad + (120\%)(\$11{,}690{,}582 + \$2{,}834)(P/A,\ 4\%,\ 5) \\ &= \$2{,}004{,}311 \end{aligned}$$

16.A.3 Maintenance Cost

$$\begin{aligned} \text{PWMC, } 75\%(4\%) &= (\$2{,}100{,}393)(P/A,\ 4\%,\ 5) - (\$172{,}360) \\ &\quad + (\$1{,}690{,}582 + (\$2{,}834)(75\%))(P/A,\ 4\%,\ 5) \\ &= \$1{,}642{,}584 \end{aligned}$$

$$\begin{aligned} \text{PWMC, } 80\%(4\%) &= (\$2{,}100{,}393)(P/A,\ 4\%,\ 5) - (\$172{,}360) \\ &\quad + (\$1{,}690{,}582 + (\$2{,}834)(80\%))(P/A,\ 4\%,\ 5) \\ &= \$1{,}641{,}953 \end{aligned}$$

$$\begin{aligned} \text{PWMC, } 95\%(4\%) &= (\$2{,}100{,}393)(P/A,\ 4\%,\ 5) - (\$172{,}360) \\ &\quad + (\$1{,}690{,}582 + (\$2{,}834)(95\%))(P/A,\ 4\%,\ 5) \\ &= \$1{,}640{,}060 \end{aligned}$$

$$\begin{aligned} \text{PWMC, } 100\%(4\%) &= (\$2{,}100{,}393)(P/A,\ 4\%,\ 5) - (\$172{,}360) \\ &\quad + (\$1{,}690{,}582 + (\$2{,}834)(100\%))(P/A,\ 4\%,\ 5) \\ &= \$1{,}639{,}430 \end{aligned}$$

$$\begin{aligned} \text{PWMC, 115\%(4\%)} &= (\$2{,}100{,}393)(\text{P/A}, 4\%, 5) - (\$172{,}360) \\ &\quad + (\$1{,}690{,}582 + (\$2{,}834)(155\%))(\text{P/A}, 4\%, 5) \\ &= \$1{,}637{,}537 \end{aligned}$$

$$\begin{aligned} \text{PWMC, 120\% (4\%)} &= (\$2{,}100{,}393)(\text{P/A}, 4\%, 5) - (\$172{,}360) \\ &\quad + (\$1{,}690{,}582 + (\$2{,}834)(120\%))(\text{P/A}, 4\%, 5) \\ &= \$1{,}636{,}906 \end{aligned}$$

16.A.4 Discount Rate (*i*)

$$\begin{aligned} \text{PWi, 95\%((95\%)(4\%))} &= (\$2{,}100{,}393)(\text{P/A}, (95\%)(4\%), 5) - (\$172{,}360) \\ &\quad + (\$1{,}690{,}582 + \$2{,}834)(\text{P/A}, (95\%)(4\%), 5) \\ &= \$1{,}649{,}654 \end{aligned}$$

$$\begin{aligned} \text{PWi, 100\%((100\%)(4\%))} &= (\$2{,}100{,}393)(\text{P/A}, (100\%)(4\%), 5) - (\$172{,}360) \\ &\quad + (\$1{,}690{,}582 + \$2{,}834)(\text{P/A}, (100\%)(4\%), 5) \\ &= \$1{,}639{,}430 \end{aligned}$$

$$\begin{aligned} \text{PWi, 155\%((115\%)(4\%))} &= (\$2{,}100{,}393)(\text{P/A}, (115\%)(4\%), 5) - (\$172{,}360) \\ &\quad + (\$1{,}690{,}582 + \$2{,}834)(\text{P/A}, (115\%)(4\%), 5) \\ &= \$1{,}609{,}292 \end{aligned}$$

16.A.5 Number of Periods (*N*)

$$\begin{aligned} \text{PWN, 75\%(4\%)} &= (\$2{,}100{,}393)(\text{P/A}, 4\%, (75\%)(5)) - (\$172{,}360) \\ &\quad + (\$1{,}690{,}582 + \$2{,}834)(\text{P/A}, 4\%, (75\%)(5)) \\ &= \$1{,}219{,}228 \end{aligned}$$

$$\begin{aligned} \text{PWN, 80\%(4\%)} &= (\$2{,}100{,}393)(\text{P/A}, 4\%, (80\%)(5)) - (\$172{,}360) \\ &\quad + (\$1{,}690{,}582 + \$2{,}834)(\text{P/A}, 4\%, (80\%)(5)) \\ &= \$1{,}304{,}924 \end{aligned}$$

$$\begin{aligned} \text{PWN, 95\%(4\%)} &= (\$2{,}100{,}393)(\text{P/A}, 4\%, (95\%)(5)) - (\$172{,}360) \\ &\quad + (\$1{,}690{,}582 + \$2{,}834)(\text{P/A}, 4\%, (95\%)(5)) \\ &= \$1{,}557{,}029 \end{aligned}$$

$$\begin{aligned}\text{PWN, }100\%\ (4\%) &= (\$2{,}100{,}393)(\text{P/A, }4\%,\ (100\%)(5)) - (\$172{,}360)\\ &\quad + (\$1{,}690{,}582 + \$2{,}834)(\text{P/A, }4\%,\ (100\%)(5))\\ &= \$1{,}639{,}430\end{aligned}$$

$$\begin{aligned}\text{PWN, }115\%(4\%) &= (\$2{,}100{,}393)(\text{P/A, }4\%,\ (115\%)(5)) - (\$172{,}360)\\ &\quad + (\$1{,}690{,}582 + \$2{,}834)(\text{P/A, }4\%,\ (115\%)(5))\\ &= \$1{,}881{,}838\end{aligned}$$

$$\begin{aligned}\text{PWN, }120\%(4\%) &= (\$2{,}100{,}393\text{P/A, }4\%,\ (120\%)(5)) - (\$172{,}360)\\ &\quad + (\$1{,}690{,}582 + \$2{,}834)(\text{P/A, }4\%,\ (120\%)(5))\\ &= \$1{,}961{,}070\end{aligned}$$

$$\begin{aligned}\text{PWN, }150\%(4\%) &= (\$2{,}100{,}393)(\text{P/A, }4\%,\ (150\%)(5)) - (\$172{,}360)\\ &\quad + (\$1{,}690{,}582 + \$2{,}834)(\text{P/A, }4\%,\ (150\%)(5))\\ &= \$2{,}420{,}484\end{aligned}$$

17 Forecasting

Forecasting is a method used for predicting the future. Within a business, marketing and sales utilize forecasting methods the most. A marketing department will commonly use forecasting for sales of both new and existing product lines while the production department will try to forecast sales for operations planning. Not all events can be accurately forecasted. For example, games of chance like those played at casinos are considered random. The house is always guaranteed to win over the long term by placing the probabilities in its favor. Evidence exists that the daily prices of stocks follow a purely random process, and studies have shown that professional traders rarely outperform stock portfolios generated purely at random. Forecasting product demand is our primary interest. Even though demand is usually random, forecasting methods can be used on the parts of the demand process that are predictable. Trends, cycles, and seasonal variation may be present within the demand process, which gives us the advantage of predicting the future demand processes.

17.1 TIME HORIZON IN FORECASTING

One dimension used to evaluate forecasting is the time horizon. The three different time horizons associated with forecasting are seen in Figure 17.1: short-term, intermediate-term, and long-term. Short-term forecasting is crucial for day-to-day planning and is typically measured in days or weeks. Examples of the uses of short-term forecasts are inventory management, production plans that may be derived from a materials requirements planning system, resource requirements planning, and shift scheduling that may require forecasts of workers' availabilities and preferences. The intermediate-term is measured in weeks or months. Typical intermediate-term forecasting problems are sales patterns for product families, requirements and availabilities of workers, and resource requirements. Long-term planning involves capacity needs. When demands are expected to increase, the firm must plan for the construction of new facilities or implementing existing facilities with new technologies.

17.2 CHARACTERISTICS OF FORECASTS

There are five important characteristics of forecasting:

1. They are usually wrong.
2. A good forecast is more than a single number.
3. Aggregate forecasts are more accurate.
4. The longer the forecast horizon, the less accurate the forecast will be.
5. Forecasts should not be used to the exclusion of known information.

To create robust forecasts, these characteristics are important to understand. Forecasts may require modifications if the forecast of demand proves to be inaccurate. The forecast must have the ability to react to unanticipated forecast errors. When creating a good forecast, a measure must be included to demonstrate forecast error. This measure can be in the form of a range or variance of the distribution of the forecast error. It is important to remember that the error made in forecasting sales for an entire product line is usually less than the error made in forecasting sales for an individual item. A particular technique may result in reasonably accurate forecasts in most circumstances, but there may be information available that will affect future demand that is not demonstrated in the past history of the series.

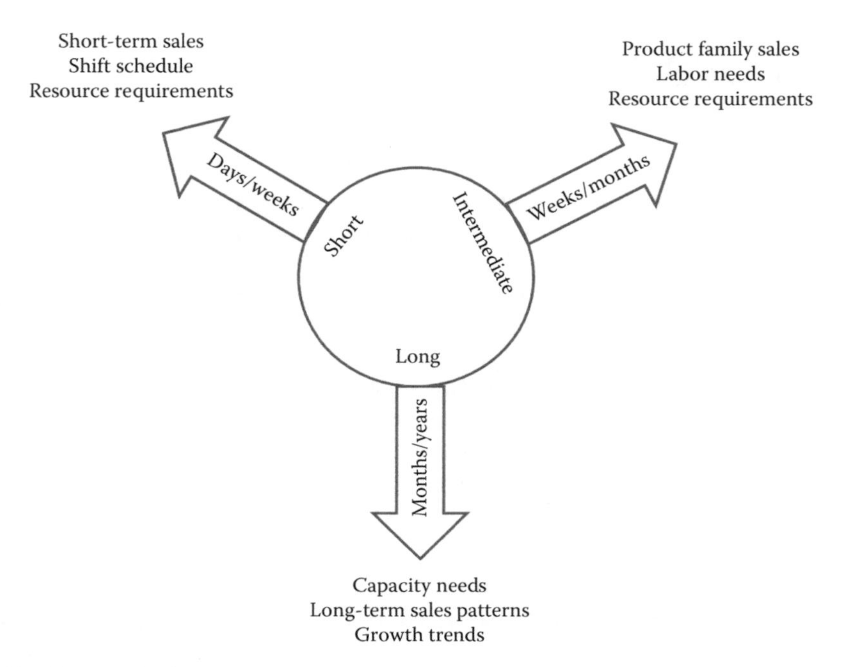

FIGURE 17.1 Forecast horizons in operations planning.

17.3 SUBJECTIVE FORECASTING MODELS

Forecasting methods have two classifications: subjective or objective. Subjective forecasting methods are based on human judgment and there are several techniques for finding out opinions needed for forecasting:

1. *Sales force composites*: Since the sales force has direct contact with consumers, they are able to notice the changes in consumer preferences. Members of the sales force can submit sales estimates of the products that they will sell in the coming year in order to develop a sales force composite. The estimates sent in might be single numbers or several numbers (pessimistic, most likely, optimistic). These composites may be inaccurate when compensation of sales personnel is based on meeting a quota because the sales force is more likely to report lower estimates.
2. *Customer surveys*: Customer surveys may be a good indicator of shifts in preference or future trends, but to be effective, these surveys must be carefully designed. They must be designed to withstand bias and must represent the entire customer base. Poorly designed surveys may result in the wrong conclusions.
3. *Jury of executive opinion*: Expert opinion may be the only source of information when there is no past history available (i.e., new product). Opinions of personnel in the functional areas of marketing, finance, and production should be obtained. Combining individual forecasts may be done in several ways. Two methods can be done in order to prepare a forecast: interview the personnel and use the results of these interviews or meet as a group and a come to a consensus.
4. *The Delphi method*: The Delphi method is similar to the previous method in that it uses expert opinions. The difference between the Delphi method and the previous method is that the Delphi method attempts to eliminate some of the inherent shortcomings of group dynamics (i.e., the personalities of some group members overshadow those of other members). The method requires a group of experts to express their opinions

by an individual sample survey. The results are then compiled and summarized. The experts are presented with these compilations and special attention is given to the opinions that are significantly different from the group averages. The experts are asked if they wish to reconsider their original opinions in light of the group response. The process is repeated until an overall group consensus is reached. The Delphi method's primary advantage is that it provides a means of assessing individual opinion without the usual concerns of personal interactions. A disadvantage is that one must be very careful with the construction of the questionnaire.

17.4 OBJECTIVE FORECASTING METHODS

Unlike subjective methods, which are based on human judgment, objective forecasting methods are derived from analyzing data. A time series method utilizes past values of the problem one is trying to predict. Causal models use other variables with values that may be connected in some way to the phenomenon being predicted.

17.4.1 Causal Models

Let Y represent the phenomenon to be forecasted and $X_1, X_2, \ldots, X_n$ be n variables that may be related to Y. The causal model is one in which the forecast for Y is some function of these variables:

$$Y = f(X_1, X_2, \ldots, X_n)$$

Econometric models are special causal models in which the relationship between Y and $(X_1, X_2, \ldots, X_n)$ is linear:

$$Y = \alpha_0 + \alpha_1 X_1 + \alpha_2 X_2 + \cdots + \alpha_n X_n$$

for some constants $(\alpha_1, \ldots, \alpha_n)$. To find estimators for the constants, the method of least squares is commonly used.

A simple example of a causal forecasting model involves a salesman who is trying to estimate his income for the succeeding year. In the past, he has found that his income is close to being proportional to the total number of product sales in his territory. He also has noticed that there has typically been a close relationship between product sales and interest rates for loans. He might construct a model for the form

$$Y_t = \alpha_0 + \alpha_1 X_{t-1}$$

where

Y is the number of sales in year t

X_{t-1} is the interest rate in year $t-1$

Based on past data, he determined the least squares estimators are currently $\alpha_0 = 385.7$ and $\alpha_1 = -1.878$. The estimated relationship between product sales and interest rates is

$$Y_t = 385.7 - 1.878 X_{t-1}$$

where X_{t-1} is the previous year's interest rate. If the current mortgage interest rate is 10%, the model would predict that the number of sales the following year in his territory would be $385.7 - 187.8 = 197.9$, or about 198 products sold.

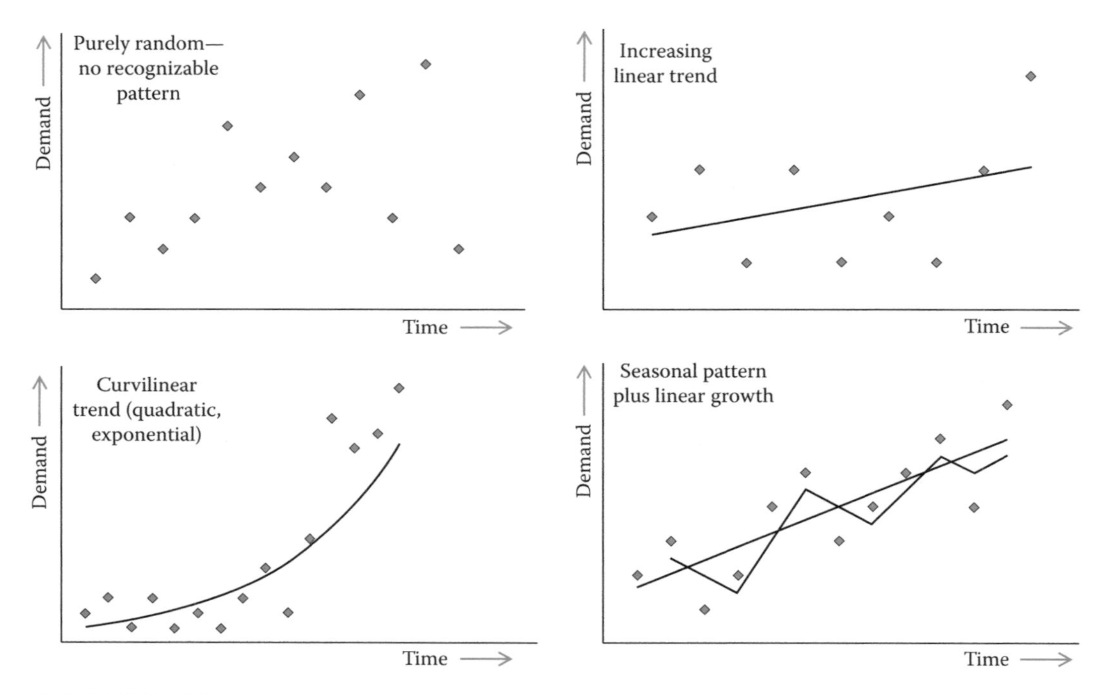

FIGURE 17.2 Time series patterns.

Causal models of this type are common for predicting economic phenomena such as the gross national product (GNP) and the gross domestic product (GDP) while time series methods are more commonly used for operations planning.

17.4.2 Time Series Methods

Unlike causal models, time series methods require no information other than the past values of the variable being predicted. A time series is a collection of observations of some economic or physical phenomenon that are drawn at discrete points in time. When conducting time series analysis, the objective is to attempt to isolate the patterns that arise most often like those seen in Figure 17.2.

Four such patterns that exist are trend, seasonality, cycles, and randomness. Trend refers to a time series that exhibits a tendency to grow or decline. The trend can be either linear or nonlinear, but when the pattern of trend is not specified, it is assumed to be linear. A seasonal pattern is one that repeats itself after fixed intervals. This repetition could occur daily, weekly, monthly, or yearly depending on the problem. An example of something that follows a seasonal pattern is clothing. Although similar to seasonality, cyclic variation demonstrates variation in the length and magnitude of the cycle. A pure random process exists when there is no recognizable pattern within the data. Although data may appear to be random, it has the possibility of having a very definite structure. Truly random data that fluctuate around a fixed mean form what is called a horizontal pattern.

17.5 NOTATION CONVENTIONS

Define $D_1, D_2, \ldots, D_t, \ldots$ as the observed values of demand during periods $1, 2, \ldots, t, \ldots$. Two assumptions will be made: (1) $(D_t, t \geq 1)$ is the time series to be predicted and (2) if period t is being forecasted, then $D_t, D_{t-1}, \ldots$ have been observed (D_{t+1} has not).

Define F_t as the forecast made for period t in period $t - 1$; therefore, it is the forecast made at the end of period $t - 1$ after having observed $D_{t-1}, D_{t-2}, \ldots$, but before observing D_t. These forecasts are one-step-ahead forecasts, which mean that they are made for the demand in the next period.

Finally, note that a time series forecast it obtained by applying some set of weights to past data (α_n):

$$F_t = \sum_{i=0}^{\infty} \alpha_n D_{t-n}$$

17.6 EVALUATING FORECASTS

Define the forecast error in period t, e_t, as the difference between the observed forecast value for that period and the actual demand for that period. For multiple-step-ahead forecasts,

$$e_t = F_{t-\tau,t} - D_t$$

and for one-step-ahead forecasts,

$$e_t = F_t - D_t$$

Let $e_1, e_2, \ldots, e_n$ be the forecast errors observed over n periods. There are two common measures of forecast accuracy during these n periods: mean absolute deviation (MAD) and the mean squared error (MSE), given by the following formulas:

$$\text{MAD} = (1/n)\sum_{i=1}^{n} e_i$$

$$\text{MSE} = (1/n)\sum_{i=1}^{n} e_i^2$$

The MSE is similar to the variance of a random sample. The MAD is often the preferred method of measuring the forecast error because it does not require squaring. When forecast errors are normally distributed, an estimate of the standard deviation of the forecast error, σ_e, is given by 1.25 times the MAD.

Another measure that can be used is mean absolute percentage error (MAPE), which is not dependent on the magnitude of the values of demand and is given by the formula

$$\text{MAPE} = \left[\left(\frac{1}{n}\right)\sum_{i=1}^{n}\left|\frac{\theta_i}{D_t}\right|\right] \times 100$$

A desirable property of forecasts is that they should be unbiased, so $E(e_i) = 0$. A method of tracking a forecast is to graph the values of the forecast error e_i over time. An unbiased method demonstrates forecast errors that fluctuate randomly above and below zero. An example is presented in Figure 17.3.

Alternatively, the cumulative sum of the forecast errors, Σe_i can be computed to check for bias. It is an indication that the forecasting method is biased if the value of this sum digresses too far from zero either above or below.

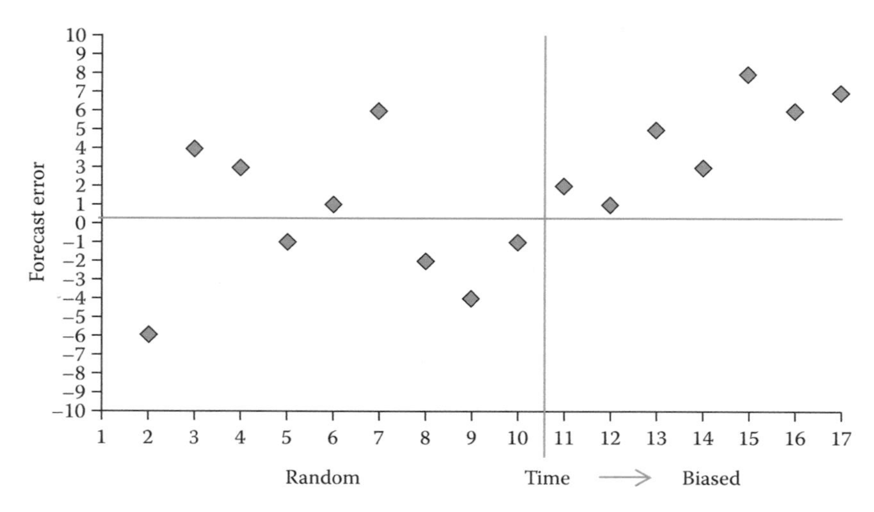

FIGURE 17.3 Forecast errors over time.

17.7 METHODS FOR FORECASTING STATIONARY SERIES

Two popular techniques for forecasting stationary time series are moving averages and exponential smoothing. A stationary time series is a time series where each observation can be represented by a constant and a random fluctuation. In symbols,

$$D_t = \mu + \varepsilon_t$$

where

μ is an unknown constant corresponding to the mean of the series

ε_t is a random error with mean zero and variance σ^2

The exponential and moving average methods are more precisely known as single or simple exponential smoothing and single or simple moving averages. Since the underlying model assumed for both moving averages and exponential smoothing is stationary, the one-step-ahead and multiple-step-ahead forecasts are the same.

17.7.1 Moving Averages

A simple but popular forecasting method is the method of moving averages. A moving average of order N is simply the arithmetic average of the most recent N observations. The F_t value, the forecast made in period $t-1$ for period t, is given by

$$F_t = (1/n)\sum_{i=t-N}^{t-1} D_i = \left(\frac{1}{N}\right)(D_{t-1} + D_{t-2} + \cdots + D_{t-N})$$

The mean of the N most recent observations is used as the forecast for the next period. MA(N) signifies N-period moving averages.

Each time a new demand observation is available, the average must be recomputed, but the recalculation of the full N-period average is not necessary every period, since

$$F_{t+1} = (1/n) \sum_{i=t-N+1}^{t} D_i = \left(\frac{1}{N}\right)\left[D_t + \sum_{i=t-n+1}^{t} D_i + D_{t-N}\right]$$

$$= F_t + (1/N)[D_t - D_{t-N}]$$

This shows that to update the forecast, it is necessary only to compute the difference between the most recent demand and the demand N periods old.

17.7.1.1 Moving Average Lags behind the Trend

Consider a demand process that contains a definite trend. Suppose that the observed demand is 2, 4, 6, 8, 10, 12, 14, 16, 18, 20, 22, 24. Consider the one-step-ahead MA(3) and MA(6) forecasts for

Period	Demand	MA(3)	MA(6)
1	2		
2	4		
3	6		
4	8	4	
5	10	6	
6	12	8	
7	14	10	7
8	16	12	9
9	18	14	11
10	20	16	13
11	22	18	15
12	24	20	17

The demand and the forecasts for the respective periods are pictured in Figure 17.4. It can be seen that the MA(3) and the MA(6) forecasts lag behind the trend.

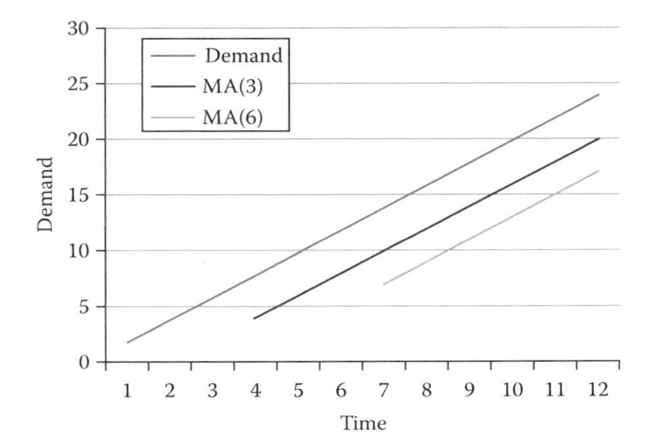

FIGURE 17.4 Moving average forecasts lag behind the trend.

MA(6) has a greater lag than MA(3), which implies that moving averages is not an appropriate forecasting method when there exists a trend in the series.

17.7.2 Exponential Smoothing

Another very popular forecasting method for stationary time series is exponential smoothing. The current forecast can be found from the weighted average of the current value of demand and the last forecast.

$$\text{New forecast} = \alpha(\text{Current observation of demand}) + (1-\alpha)(\text{Last forecast}).$$

In symbols,

$$F_t = \alpha D_{t-1} + (1-\alpha)F_{t-1}$$

where $0 < \alpha \leq 1$ is the smoothing constant, which determines the relative weight placed on the current observation of demand. Interpret $(1 - \alpha)$ as the weight placed on past observations of demand. With rearrangement and substitution, the exponential smoothing equation for F_t can be written as

$$\begin{aligned} F_t &= F_{t-1} - \alpha(F_{t-1} - D_{t-1}) \\ &= F_{t-1} - \alpha e_{t-1} \end{aligned}$$

Notice that if one forecasts high in period $t - 1$, e_{t-1} is positive and the adjustment is to decrease the forecast. Similarly, if one forecasts low in period $t - 1$, the error is negative, and the adjustment is to increase the current forecast.

Using the previous notation,

$$F_t = \alpha D_{t-1} + (1-\alpha)F_{t-1}$$

Substitution of the above formula obtains

$$F_t = \alpha D_{t-1} + \alpha(1-\alpha)D_{t-2} + (1-\alpha)^2 F_{t-2}$$

Similarly, substitution for F_{t-2} can be performed. The infinite expansion for F_t can be obtained by continuing to substitute

$$F_t = \sum_{i=0}^{\infty} \alpha(1-\alpha)^i D_{t-i-1} = \sum_{i=0}^{\infty} \alpha_i D_{t-i-1}$$

where the weights are $\alpha_0 > \alpha_1 > \alpha_2 > \ldots > \alpha_i = \alpha(1 - \alpha)^i$, and

$$\sum_{i=0}^{\infty} \alpha_i = \sum_{i=0}^{\infty} \alpha(1-\alpha)^i = \alpha \sum_{i=0}^{\infty} (1-\alpha)^i = \alpha \times \frac{1}{1[1-(1-\alpha)]} = 1$$

Therefore, exponential smoothing employs a declining set of weights to all past data.

A continuous exponential curve $g(i) = \alpha \exp(-\alpha i)$ could be fit to these weights. The smoothing constant α plays the same roles as the N value does in moving averages. A large α means that more weight is placed on the current observation of demand, thus less weight is placed on past observations. This means that the forecasts will have much greater variation from period to period, but they will react quickly to changes in the demand pattern. A small α means that more weight is placed on past data and thus the forecasts are more stable.

The derivation of the mean and variance of the forecast error can be found for both moving averages and exponential smoothing. If the assumption is made that the underlying demand process is stationary, then this mean and variance can be written in terms of the variance of each individual observation. This derivation demonstrates that both methods are unbiased and by equating the expressions for the variances and the average age of data, one can obtain the relationship between α and N. This means that if both exponential smoothing and moving averages are used to predict the same stationary demand pattern, forecast errors are normally distributed, and $\alpha = 2/(N + 1)$, then both methods will have exactly the same distribution of forecast errors.

17.7.3 Comparison of Exponential Smoothing and Moving Averages

There exist several similarities and several differences between moving averages and exponential smoothing.

17.7.3.1 Similarities

1. They both rely on the assumption that the underlying demand process is stationary.
2. Both methods depend on the specification of a single parameter. For moving averages, the parameter is N, the number of periods in the moving average, and for exponential smoothing the parameter is α, the smoothing constant. Small values of N or large values of α result in forecasts that put greater weight on current data and large values of N, and small values of α put greater weight on past data. Also, small N and large α may react quicker to changes in the demand process, but they will also result in forecast errors with higher variance.
3. If a trend exists, then both methods will lag behind.
4. When $\alpha = 2/(N + 1)$, both methods have the same distribution of forecast error, but they do not have the same forecasts. (Error affects accuracy.)

17.7.3.2 Differences

1. The moving average forecast is a weighted average of only the last N periods of data while the exponential smoothing forecast is a weighted average of all past data points (when $\alpha \leq 1$). This can be an important advantage of moving averages because an outlier will forever be part of an exponential smoothing forecast, whereas it will not be a part of moving averages forever.
2. In order to use moving averages, one must save all N past data points while one must only save the last forecast to use exponential smoothing. This is the most significant advantage of the exponential smoothing method and one reason for its popularity in practice.

17.8 TREND-BASED METHODS

As seen before, both exponential smoothing and moving average forecasts will lag behind a trend if one exists. Two forecasting methods that specifically account for trends in the data are regression analysis and Holt's method. Regression analysis is a method that fits a straight line to a set of data. Holt's method is a type of double exponential smoothing that allows for simultaneous smoothing on the series and on the trend.

17.8.1 Regression Analysis

Let $(x_1, y_1), (x_2, y_2), \ldots, (x_n, y_n)$ be n paired data points for the two variables X and Y. Assume that y_i is the observed value of Y when x_i is the observed value of X. Refer to Y as the dependent variable and X as the independent variable. A relationship exists between X and Y, which can be represented by the straight line

$$\hat{Y} = a + bX$$

Interpret Y as the predicted value of Y. The objective is to find the values of a and b so that the line $\hat{Y} = a + bX$ gives the best fit of the data. The values of a and b are chosen so that the sum of the squared distances between the regression line and the data points is minimized (see Figure 17.5).

When applying regression analysis to the forecasting problem, the independent variable often corresponds to time and the dependent variable to the series to be forecast. Assume that $D_1, D_2, \ldots, D_n$ are the values of the demand at times $1, 2, \ldots, n$. The optimal values of a and b are given by

$$b = \frac{S_{xy}}{S_{xy}}$$

and

$$a = \bar{D} - \frac{b(n+1)}{2}$$

where

$$S_{xy} = n\sum_{i=1}^{\infty} iD_i - \frac{n(n-1)}{2}\sum_{i=1}^{n} D_i$$

$$S_{xx} = \frac{n^2(n+1)(2n+1)}{6} - \frac{n^2(n+1)^2}{4}$$

and $\bar{D}$ is the arithmetic average of the observed demands during periods $1, 2, \ldots, n$.

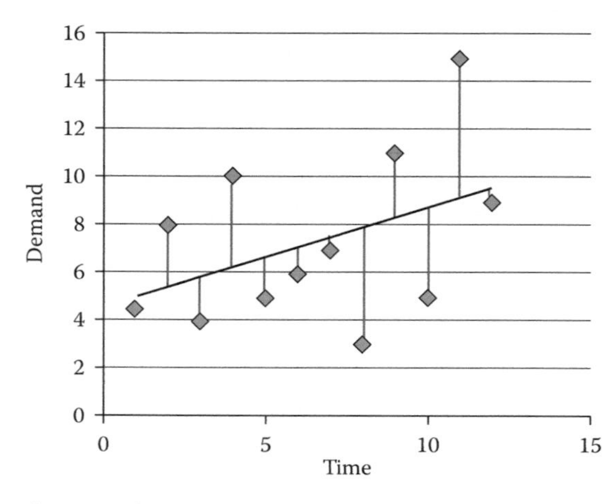

FIGURE 17.5 Example of a regression line.

17.8.2 Double Exponential Smoothing Using Holt's Method

Holt's method is a type of double exponential smoothing designed to track time series that demonstrate a linear trend. The method requires the specification of two smoothing constants, α and β, and uses two smoothing equations: one for the value of the series (the intercept) and one for the trend (the slope). The equations are

$$S_t = \alpha D_t + (1-\alpha)(S_{t-1} + G_{t-1})$$

$$G_t = \beta(S_t - S_{t-1}) + (1-\beta)G_{t-1}$$

The S_t value represents the intercept at time t and the G_t value is the slope at time t. The first equation is very similar to that used for simple exponential smoothing. When the most current observation of demand, D_t, becomes available, it is averaged with the prior forecast of the current demand, which is the previous intercept, S_{t-1}, plus 1 times the previous slope, G_{t-1}. The second equation can be explained as follows. Our new estimate of the intercept, S_t, causes us to revise our estimate of the slope to $S_t - S_{t-1}$. This value is then averaged with the previous estimate of the slope, G_{t-1}. The smoothing constants may be the same, but for most applications, more stability is given to the slope estimate (implying $\beta \leq \alpha$).

The τ-step-ahead forecast made in period t, which is denoted by $F_{t,t+\tau}$, is given by

$$F_{t,t+\tau} = S_t + \tau G_\tau$$

17.9 ADVANCED METHODS

Several other models exist for more complex forecasting. These models can be used when seasonality exists, trends exist, or both. One such method is the seasonal factors for stationary series. This method is very simple for computing factors for a time series with seasonal variation and no trend. It requires a minimum of two seasons of data. Another method is the seasonal decomposition using moving averages. This method is more accurate when a trend exists, and it requires a minimum of two seasons of data. It is a slightly more complex method for estimating seasonal factors and requires the computation of N-period moving averages, where N is the length of the season. The moving average method described can be used to predict a seasonal series with or without a trend. Unfortunately, as new data become available, the previous method requires that all seasonal factors be recalculated from scratch. Winter's method is a type of triple exponential smoothing; therefore, it is easy to update as new data become available. These models all require more complex calculations and should be used accordingly.

Determining the proper model depends on the characteristics of the history of observations and on the context in which the forecasts are required. When historical data are available, they should be examined carefully in order to determine if obvious patterns exist, such as trend or seasonal fluctuations. Usually, these patterns can be spotted by graphing the data. Statistical tests, such as significance of regression, can be used to verify the existence of a trend, for example. Identifying complex relationships not seen in graphical methods requires more sophisticated method like the *sample autocorrelation function*, which can reveal intricate relationships. Another method is the Box–Jenkins, which determines the appropriate model from an examination of the autocorrelation structure. Once a model has been chosen, forecasts should be monitored regularly to see if the model is appropriate or if some unforeseen change has occurred in the series. As indicated, a forecasting method should not be biased.

18 Manpower Planning

Aggregate planning addresses the problem of deciding how many employees the firm should retain or the quantity and the mix of products to be produced. Some firms utilize aggregate planning to control costs within the company. To determine proper planning, a firm must evaluate how responsive it can be to anticipated changes in the demand. The aggregate planning methodology is based on the assumption that demand is deterministic, which means that the demand is known in advance. Aggregate planning involves competing objectives. One objective is to react quickly to the anticipated changes in demand while another is maintaining a stable workforce to make sure they have enough workers. The final objective of aggregate planning is developing a plan that maximizes profit, which is subject to several constraints. Production planning can be seen as a hierarchical process that can be utilized at several levels in the firm to make purchasing, production, and staffing decisions. Although aggregate planning relies on managing groups of items, it may be applied at any level.

18.1 AGGREGATE UNITS OF PRODUCTION

The aggregate planning approach utilizes an aggregate unit of production. This unit can be considered an average item if the items are similar, but if there are many different types of items, it is more appropriate to consider aggregate units in terms of weight (tons of aluminum), volume (gallons of gasoline), amount of work required (worker-years of military training), or dollar value (value of inventory in dollars). The appropriate scheme to use is not always obvious and depends on the context of the problem as well as the level of aggregation required.

Example 18.1

A plant manager working for a large military guns manufacturer is considering implementing an aggregate planning system in order to determine the workforce and production levels in his plant. In his plant, they produce six weapon models. The characteristics of the guns are as follows:

Model Number	Number of Worker-Hours Required to Produce	Selling Price
MK123	4.2	$285
AM545	4.9	345
BP435	5.1	395
BP800	5.2	425
RF500	5.4	525
BK160	5.8	725

He must decide on which specific aggregation scheme he must use. One possible scheme involves defining an aggregate unit as $1 of output, but the selling prices of the different models produced are not consistent with the worker-hours required to produce them. While evaluating the sales, the manager notices that the percentages of the total number of sales for these six models have been fairly constant, with values of 32% for MK123, 21% for AM545, 17% for BP435, 14% for BP800, 10% for RF500, and 6% for BK160. With this knowledge, he decides to define an aggregate unit of production as a fictitious gun requiring (0.32)(4.2) + (0.21)(4.9) + (0.17)(5.1) + (0.14)(5.2) + (0.10)(5.4) + (0.06)(5.8) = 4.856 h of labor time. Sales forecasts can be obtained in a similar fashion. He can obtain sales forecasts for aggregate production units in essentially the same way by multiplying the appropriate fractions by the forecasts for unit sales of each type of machine. ■

The approach in Example 18.1 was possible because the products were similar, but this is not always the case, which makes defining the aggregate unit a little more difficult. When a firm produces a large variety of products, the natural aggregate unit is sales dollars, which will provide a good approximation for planning at the high level.

Aggregate planning is closely related to Hax and Meal's (1975) hierarchical production planning (HPP). This type of planning not only considers planning at the top level but also considers workforce sizes and production rates at a variety of levels of the firm. Hax and Meal recommend the follow hierarchy for aggregate planning purposes:

1. *Items*: Final products to be delivered to the customer and are referred to as an SKU (for stock-keeping unit).
2. *Families*: Group of items that share a common manufacturing setup cost.
3. *Types*: Groups of families with production quantities that are determined by a single aggregate production plan.

The items in the above example would be individual gun models, while a family would include all gun models. Finally, a type level example would be weapons. In Figure 18.1, we present a schematic of the aggregate planning function and its place in the hierarchy of production planning decisions.

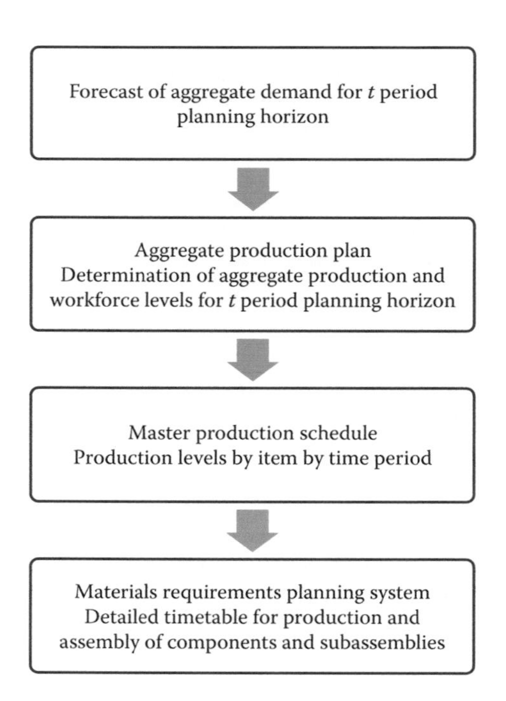

FIGURE 18.1 The hierarchy of production planning decisions.

18.2 OVERVIEW OF THE AGGREGATE PLANNING PROBLEM

After defining an aggregate unit, assume that there exists a forecast of the demand for a specified planning horizon expressed in terms of aggregate production units. The demand forecasts $D_1, D_2, \ldots, D_T$ occur over the next T planning periods. In most cases, a planning period is a month, (other periods can be used as well). These demand forecasts are used as known constants.

The objective of aggregate planning is to find the aggregate production quantities and the levels of resources required to achieve these production goals. Basically, aggregate planning determines the number of workers that should be employed and how much these workers must produce in each planning period (1, 2, …, T). Aggregate planning balances the positives of meeting demands against the negatives of changing the levels of production and/or the workforce levels.

The primary issues related to the aggregate planning problem include smoothing, bottleneck problems, planning horizon, and treatment of demand. Smoothing refers to costs that result from changing production and workforce levels from one period to the next. Two of the key components of smoothing costs are the costs that result from hiring and firing workers. Firing workers can be difficult to calculate because it is hard to discern the total effects of this action; therefore, it is hard to accurately calculate these costs. Companies that fire frequently give a poor image, which in turn may discourage potential employees from joining the company. Also, employees that are fired will look for other jobs instead of waiting around for the firm to rehire them when business picks up. Finally, most companies are not able to hire and fire at free will because labor agreements monitor these decisions.

Bottleneck problems occur when capacity restrictions are unable to respond to sudden changes in demand. This could occur in a plant that may all of sudden get an unusually high order that was not planned for; therefore, the plant is unable to meet the demand. Bottlenecks can also occur within different machines in a plant.

The planning horizon is another issue of aggregate planning. The number of periods for which workforce and inventory levels are to be determined, known as the period to be forecasted, must be specified in advance. The choice of the forecast horizon, T, affects whether the aggregate plan is useful. When T is chosen to be too small, the decisions made by the plan can be incorrect. The *end-of-horizon* effect is another important issue. In order to minimize holding costs, the aggregate plan might recommend that the inventory at the end of the horizon be drawn to zero, which could result in a poor strategy. To address this issue, rolling schedules are used, which means that at the time of the next decision, the old forecast is updated to reflect new information.

Treatment of demand is the final important issue to aggregate planning. Since aggregate planning methodology is based on the assumption that demand is known with certainty, the approach is both weak and strong. It is weak because it ignores the possibility of forecast errors; therefore, it does not provide a buffer for forecast errors. This approach is strong because by assuming deterministic demand, the effects of seasonal fluctuations and business cycles can be integrated into the planning function.

18.3 COSTS IN AGGREGATE PLANNING

Since the goal of aggregate planning is to minimize cost, one must identify all the costs that are affected by planning decisions including smoothing costs, holding costs, shortage costs, regular time costs, overtime and subcontracting costs, and idle time costs.

Smoothing costs are costs that accumulate as a result of changing the production levels from one period to the next like changing the size of the workforce. When increasing the size of the workforce, several different costs are incurred. For example, one must account for the time and expenses for advertising for new positions, interviewing these employees, and training new employees. When the workforce size decreases, one must consider costs such severance pay. Other costs are harder to measure such as the costs of a decline in worker morale that may result. In most of the models

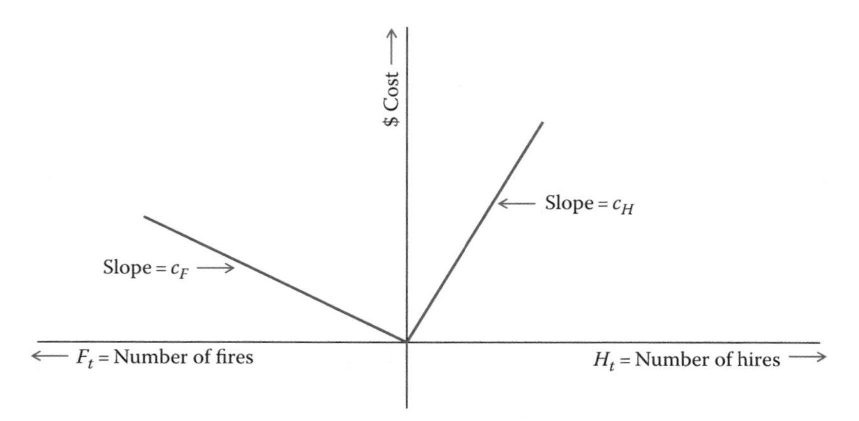

FIGURE 18.2 Cost of changing the size of the workforce.

considered, these costs are considered linear functions of the number of employees that are hired or fired, which means a constant dollar amount is charged for each employee hired or fired. The assumption of linearity is probably reasonable up to a point. A typical cost function for changing the size of the workforce appears in Figure 18.2.

Holding costs are the costs that result from having capital tied up in inventory. This cost can be considered rent on inventory since one must pay to hold the inventory from month to month. Holding costs are usually assumed to be linear in the number of units being held at a particular point in time. For aggregate planning, the assumption is made that holding cost is expressed in terms of dollars per unit held per planning period. Holding costs can either be charged against the inventory remaining on hand at the beginning or the end of the planning period. Shortage costs are the costs that occur when forecasted demand exceeds the capacity of the production facility or when demands are higher than anticipated. As with holding costs, shortage costs are generally assumed to be linear. Convex functions also can accurately describe shortage costs, but linear functions seem to be the most common. Figure 18.3 shows a typical holding/shortage cost function.

Regular time costs are the costs of producing one unit of output during regular working hours. Some costs included in regular time costs are payroll costs of regular employees working on regular time, the direct and indirect costs of materials, and other manufacturing expenses. Regular payroll costs become a "sunk cost" when all production is carried out on regular time. This is due to the fact that the number of units produced must equal the number of units demanded over any planning

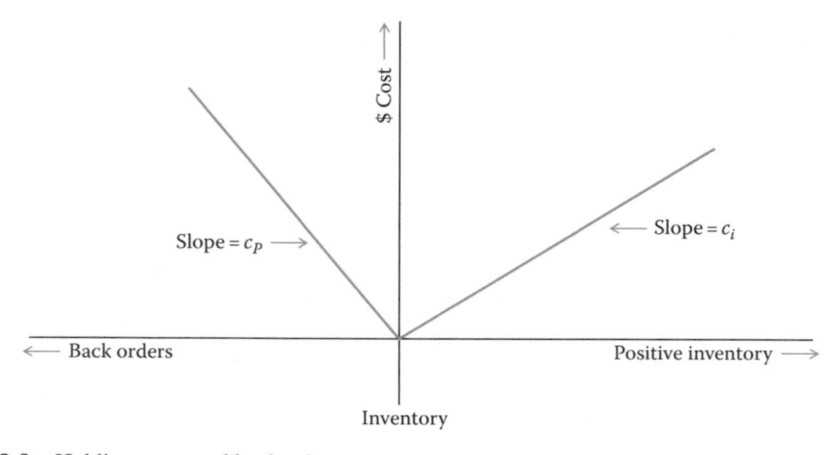

FIGURE 18.3 Holding costs and backorder costs.

horizon of adequate length. When there is no overtime or worker idle time, then regular payroll costs do not have to be included.

Overtime and subcontracting costs occur when units are not produced on regular time. Overtime will occur when employees must work during hours beyond the normal hours of a workday. Subcontracting refers to costs accrued from hiring outside suppliers to produce items. These costs are also considered linear.

Idle time costs occur when the workforce is underutilized, which means that workers are idle. Most times, idle time cost is 0 because the direct costs of idle time would be taken into account in labor costs and lower production levels. Idle time may have other consequences for the firm (i.e., if the aggregate units are input to another process, idle time on the line may result in higher costs to the following process). In these cases, a positive idle cost would be included.

Many things must be considered when creating an aggregate plan at a high level of the firm. An optimal solution may be found with a mathematical model, but one must consider things such as prior contract agreements or the negative effects on the firm's public image.

18.4 PROTOTYPE PROBLEM

To obtain the solutions to aggregate problems, one can either solve these problems by hand, graphically, or by linear program. The following example explores the different techniques for obtaining optimal solutions.

Example 18.2

RFMark, a firm that produces a line of RFID readers, is going to plan workforce and production levels for the 6 month period from January to June. Forecast demands over the next 6 months for a particular line of readers produced in the Lincoln, Nebraska plant are 1280, 640, 1200, 2000, and 1400. There are currently 300 workers employed in the Lincoln plant at the end of December. The ending inventory in December is expected to be 500 units, and the firm has decided that they would like to have 600 units on hand at the end of June.

To incorporate the starting and ending inventory constraints into the formulation, the easiest way is to modify the values of the predicted demand; therefore, in period 1, the net predicted demand is predicted demand minus initial inventory. If there is a minimum ending inventory constraint, this amount should be added to the demand in period T. Actual ending inventories should be computed using the original demand pattern, however.

In Example 18.2, net predicted demand for January is 780(1280 − 500) and the net predicted demand for June is 2000(1400 + 600). Net demand is based on the assumption that starting and ending inventory are both zero. For the 6 months, the net predicted demand and the net cumulative demand are as follows:

Month	Net Predicted Demand	Net Cumulative Demand
January	780	780
February	640	1420
March	900	2320
April	1200	3520
May	2000	5520
June	2000	7520

The cumulative net demand is pictured in Figure 18.4 as well as the feasible production plan, which is the specification of the production levels for each month. Cumulative production must be equal to or greater than cumulative demand for each period if shortages are not permitted.

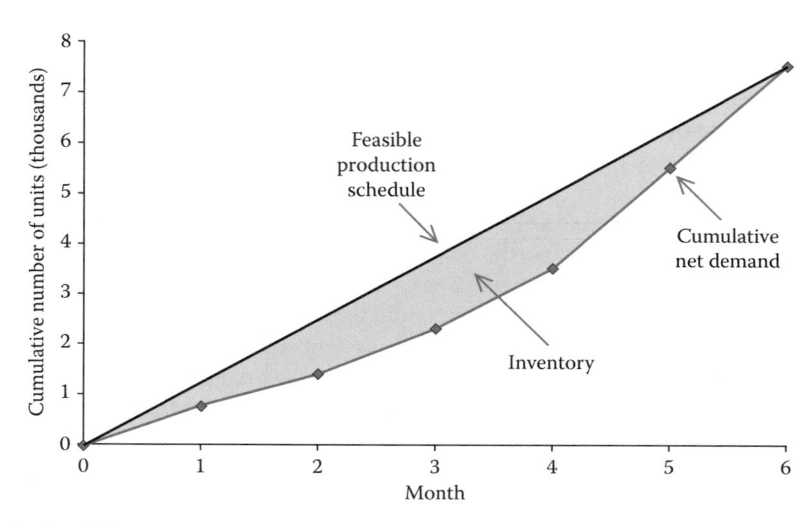

FIGURE 18.4 Feasible aggregate plan for RFMark.

To evaluate cost trade-offs, three costs are considered: cost of hiring workers, cost of firing workers, and cost of holding inventory:

- C_H = Cost of hiring one worker = \$500
- C_F = Cost of firing one worker = \$1000
- C_I = Cost of holding one unit of inventory for 1 month = \$80

To translate aggregate production in units to workforce levels, K will be evaluated as

- K = Number of aggregate units produced by one worker in 1 day

In the past, the plant manager observed the firm produced 245 RFID readers over 22 working days, with the workforce level constant at 76 workers. On average, the production rate was 245/22 = 11.1364 readers per day at a constant 76 workers; therefore, one worker produced an average of 11.1364/76 = 0.14653 readers in 1 day. The K value is equal to 0.14653 for this example. ■

This problem can be evaluated using two other strategies: zero inventory plan and the constant workforce plan. The zero inventory plan, also known as the chase strategy, changes the workforce each month in order to produce enough units to closely match the demand pattern. The constant workforce plan uses a strategy that maintains the minimum constant workforce necessary to satisfy the net demand.

18.4.1 Evaluation of a Chase Strategy (Zero Inventory Plan)

The chase strategy develops a production plan for RFMark, which minimizes the levels of inventory the firm must hold during the 6 month planning horizon. The numbers for the calculations along with the minimum number of workers required in each month is shown Table 18.1.

The final column in Table 18.1 is found by dividing the forecasted net demand by the number of units produced per worker. This ratio is rounded up to the next highest integer to guarantee that shortages will not occur (i.e., January, 780/2.931 gives 266.12, which is rounded up to 267). The number of working days each month may depend on a variety of factors, such as paid holidays and worker schedules.

TABLE 18.1
Initial Calculations for Zero Inventory Plan for RFMark

A	B	C	D	E
Month	Number of Working Days	Number of Units Produced per Worker (B × 0.14653)	Forecast Net Demand	Minimum Number of Workers Required (D/C Rounded up)
January	20	2.931	780	266
February	24	3.517	640	182
March	18	2.638	900	341
April	26	3.810	1200	315
May	22	3.224	2000	620
June	15	2.198	2000	910

Recall that the number of workers employed at the end of December is 300. Hiring and firing workers each month to match forecast demand as closely as possible results in the aggregate plan given in Table 18.2.

The number of units produced each month (column F in Table 18.2) is found by using the formula as follows:

Number of units produced = Number of workers × Average number of aggregate units produced in a month by a single worker

rounded to the nearest integer.

The totals at the bottom of Table 18.2 are multiplied by their respective costs to find the total cost of the production plan. The total cost of hiring, firing, holding is (755)(500) + (145)(1,000) + (30)(80) = $524,900. This cost is adjusted to include the holding cost for ending inventory of 600 units in June; therefore, the total cost of this plan is 524,900 + (600)(80) = $572,900.

There is always some remaining inventory at the end of each period because it is impossible to employ a fractional number of workers; therefore, zero inventory is hardly ever achieved. It is usually impossible to achieve zero inventory at the end of each planning period because it is not possible to employ a fractional number of workers.

TABLE 18.2
Zero Inventory Aggregate Plan for RFMark

A	B	C	D	E	F	G	H	I
Month	Number of Workers	Number Hired	Number Fired	Number of Units per Worker	Number of Units Produced (B × E)	Cumulative Production	Cumulative Net Demand	Ending Inventory (G − H)
January	267		33	2931	783	783	780	3
February	182		85	3517	640	1423	1420	3
March	342	160		2638	902	2325	2320	5
April	315		27	3810	1200	3525	3520	5
May	621	306		3224	2002	5527	5520	7
June	910	289		2198	2000	7527	7520	7
Totals		755	145					30

18.4.2 Evaluation of the Constant Workforce Plan

The goal of the constant workforce plan is to eliminate hiring and firing during the planning horizon. To do this, the minimum workforce required for every month in the planning horizon needs to be computed. The minimum workforce for January is found to be 267 using the net cumulative demand and units produced per worker. In January and February combined, there are 2931 + 3517 = 6448 units produced per worker and the cumulative demand is 1420. By calculating the ratio, 1420/6448 = 220.22 ≈ 221 workers are required for both months. Table 18.3 shows the ratios. The maximum of these numbers is the minimum number of workers needed for the 6 month planning period (411).

The inventory levels resulting from a constant workforce of 411 workers are found in Table 18.4. Column C displays the monthly production levels, which are found by multiplying the number of units produced per worker each month by the fixed workforce size of 411 workers. The total of the ending inventory levels is 5962 + 600 = 6562 (600 is the ending inventory in June). The total inventory cost of the plan is (6.562)(80) = $524,960. The cost of increasing the workforce from 300 to 411 in January, which is (111)(500) = $55,500 must be added to the previous number to find that the total cost is $580,460. This is slightly higher than the cost found by the zero inventory plan ($569,540). Because costs of the two plans are close, it is likely that

TABLE 18.3
Computation of the Minimum Workforce Required by RFMark

A	B	C	D
Month	Cumulative Net Demand	Cumulative Number of Units Produced per Worker	Ratio B/C (Rounded Up)
January	780	2.931	267
February	1420	6,448	221
March	2320	9,086	256
April	3520	12,896	273
May	5520	16,120	243
June	7520	18,318	411

TABLE 18.4
Inventory Levels for Constant Workforce Schedule

A	B	C	D	E	F
Month	Number of Units Produced per Worker	Monthly Production (B × 411)	Cumulative Production	Cumulative Net Demand	Ending Inventory (D – E)
January	2931	1205	1205	780	425
February	3517	1445	2650	1420	1230
March	2638	1084	3734	2320	1414
April	3810	1566	5300	3520	1780
May	3224	1325	6625	5520	1105
June	2198	903	7528	7520	8
Total					5962

the firm would choose the constant workforce plan so they can avoid the costs that cannot be measured for frequently changing their workforce numbers.

18.5 SOLUTION OF AGGREGATE PLANNING PROBLEMS BY LINEAR PROGRAMMING

Linear programming is a way to solve optimization problems. The objective of linear programming is to determine the values of n nonnegative real variables that maximize or minimize a linear function of these variables and is subject to m linear constraints of these variables. The main advantage to formulating a problem as a linear program is that it is very efficient to find optimal solutions through simplex method.

When all cost functions are linear, there is a linear programming formulation of the general aggregate planning problem. Optimal solutions of large problems can be found using computer software.

18.5.1 Cost Parameters and Given Information

The following values are assumed to be known:

- c_H = Cost of hiring one worker
- c_F = Cost of firing one worker
- c_I = Cost of holding one unit of stock for one period
- c_R = Cost of producing one unit of stock for one period
- c_O = Incremental cost of producing one unit on overtime
- c_U = Idle cost per unit of production
- c_S = Cost to subcontract one unit of production
- n_t = Number of production days in period t
- K = Number of aggregate units produced by one worker in 1 day
- I_0 = Initial inventory on hand at the start of the planning horizon
- D_t = Forecast of demand in period t

18.5.2 Problem Variables

The following are the problem variables:

- W_t = Workforce level in period t
- P_t = Production level in period t
- I_t = Inventory level in period t
- H_t = Number of workers hired in period t
- F_t = Number of workers fired in period t
- O_t = Overtime production in units
- U_t = Worker idle time in units ("undertime")
- S_t = Number of units subcontracted from outside

The overtime and idle time variables are determined in the following way. The number of units produced by one worker term in period t is represented by the term Kn_t; therefore, the number of units produced by the entire workforce in period t is represented by Kn_tW_t. The term Kn_tW_t does not have to equal P_t. For example, if P_t is greater than Kn_tW_t, then it means that the number of units produced exceeds what the workforce can produce over a regular time period; therefore, overtime is the difference of these values ($O_t = P_t - Kn_tW_t$). When P_t is less than Kn_tW_t, then this means that the workforce is producing less than it should be over a regular time period. Idle time is what occurs when this happens and measured in production units ($U_t = Kn_tW_t - P_t$).

18.5.3 Problem Constraints

Three sets of constraints are required for the problem, and they are included to ensure that the labor and units are conserved:

1. Conservation of workforce constraints

$$\underbrace{W_t}_{\text{Number of workers in } t} = \underbrace{W_{t-1}}_{\text{Number of workers in } t-1} + \underbrace{H_t}_{\text{Number hired in } t} - \underbrace{F_t}_{\text{Number fired in } t} \quad \text{for } 1 \le t \le T$$

2. Conservation of units constraints

$$\underbrace{I_t}_{\text{Inventory in } t} = \underbrace{I_{t-1}}_{\text{Inventory in } t-1} + \underbrace{P_t}_{\text{Number of units produced in } t} - \underbrace{S_t}_{\text{Number of units subcontracted in } t} - \underbrace{D_t}_{\text{Demand in } t} \quad \text{for } 1 \le t \le T$$

3. Constraints relating production levels to workforce levels

$$\underbrace{P_t}_{\substack{\text{Number of} \\ \text{units produced in } t}} = \underbrace{Kn_tW_t}_{\substack{\text{Number of units produced by} \\ \text{regular workforce in } t-1}} + \underbrace{O_t}_{\substack{\text{Number of units produced} \\ \text{on overtime in } t}} - \underbrace{U_t}_{\substack{\text{Number units of} \\ \text{idle production in } t}} \quad \text{for } 1 \le t \le T$$

All problem variables must be nonnegative. The formulation is subject to these constraints and the nonnegativity constraints.

To formulate the linear program, the initial inventory, I_0, and the initial workforce, W_0, must be specified.

The objective function involves all cost constants. The linear programming solution will choose values of the problem variables W_t, P_t, I_t, H_t, F_t, O_t, U_t, and S, that will minimize

$$\sum_{t=1}^{T} \left(c_H H_t + c_F F_t + c_I I_t + c_R P_t + c_o O_t + c_S S_t \right)$$

subject to

$$W_t = W_{t-1} + H_t - F_t \quad \text{for } 1 \le t \le T \text{ (conservation of workforce),} \tag{18.1}$$

$$P_t = Kn_tW_t + H_t - F_t \quad \text{for } 1 \le t \le T \text{ (production and workforce),} \tag{18.2}$$

$$I_t = I_{t-1} + P_t + S_t - D_t \quad \text{for } 1 \le t \le T \text{ (inventory balance),} \tag{18.3}$$

$$H_t, F_t, I_t, O_t, U_t, S_t, W_t, P_t \ge 0 \text{ (nonnegativity),} \tag{18.4}$$

plus any additional constraints that may define the values of starting inventory, ending workforce, or any other fixed variables.

18.6 SOLVING AGGREGATE PLANNING PROBLEMS BY LINEAR PROGRAMMING: AN EXAMPLE

The linear programming can be used to find the optimal solution to the example presented in Section 18.4. Assume the cost coefficients are constant with respect to time, so the objective function is

$$\text{Minimize}\left(500\sum_{t=1}^{6} H_t + 1000\sum_{t=1}^{6} F_t + 80\sum_{t=1}^{6} I_t\right)$$

Additional constraints for each boundary condition are added to represent the initial inventory of 500 units, the initial workforce of 300 workers, and the ending inventory of 600 units.

Substituting $t = 1,\ldots,6$ into Equations 18.1 through 18.3 obtains the constraints. The full set of constraints expressed in standard linear programming format is as follows:

$$\begin{aligned} W_1 - W_0 - H_1 + F_1 &= 0,\\ W_2 - W_1 - H_2 + F_2 &= 0,\\ W_3 - W_2 - H_3 + F_3 &= 0,\\ W_4 - W_3 - H_4 + F_4 &= 0,\\ W_5 - W_4 - H_5 + F_5 &= 0,\\ W_6 - W_5 - H_6 + F_6 &= 0; \end{aligned} \tag{18.5}$$

$$\begin{aligned} P_1 - I_1 - I_0 &= 1280,\\ P_2 - I_2 - I_1 &= 640,\\ P_3 - I_3 - I_2 &= 900,\\ P_4 - I_4 - I_3 &= 1200,\\ P_5 - I_5 - I_4 &= 2000,\\ P_6 - I_6 - I_5 &= 1400; \end{aligned} \tag{18.6}$$

$$\begin{aligned} P_1 - 2.931W_1 &= 0,\\ P_2 - 3.517W_2 &= 0,\\ P_3 - 2.638W_3 &= 0,\\ P_4 - 3.810W_4 &= 0,\\ P_5 - 3.224W_5 &= 0,\\ P_6 - 2.198W_6 &= 0; \end{aligned} \tag{18.7}$$

$$W_1,\ldots,W_6, P_1,\ldots,P_6, I_1,\ldots,I_6, H_1,\ldots,H_6 \geq 0; \tag{18.8}$$

$$\begin{aligned} W_1 &= 300,\\ I_0 &= 500,\\ I_6 &= 600, \end{aligned} \tag{18.9}$$

TABLE 18.5
Aggregate Plan for RFMark Obtained from Rounding the Linear Programming Solution

A	B	C	D	E	F	G	H	I
Month	Number of Workers	Number Hired	Number Fired	Number of Units per Worker	Number of Units Produced (B × E)	Cumulative Production	Cumulative Net Demand	Ending Inventory (G − H)
January	273		27	2931	800	800	780	20
February	273			3517	960	1760	1420	340
March	273			2638	720	2480	2320	160
April	273			3810	1040	3520	3520	0
May	738	465		3224	2379	5899	5520	379
June	738			2198	1622	7521	7520	1
Totals		465	27					900

Using computer software, the linear program can be solved to find the optimal solution of the objective function at $379,320.90. This value is found with fractional values of the variables, which is not feasible with units, so after rounding, the actual cost will be slightly higher.

By rounding all the values of W_t to the next higher integer, the values become $W_1 = \ldots = W_4 = 273$ and $W_5 = W_6 = 738$; therefore, the firm should fire 27 workers in January and hire 465 workers in May. Table 18.5 shows the complete solution.

Since column H in Table 18.5 corresponds to net demand, 600 units of ending inventory must be added in June, giving a total inventory of 900 + 600 = 1500 units. The total cost of this plan is then (500)(465) + (1,000)(27) + (80)(1,500) = $379,500. This value is much lower than the zero inventory plan and the constant workforce plan.

The linear programming analysis suggests that the optimal strategy is to decrease the workforce in January and build it back up again in May. Considering these results, a reasonable alternative might be to not fire the 27 workers in January and also hire fewer workers in May. Using this alternative plan, the most efficient method for finding the correct number of workers to hire in May can be found by resolving the linear program without the variables $F_1, \ldots, F_6$, as no firing of workers means that these variables are forced to zero. If no workers are fired in January, the optimal number of workers to hire in May turns out to be 374, and the cost of the plan is seen to be approximately $386,120, which is only slightly more expensive than the optimal plan. This suggestion shows that one must consider not only the results of the output, but also the impact the decisions have on the company.

18.7 ADVANTAGES AND DISADVANTAGES

Aggregate planning can be very helpful in production planning and manpower levels of a firm. The objective of aggregate planning is to absorb demand fluctuations by smoothing workforce and production levels. There are several advantages to aggregate planning. One advantage is that aggregate planning is much less expensive than the cost of preparing forecasts and determining productivity and cost parameters on an individual item basis. Another advantage is aggregate planning improves forecast accuracy; aggregate forecasts are usually more accurate than individual forecasts.

Although there are several advantages, there are also a few drawbacks to aggregate planning. To begin with, it is difficult to define an aggregate unit of production. Second, it is difficult to obtain accurate cost and demand information for aggregate units. Another disadvantage is that it is not that easy to change workforce levels. Finally, managers have trouble relying on mathematical models to address workforce levels.

19 Production Planning and Scheduling

19.1 INTRODUCTION

Production scheduling is very similar to aggregate planning. While aggregate planning determines the amount of resources necessary to produce a certain number of units, production scheduling determines the development of a schedule of how long it takes to make a certain number of products. Production scheduling answers the following questions:

1. When are the products to be produced?
2. How much of each product will be produced?
3. For how long will the products be produced?

In this chapter, we focus on what is necessary to generate a schedule that meets demand. A firm's success is dependent on having an efficient production system; therefore, a well-thought-out production schedule must be made to ensure production meets the customer's demand.

19.2 SCHEDULE DESIGN

Schedule design decisions provide answers to questions involving how much to produce and when to produce. Lot-size decisions, determining when to produce, are referred to as production scheduling and are heavily based on the aggregate planning strategies determined using methods as those described in the previous chapter. In addition to how much and when, it is important to know how long production will continue, which is based on the forecasts of market demand.

The impact of schedule design decisions is seen throughout the following areas: machine selection, number of machines, number of shifts, number of employees, space requirements, storage equipment, material handling equipment, personnel requirements, storage policies, unit load design, building size, and so on.

To plan a production schedule, information is needed concerning production volumes, trends, and the predictability of future demands for the products to be produced, which can be obtained utilizing the methods described in Chapter 19. The more specific the inputs from product, process, and schedule designs, the greater the likelihood of optimizing the facility and meeting the needs of manufacturing.

19.2.1 Process Requirements

Process design determines the specific equipment types required to produce the product. Schedule design determines the number of each equipment type required to meet the production schedule.

The determination of process requirements occurs in three phases:

- Phase 1 determines the quantity of components that must be produced, including scrap allowance, in order to meet the market estimate.
- Phase 2 determines the equipment requirements for each operation.
- Phase 3 combines the operation requirements to obtain overall equipment requirements.

19.2.2 Scrap Estimates

It is important to take into account the scrap estimates in order to create the most efficient production schedule. The market estimate found by forecasting and aggregate planning specifies the annual volume to be produced for each product. To produce the required amount of product, the number of units scheduled through production must equal the market estimate plus a scrap estimate; therefore, production capacity must be planned for the production of scrap. The definition of scrap is the material waste generated in the manufacturing process.

An estimate of the percentage of scrap to be incurred from each operation must be made. This may be based on historical data or estimated from similar operations. Let P_k represent the percentage of scrap produced on the kth operation, O_k the desired output of non-defective product from operation k, and I_k the production input to operation k. It follows that, on average,

$$O_k = I_k - P_k I_k \tag{19.1}$$

or

$$O_k = I_k(1 - P_k)$$

so,

$$I_k = \frac{O_k}{1 - P_k} \tag{19.2}$$

Therefore, the expected number of units to start into production for a part having n operations is

$$I_1 = \frac{O_n}{(1 - P_1)(1 - P_2)\cdots(1 - P_n)} \tag{19.3}$$

where, in this case, O_n is the market estimate.

19.3 TOOLS FOR PRODUCTION SCHEDULING

Some tools that are frequently used by quality experts are helpful in production scheduling. More recently, the seven management and planning tools (*affinity diagram*, the *interrelationship digraph*, the *tree diagram*, the *matrix diagram*, the *contingency diagram*, the *activity network diagram*, and the *prioritization matrix*) have gained acceptance as a methodology for improving planning and implementation efforts in general. The tools that are most likely to help in production scheduling are the affinity diagram, interrelationship digraph, and activity network diagram.

19.3.1 Affinity Diagram

The affinity diagram is used to gather language data, such as ideas and issues, and organize it into groupings. Suppose we are interested in generating ideas for reducing manufacturing leadtime. In a brainstorming session, the issues are written down on "post-it" notes and grouped on a board or wall. Each group then receives a heading. An affinity diagram for reducing manufacturing lead time is presented in Figure 2.19. The headings selected were facilities design, equipment issues, quality, setup time, and scheduling.

19.3.2 Interrelationship Digraph

The interrelationship digraph is used to map the logical links among related items, trying to identify which items impact others the most. The term digraph is employed because the graph uses directed arcs. This digraph helps us understand the logical sequence of steps followed by a process in order to produce a unit. The efforts must be initiated with the formation of product families.

19.3.3 Activity Network Diagram

The activity network diagram is used to develop a work schedule that details the pessimistic, expected, and optimistic times that a process will take to finish production. This diagram is synonymous to the critical path method graph. It can also be replaced by a Gantt chart and if a range is defined for the duration of each activity, the Program Evaluation and Review Technique chart can also be used. The important message is that a well-thought-out time table is needed to understand the length of a production project.

19.4 SUMMARY

As can be seen, forecasting, aggregate planning, and production scheduling are all intertwined. Without forecasting and aggregate planning, it would be impossible to determine a sufficient production schedule that would allow a firm to meet their customers' demand. Utilization of tools aids production planners to determine the amount needed to meet demand and also leads to more efficient scheduling.

20 RFID in Logistics

20.1 INTRODUCTION

In this chapter, we discuss how radio frequency identification (RFID) supports

- Information use in the supply chain
- Open and closed-loop systems as an intelligent agent
- Real-time item visibility that facilitates inventory control
- Supply chain planning and operational optimization
- Organizational improvement with best practices

20.2 RFID SUPPORTS INFORMATION USE IN THE SUPPLY CHAIN

In this text, we discuss how RFID-captured information influences supply chain performance with data analysis, inventory management, transportation, and supply chain visibility.

1. *Data analysis* includes the evaluations concerning facilities, inventory, transportation, costs, prices, and customers throughout the supply chain.
2. *Inventory management* includes raw materials, work in process, and finished goods within a supply chain. Changing inventory policies can affect the supply chain's efficiency and responsiveness.
3. *Transportation visibility* provides information about inventory transported from point to point in the supply chain. Transportation includes many combinations of modes and routes, each with its own performance characteristics. Transportation modes have a large impact on supply chain responsiveness and cost.
4. *Supply chain visibility* provides information including status updates that allow for evaluation of the physical locations in the supply chain network where product is stored, assembled, or fabricated. The two major types of facilities are production sites and storage sites. The location, capacity, and flexibility of facilities have a significant impact on the supply chain's performance.

In this section, we discuss the role that information plays in the supply chain, as well as key information-related decisions that supply chain managers must make.

20.2.1 Data Analysis and Information Gathering

Data analysis and information gathering affects every part of the supply chain. Information gathering affects a supply chain in many different ways. Consider the following:

1. Information serves as the connection between various stages of supply chain, allowing them to coordinate and maximize total supply chain profitability.
2. Information is important to the daily operations of each stage in the supply chain. Consider a production scheduling system that uses information on demand to create schedules that allow a factory to produce the right products at the right time. A warehouse management system (WMS) uses information to create visibility of the warehouse's inventory. The company can then use this information to determine whether new orders can be filled.

3. Information is an important driver that companies have used to become more responsive and efficient. The growth of the importance of information technology is due to the fact that it has effectively improved business. Though information can support efficiencies if integrated too aggressively, it can result into costly decisions such expensive software system implementations. Organizations should decide what information is necessary for reducing cost and improving their responsiveness within a supply chain.

20.2.1.1 Push and Pull Operational Strategies

Different types of operational strategies require different types of information. Push systems are associated with *material requirements planning* systems that use master production schedules to create schedules for suppliers with part types, quantities, and delivery dates.

Pull systems are associated with kanban systems which fulfill only the necessary requirements from actual customer demand.

For practical use, they require the latest information on actual demand. Modern execution systems such as WMS, and transportation management systems, and related execution modules in enterprise resource systems (ERP) use these strategies in their programmed logic. The effectiveness of these types of systems is predicated on timeliness of the collected information. RFID, bar codes, and other automatic identification (auto ID) systems allow these types of systems to effectively reduce operational costs.

20.2.1.2 Supply Chain Coordination

Data collection allows supply chain coordination to occur when all stages of a supply chain work toward the objective of maximizing total supply chain profitability based on shared information. Lack of coordination can result in a significant loss of supply chain profit. Coordination among different stages in a supply chain requires each stage to share appropriate information with other stages.

20.2.1.3 Forecasting

Timely information enables more accurate forecasting about what future demand and conditions will be. Obtaining forecasting information frequently means using sophisticated techniques to estimate future sales or market conditions. Managers must decide how they will make forecasts and to what extent they will rely on forecasts to make decisions. Companies often use forecasts both on a tactical level to schedule production and on a strategic level to determine whether to build new plants or even whether to enter a new market.

Forecasting is a method used by firms to predict the future. Marketing and sales utilize forecasting methods to predict product demand. It is difficult to accurately forecast events because events are often random. Events such as tossing a coin or rolling a dice are considered random events. The main focus of forecasting is product demand. Forecasting methods can be used on parts of the demand process even though product demand is usually random. This is because parts of the demand process are predictable. Trends, cycles, and seasonal variation may be present within the demand process, which gives us the advantage of predicting the future demand processes.

20.2.1.4 Aggregate Planning

Once a company creates a forecast, the company needs a plan to act on the forecast. Aggregate planning transforms forecasts into plans of activity to satisfy the projected demand. A key decision manager's face is how to collaborate on aggregate planning throughout the entire supply chain. The aggregate plan becomes a critical piece of information to be shared across the supply chain because it affects both the demand on a firm's suppliers and the supply to its customers.

Aggregate planning is used when demand is assumed to be deterministic. A demand that is deterministic is known in advance. Production planning can be seen as a hierarchical process that can be utilized at several levels in the firm to make purchasing, production, and staffing decisions. Aggregate planning may be applied at any level but it is mainly relying on managing groups of items. With the aggregate planning methodology, a firm can evaluate its responsiveness to anticipated changes. There are several goals of aggregate planning, which are often conflicting. To begin with, an aggregate plan must react quickly to the anticipated changes. The second goal that conflicts with the first one is that the firm must maintain enough people to ensure that there are enough workers to produce enough to meet the demand. The final objective is that an aggregate plan must maximize profit when subjected to several constraints.

20.2.2 RFID and Other Enabling Technologies

Many technologies along with RFID exist to share and analyze information in the supply chain. Some of these technologies include the following.

20.2.2.1 EDI Business Transmissions

Electronic data interchange (EDI) refers to the electronic transmission of standard business documents in a predetermined format from one company's business computer to its trading partners' computer. EDI relies on two standards, ANSI and EDIFACT, to ensure standardize business communication. EDI allows a firm to transmit information, such as point-of-sale (POS) demand for information, purchase orders, and inventory status information, to users within the firm and to customers and trading partners. EDI systems have been implemented generally by larger firms because of the expense it requires for dedicated software and advanced hardware.

The Internet has critical advantages over EDI with respect to information sharing. The Internet conveys much more information and therefore offers much more visibility than EDI. Better visibility improves decisions across the supply chain. Internet communication among stages in the supply chain is also easier because a standard infrastructure (the World Wide Web) already exists. Thanks to the Internet, e-commerce has become a major force in the supply chain.

20.2.2.2 Web-Based Application Systems

The use of the web for both business to consumer (B2C) and business to business (B2B) transactions is growing quickly. The web will have significant implications for supply chain management in the coming years. Thomas Freidman, a leader on political thought, mentions that the web is allowing for small companies to compete with large companies in both B2C and B2B transactions using the web. Unlike the failed dot-bomb companies in the early twentieth century, now well-thought-out small businesses are able to compete, leveraging web-based transactions.

Many software firms offer web-based systems. Some advantages that web-based supply chain systems provide include

1. Unlimited web access
2. Common platform unlike the complexity of EDI
3. Cost-effective implementation

Because many firms are concerned with Internet security, some prefer the complex EDI protocols that provide more extensive security than do web-based systems. However, given the new web-based EDI protocol standards and the cost implications, the web-based systems will begin to replace EDI systems in the coming years.

20.2.2.3 Business Operations Systems

ERP systems provide the transactional tracking and global visibility of information from within a company and across its supply chain. The real-time information (RTI) helps a supply chain to improve the quality of its operational decisions. ERP systems keep track of the information, whereas the Internet provides one method with which to communicate this information.

Supply chain management (SCM) software uses the information in ERP systems to provide analytical decision to support in addition to the visibility of information. ERP systems show a company what is going on, while SCM systems help a company decide what it should do.

20.2.2.4 Overall Trade-Off: Responsiveness versus Efficiency

Good information can help a firm improve both its responsiveness and efficiency. The information driver is used to improve the performance of other drivers, and the use of information is based on the strategic position the other drivers support. Accurate information can help a firm improve efficiency by decreasing inventory and transportation costs. Accurate information can improve responsiveness by helping a supply chain better match supply and demand.

Common literature suggests that we are living in the "information age." The availability of information from many sources appears in academic publications, trade journals, magazines, newsletters, blogs, e-magazines, and so on. The explosion of information availability on the web due to web search companies such as Yahoo, Google, and Microsoft allow people to perform web searches for information on almost anything.

Supply chains information provides the organization strategic advantage over competition and is the key to running a business efficiently and effectively in an ever changing and more complex environment. Information plays a key role in the management of the supply chain as evidenced in such uses for forecasts, aggregate manpower planning, and customer inquires.

Concepts such as just-in-time (JIT) manufacturing and delivery, vendor managed inventory (VMI), and cross-docking require timely information within the supply chain. JIT uses timely information to optimize the scheduling of deliveries or manufacturing in such a way to minimize inventories. The VMI concept allowed vendors to review information from a customer and order only what is needed in specific time period, which allowed them to reduce ordering excess inventory. These concepts leverage the concept of information in the supply chain. In the next section, we describe some of the concepts for leveraging information in the supply chain. The section will introduce the bullwhip effect, which is a general term that describes the inefficiencies realized in supply chain operations with imperfect information. We will discuss the determinants and current ways information can offset these inefficiencies, including electronic commerce, web-based systems, and RFID.

20.2.2.5 e-Commerce and Technology

Electronic commerce or e-commerce refers to a technology that allows businesses to operate a common transaction that was traditionally performed on a paper-based system and now is performed electronically. These technologies include EDI, e-mail, electronic funds transfers, electronic publishing, image processing, electronic bulletin boards, blogs, Internet voice mail, Internet video meetings, mp3 sharing, shared data bases, POS bar code systems in supermarkets, and all manner of web-based business systems.

Some well-known companies such as General Electric Corporation, one of the world's largest diversified manufacturers of a wide variety of products, uses web-based transaction systems and EDI as a regular part of its business practices in most divisions. Other companies use EDI, Internet-based systems, electronic forecasting, and WMS to gain competitiveness.

During the last few years of the twentieth century, IPOs of the "dot-com" companies were occurring almost everyday and their share prices rose steadily, even though many of these companies had few customers. These "pure play" e-tailers, which represented Internet-based retailers, without traditional brick and mortar operations have all but disappeared. One that survived and thrived is Amazon.com. Amazon has significantly expanded their product line, and is one of the few successful "pure play" e-tailers that survived the dot-com bust. One of the authors, who was a consultant during the time of the e-tailer craze, recalls that the main failure was the real lack of a business plan and focusing on developing a "killer ap" or creative front-end software and Web sites as their major objective. Some of the primary portals (e.g., Yahoo and Google) require big money to allow direct access to an e-commerce site. These failed e-tailers were also referred to "dot-bomb" companies instead of "dot-com" companies.

A new phenomenon that was derived from these companies' failures and was recently documented by Thomas Friedman in his landmark text "The World is Flat" is that smarter versions of these e-businesses initiatives are reappearing and competing against larger companies in local markets, domestically and internationally. Most profits for Internet activities are B2B web-based systems account. They represent a much greater share of the e-commerce marketplace than web-based retailers.

20.2.2.6 RFID as Part of the Information Supply Chain

RFID tags are emerging as the bar codes of the future. As we discussed in the earlier chapters, bar codes, which have become common in retailing, were only accepted en masse in 1985. The expectation is that though active RFID tags were commercially viable in 1973 and utilized in toll roads and animal tracking in the mid-1980s, and passive tags were arguably commercially viable in 2005, mass acceptance in logistics is expected by 2010. This emerging technology may have one of the fastest technology acceptance rates in history.

Common applications include (1) EZ Pass for paying bridge or highway tolls, (2) tagging of library books in some libraries, and (3) tagging of cargo containers at most of the world's ports. Reconciling shipments against bill of landings or packing and customer orders can be performed succinctly and accurately, thus eliminating the need to perform these functions manually. Beyond the supply chain, RFID technologies have broader applications such as emergency human identification for finding abducted children in Mexico or mountain climbers in Colorado who may become lost in an avalanche. Such applications of RFID technology benefits are taunted, yet fiercely debated as to how they may threaten individual rights and privacy. In the application chapters, details of these applications and their challenges are described. We now further detail how RFID technologies provide information strategically to allow organizations to improve operational effectiveness.

20.2.3 RFID as an Intelligent Agent System

Because of the differing auto-ID technologies such as bar code, RFID passive, RFID active, SAW tags, and sensor tags, there must be exploration of how they can be integrated for use in the supply chain. We suggest using RFID technologies as an intelligent agent system that supports real-time decision systems as solution for this integration challenge.

The idea of an intelligent agent is pervasive control system frameworks. Control frameworks can be classified as *hierarchical, heterarchical*, and *hybrid*. We consider RFID tags that have a master–slave relationship that exists between higher and lower levels in a *hierarchical* auto-ID framework. An operational example would be using bar codes affixed to cartons to write information to passive RFID pallet tags. The information is passed to the next highest unit load; case to pallet level with each technology acting as independent systems. This is similar to the control system concept in which response to input data is passed up the chain of command, higher level controllers pass down command data for execution by the lower level controllers. In control systems, this theory works well when there is little interference between the technologies; RFID integration does present this problem. We will refer to this type of system as open system. A *heterarchical* framework is present when there exists interactions between the lower level controllers, and it permits these to engage in one-on-one communication assuming there is no hierarchy or higher level controller. For RFID systems, this represents the use of reading tags on a common protocol such as the EPCglobal passive standard where multiple readers can read standardized tags. We refer to this type of system in our text as a closed system. *Hybrid* frameworks discuss how these frameworks capture the benefits of hierarchical and heterarchical frameworks while avoiding their pitfalls. For RFID technologies to work with other auto-ID technologies in the short run, this type of approach will be necessary to realize organizational savings.

A practical example of how the integration of multiple RFID technologies that operate at different frequencies can be modeled in control frameworks so that RTI can be used to determine an inventory policy is given as follows: a high frequency (HF) 13.56 passive tag is used to track retail over-the-counter drugs at the item level, ultrahigh frequency (UHF) 915 MHz passive RFID tags can be used to track inventory at the case, and pallet level inventory, and UHF 433 MHz active tags track the status of inventory on tractor trailers. Popular industrial literature assumes that linking information with relational databases provides RTI on the status at the item level (i.e., the active tag can show the status of the drugs because the tags were relationally linked as they moved up in container level).

The flaws in this assumption may be that different technologies have different error rates in scanning validation, human error of integrating these relations in database programming, and the technologies do not have common standards. The current mandate from Walmart encompasses only one standard, the electronic product code (EPC) global standard for Generation 1 and 2 UHF 856–915 MHz passive tags.

Furthermore, this EPCglobal standard is currently accepted in the United States but has not been completely adopted by other countries. Also, current FDA initiatives for over-the-counter drug tracking incorporate the 13.56 MHz RFID tags. The lack of understanding how the mixed RFID technologies will have negative impacts, such as higher error rates and lower productivity, provides a gap that I seek to investigate during this research project. One of the authors identified this gap when testing technologies, NASA ISS. The most operationally valid solutions included multiple RFID technologies.

20.2.4 Summary of RFID and Information Enablers

This section provides understanding of key technologies and how all the technologies differ and how they can be integrated to work for operational effectiveness. This will allow for better interpretation of WMS algorithms such as "bucket brigades" calculations, picking route optimization, and other effective system updates that will improve operations. Further insights into how safety stock minimization, customer order optimization, and pick/stock labor minimization will be affected are given later in the text.

20.2.5 RFID Provides Timely Visibility in Logistics

RFID supports information in the supply chain by enabling visibility. The concept of visibility describes the ability for anyone including customers to have access to inventory, orders, raw materials, and delivery points at any time. Visibility is currently provided by a mixture of auto-ID technologies such as bar codes, smart labels, ISBN, and UPC codes, along with others. The opportunity for RFID is that its non-line of sight scanning, the integration of the aforementioned auto-ID identifiers into RFID nomenclature, and push for standardized technology protocols will provide large supply chain savings.

The real-time nature of RFID is considered a benefit and currently a challenge. The benefit is that you have the latest information to make the best decisions; the drawback is that the amount of data currently presents a data storage and handling problem for operational systems.

Better visibility provides reduced inventory, labor and assets management using inventory policies, scheduling, and decision support system (DSS) information. This is exemplified by

- RFID supports reduced inventory costs with more effective labor policies
- RFID supports labor reduction with more effective scheduling
- RFID supports the reduction of expensive assets such as facilities, trucks, containers, and railroad time because of more accurate information in DSS

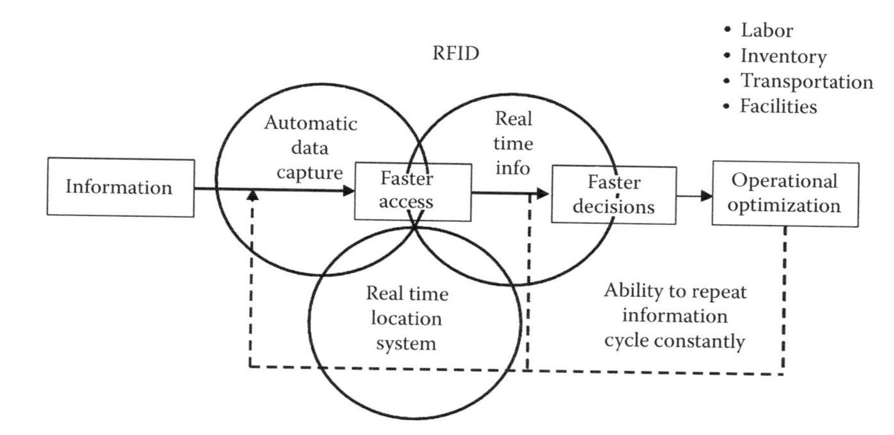

FIGURE 20.1 RFID enabling technologies.

The ability for RFID to provide timely information and visibility into the supply chain are based on three components of RFID technologies. They are

1. Automatic data capture
2. Real-time information
3. Real-time location system

The RFID enabling technologies diagram represents these components as interconnecting orbits in Figure 20.1.

Figure 20.1 also shows how RFID supports timely information in the supply chain by enabling information to be accessed faster. This implies faster decisions can be made, which produces operational optimization that can be effectively repeated. In Figure 20.1, the boxes represent the RFID information flow. RFID shows future promise in the ability to allow resident information collected automatically in real time that leads to faster, more effective decisions. Business costs are reduced as operations become more productive by reducing labor, transportation, and facility cost of moving inventory in the supply chain.

Many organizations see the benefit of using RFID as the ability to effectively manipulate inventory. Inventory exists in the supply chain because of the variance between supply and demand. This variance is necessary for manufacturers when it is economical to manufacture in large lot quantities and then store these quantities for future sales. The variance is also present in retail stores where inventory is held for future customer demand. Oftentimes, businesses suggest that inventory is a marketing vehicle creating demand by passing customers. The main role for inventory is to satisfy customer demand by providing product available when the customers want it. Another significant role that inventory plays is to reduce cost by exploiting economies of scale that may exist during production and distribution. Given that RFID will have such a large impact on inventory, we will present some relevant information of inventory in the supply chain.

20.3 INVENTORY IN THE SUPPLY CHAIN

Inventory is held throughout the supply chain in the form of raw materials, work-in-process, and finished goods. Inventory is a major source of costs in a supply chain and impacts customer responsiveness and eventually customer satisfaction. Inventory also has a significant impact on the material flow time in a supply chain. Material flow time is the time that elapses between the points at which material enters the supply chain to the point at which it exits. For a supply chain, throughput is the rate at which sales occur. If inventory is represented by I, flow time by T, and throughput by D, the three can be related using Little's law as follows:

$$I = DT$$

For example, if the flow time of an auto assembly process is 10h and the throughput is 50 units an hour, Little's law tells us that the inventory is 50 × 10 = 500 units. If we were able to reduce inventory to 250 units while holding throughput constant, we would reduce our flow time to 5h (250/50). We note that in this relationship, inventory and throughput must have constant units. One can see that those inventory and flow times are related and that throughput is often determined by customer demand. The goal of many operations is to reduce amount of inventory needed without increasing cost or reducing responsiveness.

20.4 BUSINESS RESPONSIVENESS

Inventory plays a significant role in a firm's responsiveness. Inventory decisions may require strategies that locate inventory close to the customer or by locating a main warehouse centrally, stocking distribution centers seasonally using a centralized stocking concept. Each strategy has trade-offs that need to be evaluated by each organizations goals. Some of the goals involve

- Cycle inventory
- Safety inventory
- Seasonal inventory
- Level of product availability
- Inventory-related metrics

20.4.1 Cycle Inventory

Cycle inventory is defined as the average amount of inventory used to satisfy demand between receipts of supplier shipments. The size of the cycle inventory is a result of the production, transportation, or purchase of material in large lots. Companies produce or purchase in large lots to exploit economies of scale in the production, transportation, or purchase process. With the increase in lot size, however, also comes an increase in carrying costs.

20.4.2 Safety Inventory

Safety inventory is inventory held in case demand exceeds expectations; it is held to counter uncertainty. Because demand is uncertain and may exceed expectations, companies hold safety inventory to satisfy an unexpectedly high demand. If a company does not have enough inventory, they may lose sales and profit; thus, choosing safety inventory involves making a trade-off between the costs of having too much inventory and the costs of losing sales due to not having enough inventory.

20.4.3 Seasonal Inventory

Seasonal Inventory is additional inventory stored to counter predictable variability in demand due to a given repeatable period. Companies using seasonal inventory build up inventory in periods of low demand and store it for periods of high demand when they will not have the capacity to produce all that is demanded. The trade-off for organizations is determining how much seasonal inventory to build—in other words the cost of carrying the additional seasonal inventory versus the cost of having a more flexible production rate.

20.4.4 Level of Product Availability

Level of product availability is the amount of demand that is available from products currently in non-committed inventory. A high level of product availability provides a high level of responsiveness but increases cost because inventory has to be held with no prior commitment or order and oftentimes this excess inventory is held but rarely used. In contrast, a low level of product availability lowers inventory holding cost but results in customer failure and loss of current and future sales. The basic trade-off when determining the level of product availability is between the cost of inventory to increase product availability and the loss from not satisfying customers.

20.4.4.1 Inventory-Related Metrics

Often, given the importance of inventory, data are collected and assessed in order to ensure proper management. Some common metrics are described as follows:

- *Average inventory* measures the average number of inventory in dollars or units over a time period such as days, months, and years.
- *Obsolete inventory* products with more than a specified number of days of inventory identify the products for which the firm is carrying a high level of inventory.
- *Average safety inventory* measures the average amount of inventory on hand when a replenishment order arrives. Average safety inventory should be measured by SKU in both units and days of demand. It can be estimated by averaging over time the minimum inventory on hand in each replenishment cycle.
- *Seasonal inventory* measures the amount of both cycle and safety inventory that is purchased solely due to seasonal changes in demand.
- *Fill rate* measures the fraction of orders/demand that were met on time form inventory.
- *Percent of time out of stock* measures the fraction of time that a particular SKU had zero inventory.

There is underlying trade-off that organizations make with regards to inventory decisions between responsiveness and inventory costs. Increasing inventory generally makes the supply chain more responsive to the customer. A higher level of inventory also facilitates a reduction in production and transportation costs because of improved economies of scale in both functions. This choice, however, increases inventory holding cost. Moreover, these inventory costs in the supply chain can be greatly affected by a lack of supply chain coordination, commonly referred to as the bullwhip effect.

20.4.5 Bullwhip Effect

The bullwhip effect has been evaluated by both practitioners and academics. Chopra (2006) provides a brief history of the bullwhip effect. The problem was identified when Proctor & Gamble (P&G) were studying replenishment patterns for one of their best-selling products. They recognized there was greater variability between (1) orders placed by distributors against retail stores sales and (2) against requested materials from suppliers. Given that the product had consistent demand over the years, the large discrepancy was not expected. P&G coined the term "bullwhip" effect for this phenomenon. It also has been referred to as the "whiplash" or "whipsaw" effect. Other organization such as HP experienced the bullwhip effect in patterns of sales for products such as printers.

An example of this "bullwhip" pattern of increasing variance as you move up the supply chain is shown in Figure 20.2.

Many researchers and practioners have attempted to discover the origins of this effect due to the fact that it creates excess cost in the form of inventory in the supply chain. Some believe, when working with demand that is constant and highly predictable, that effect is produced when companies order products in batch quantities at operational supply chain levels is what creates the effect.

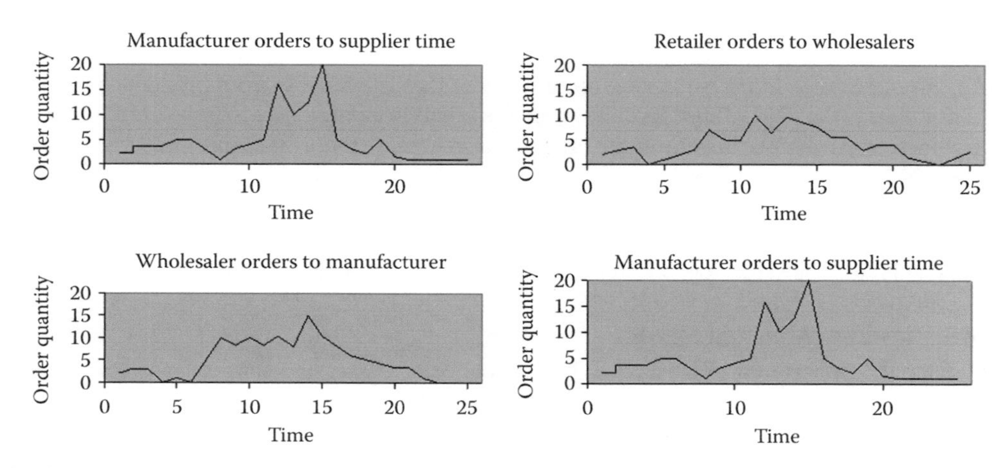

FIGURE 20.2 Increasing variability of orders up the supply chain.

The grocery industry, which exhibits this type of demand, sought to reduce the bullwhip effect with an efficient consumer response (ECR) initiative in which the food delivery supply chain would reduce a projected excess of 100 days of inventory from the supply chain. The stated goal of the ECR initiative was to save $30 billion annually by reducing the bullwhip effect in food deliveries logistics.

Some causes solution that may reduce the effects of the bullwhip affect on demand forecasting, order batching, price fluctuations, and creative order gaming are suggested by researchers.

They include sharing of information, supplier alignment, stable price structure, and incentives to prevent gaming. First, the sharing of information from all parties from common data such as a POS data and create forecasts on these data. Other techniques include electronic data interchange or EDI as it is commonly described. EDI and other web-based exchange formats provide a means for integrating information between company software platforms.

Second, supplier alignment allows for the coordination of pricing, transportation, inventory planning efforts operations in the supply chain.

Fixed costs that create order batching behavior and economies of scale costs such as transportation can be limited with RTI. Things such as smaller batch lot quantities and effective transportation scheduling is allowed by information that provides alignment. Another trend encouraging small batch ordering is the outsourcing of logistics to third parties. Logistics companies can consolidate loads from multiple suppliers. Logistics outsourcing to companies such as UPS supply chain solutions is expanding rapidly.

Third, supplier pricing is designed to motivate customers to buy in large batches and store items for future use. This behavior is called price stabilization and is designed to reduce sales demand variation. This is evident in comparing a retailer that runs frequent promotions in comparison of warehouse stores that offer everyday low pricing. The warehouse stores have more stable demand than do department stores in which promotional sales account for most of their business. Finally, exaggeration of orders to manipulate pricing is often referred to as gaming and can be affected dramatically by information. These order forecasts can be smoothed using past demand not sales forecasting.

20.5 SUMMARY

In summary, the effective use of information to manage inventory can counteract the bullwhip effect created by partners in the supply chain acting in their own best interests. The need for information timely can be supported by RFID technologies.

RFID technologies provide an opportunity to reduce the uncertainty leading to the bullwhip effect through more RTI. Given the costs of holding excess inventory in capital, obsolescence (or spoilage), handling costs, occupancy costs, pilferage, damage, taxes, and insurance, it may be worthwhile to use RTI in evaluating inventory reduction.

QUESTIONS

20.1 What are the five components of the EPC air interface standard?

20.2 What are three advantages of the Gen 2 protocol for RFID tags?

20.3 What are the three planning levels for RFID in logistics?

20.4 What are the four types of participants that make up a supply chain?

20.5 At what level is inventory control in RFID logistics planned?

20.6 What are the three components of inventory control for RFID logistics?

20.7 What does EOQ represent?

20.8 What two forms of inventory are buffered with continuous review models?

20.9 Which buffer does EOQ address?

20.10 Which buffer does safety stock address?

20.11 Weekly demand for Lego at a Walmart store is normally distributed with a mean of 2500 boxes and a standard deviation of 500. The replenishment lead time is 2 weeks. Assuming a continuous review replenishment policy, evaluate the safety inventory that the store should carry to achieve a CSL of 90%.

20.12 Carbon fiber seat posts are consumed by a bicycle manufacturing factory at a fairly steady rate of 100 per week. The seat posts cost the factory $35.00 each. It costs the plant $125 to initiate an order, and holding costs are based on an annual interest rate of 20%. Determine the optimal number of seat posts for the plant to purchase and the time between orders. (Use the EOQ formula.)

20.13 Weekly demand for Lego at a Walmart store is normally distributed with a mean of 2500 boxes and a standard deviation of 500. The replenishment lead time is 2 weeks and the store manager has decided to review inventory every 4 weeks. Assuming a periodic review replenishment policy, evaluate the safety inventory that the store should carry to provide a CSL of 90%. Evaluate the OUL for such a policy.

20.14 What two replenishment policies are associated with inventory control?

20.15 What are two impacts of RFID on inventory control?

20.16 What are two impacts of RFID on transportation?

20.17 Give two reasons to use the anechoic chamber for RFID?

20.18 What are the steps of RFID DFSSR?

20.19 Describe how order scheduling impacts labor and the effect RFID can have on minimizing this process.

21 Inventory Control Basics

21.1 INTRODUCTION

Inventory amount and size decisions are traditionally driven by the costs of maintaining inventories and costs of being out of stock. Operations managers seek to maintain inventory levels that minimize the total cost of both.

In this section, we discuss closed-loop inventory control. The approaches discussed here are relatively unsophisticated. There is a large amount of information available in other academic texts along with consultative materials available in industry. We will discuss certain aspects of inventory control theory so that the impact that RFID technologies can be recognized.

21.2 INVENTORY CARRYING COSTS

Inventory carrying costs fall into several categories. They include the following:

1. Storage costs are the costs associated with occupying space in a storeroom, warehouse, or distribution center. Inventory costs such as insurance for fire, flood, and theft are included in the expense of storing goods.
2. Theft or inventory shrinkage identifies when more items are recorded entering warehouses than leaving.
3. Obsolescence describes when items in an inventory eventually become out of date.
4. Depreciation or deterioration of inventory as a function of time, not usage.
5. Interest refers to the interest charges for the money invested in inventories. Oftentimes, this represents the investment into company inventories as opposed to money that can be invested in other investments.
6. Taxes refer to when inventories are taxed. Traditionally, the tax is derived on the basis of the inventory on hand on a certain date. Most companies make a concentrated effort to have inventory present on that day to be as low as possible.
7. Carrying costs include inventory tax, and costs associated with avoiding or evading the inventory taxes.
   a. Consider products such as fresh produce which may deteriorate in only a few days. The depreciation portion of a produce company's carrying costs might be as high as 50% per day. Other products depreciate completely given their expiration dates including products such as dairy products, drugs, bread, some soft drinks, and camera film. For these products, the rate of depreciation can be calculated because expired products that are unsold must be removed from the shelf.
   b. Specialized inventory costs are related to pets and livestock that have costs related to being watered and fed. Security cost for high-value items such as computer chips may increase inventory carrying costs.
   c. Inventory carrying costs are expressed as a percentage of the inventory's value and widely cited estimate is that carrying costs approximate 25% per year of a product's value.

Opportunity costs are not traditionally included in most carrying costs calculations. Most companies must consider the trade-off of holding inventory against having inventory to meet the fluctuations of customer demand.

21.3 STOCK-OUT COSTS

Stock-out refers to the event that occurs when an item is out of stock when a customer wants to buy the item. Stock-out costs are difficult to determine and oftentimes effect customer satisfaction. The difficulty of determining cost that is lost due to stock-outs requires a good understanding of company customer behavior. Customer can have many varied reactions to stock-outs. We suggest that the responses can be placed into three categories:

- Future sale
- Lost sale
- Loss of customer

Consider a set of 500 customers who experienced stock-outs for a given product. The three types of customer responses may suggest that of the 500 customers 50 will return as a future sell, 325 customers may go to another store which represents a lost sell, and 125 customers may never return to the company. The percentages represented by future sale, lost sale, and loss of the customer are 10%, 65%, and 25%, respectively. These percentages can be considered probabilities of the events taking place and can be used to determine the average cost of a stock-out.

Table 21.1 illustrates the procedure. Each cost is multiplied by the likelihood that it will occur, and the results are added. A delayed sale has no cost because the customer is brand loyal and purchases the product when it is again available. The lost sale alternative results in loss of the profit that would have been made on the customer's purchase. The lost customer situation is the worst. The customer tries the competitor's product and prefers it to the product originally requested. The customer is lost, and the cost involved is that of developing a new brand-loyal customer. These costs are usually determined by a firm's marketing department but we use the suggested numbers for demonstration purposes.

21.4 SAFETY STOCKS

Firms usually maintain *safety stocks* or excess inventory in order to prevent an excessive number of stock-outs. Analysis is required in order to minimize the amount of safety stock and to determine the optimum level of safety stocks. This is illustrated in Table 21.2.

TABLE 21.1
Component Breakdown of the 25% Figure

Insurance	0.25%
Storage facilities	0.25
Taxes	0.50
Transportation	0.50
Handling costs	2.50
Depreciation	5.00
Interest	6.00
Obsolescence	10.00
Total	25.00%

Source: Adapted from Alford, L.P. and Bangs, J.R. eds., *Production Handbook,* Ronald, New York, 1955, pp. 396–397.

TABLE 21.2
Determination of the Average Cost of a Stock-Out

Alternative	Loss	Probability	Average Cost
1. Brand-loyal customer	$0.00	0.10	$0.00
2. Switches and comes back	$37.00	0.65	$24.05
3. Lost customer	$1200.00	0.25	$300.00
Average cost of a stock-out		1.00	$324.05

We consider this example to demonstrate safety stock analysis. Consider goods must be ordered from a wholesaler in multiples of 10. The carrying cost of an additional or marginal 10 units is $1200. However, by stocking an additional 10 units of safety stock and maintaining it throughout the year, the firm is able to prevent 20 stock-outs. The average cost of a stock-out has already been determined to be $324.05. We derive that saving 20 stock-outs saves the firm $6481.00 ($324.05 × 20). In this case, the savings justify the investment costs. Next, we consider an alternative that maintains a safety stock throughout the year of 20 units. This adds $1200 to the costs but prevents 16 additional stock-outs from occurring, thereby saving $5184.80.

The optimum quantity of safety stock is 60 units. With this quantity, the carrying cost of 10 additional units is $1200, but $1296.20 is saved. If the safety stocks are increased from 60 to 70 units, the additional carrying cost is again $1200, while the savings are only $972.15. We conclude that the firm would be more profitable by permitting three stock-outs to occur each year. Note that these concerns determine a level of customer service.

Safety stocks indicate that a firm will attempt to meet customer demand for out-of-stock items. Many firms choose not to maintain safety stock due to the high carrying cost for inventory. Some mass merchandisers do not replace many items given their profit margins and the fact that customers are not loyal to buying at that firm. In these situations, customer behavior is to buy a complete set of items and/or fixtures needed to complete a project. They understand that the merchandiser may not have that product in the future. This is evidenced in popular "close out" stores such as Big Lots, and Hobby Lobby in which the firm buys large quantities of a product and sells it at a discount. When the product is sold out, there is no expectation of that product appearing at the store in the future.

21.5 ECONOMIC ORDER QUANTITY

Safety stock level is the minimum inventory a firm tries to keep on hand. Commonly, determining the inventory level, how they should be reordered, and how much should be ordered each time is determined by the economic order quantity (EOQ). We will provide a brief overview of EOQ; further reading is available in academic texts that discuss operation and production planning.

Given that the typical inventory order size problem deals with calculating the proper order size based on minimizing the total of two costs: (1) the costs of carrying the inventory, which are in direct proportion to the size of the order that will arrive, and (2) the costs of ordering, which mainly involve the paperwork associated with handling each order, irrespective of its size. Consider if there were no inventory carrying costs, customers would hold inventory and avoid reordering. If there were no costs associated with ordering, one would place orders continually and maintain no inventory at all, aside from safety stocks. Figure 21.1 shows the two costs on a graph and indicates the point at which they are minimized.

Mathematically, the EOQ is determined using this formula:

$$\text{EOQ} = \sqrt{\frac{2AB}{I}}$$

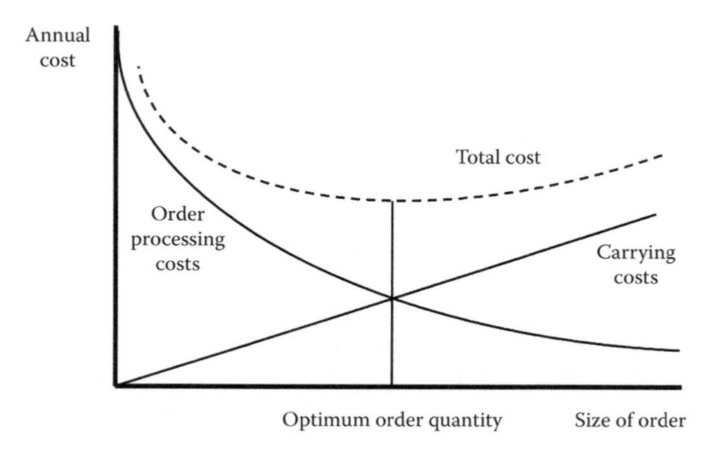

FIGURE 21.1 Determining EOQ by use of a graph.

where

EOQ is the most economic order size, in dollars
A is the annual usage, in dollars
B is the company costs per order of placing the order
I is the carrying costs of the inventory (expressed as an annual percentage)

If \$1000 of an item is used each year, if the order costs are \$25 per order submitted, and if carrying costs are 20%, what is the EOQ?

$$\text{EOQ} = \sqrt{\frac{2 \times 1,000 \times 25}{0.20}} = \sqrt{250,000} = \$500 \text{ order size}$$

Because of the assumption of even outward flow of goods, inventory carrying costs are applied to one-half the order size, which would be the average inventory on hand as illustrated in Table 21.3.

EOQs, once calculated, may not be the same as the lot sizes that the product is bought and sold at a company. EOQs can also be calculated in terms of the number of units that should be ordered. The formula is

$$\text{EOQ} = \sqrt{\frac{2(\text{annual use in number of units})(\text{cost of placing an order})}{\text{annual carrying cost per item per year}}}$$

TABLE 21.3
Safety Level Analysis

Number of Units of Safety Stock	Total Value of Safety Stock (\$480 per Unit)	25% Annual Carrying Cost	Carrying Cost of Incremental Safety Stock	Number of Additional Orders Filled	Additional Stock-Out Costs Avoided
10	\$4,800	\$1200	\$1200	20	\$6481.00
20	9,600	2400	1200	16	5184.80
30	14,400	3600	1200	12	3888.60
40	19,200	4800	1200	8	2592.40
50	24,000	6000	1200	6	1944.30
60	28,800	7200	1200	4	1296.20
70	33,600	8400	1200	3	972.15

TABLE 21.4
EOQ Calculations

Number of Orders per Year	Order Size	Ordering Cost	Carrying Cost of Average Inventory in Stock	Total Cost
1	$1000	$125	$100	$125
2	500	50	50	100
3	333	75	33	108
4	250	100	25	125
5	200	125	20	145

Assume that an item in Table 21.4 example costs $5. Substituting numbers in the new formula yields

$$\text{EOQ} = \sqrt{\frac{2 \times 1{,}000 \times 25}{0.20}} = \sqrt{\frac{10{,}000}{1}} = 100\,\text{units}$$

The earlier EOQ formula and Table 21.4 showed that $500 was the best order size, and because the product is priced at $5.00 per unit, the answer is the same.

The simple EOQ formulation just given does not take into account large volume discounts. We can review Table 21.4 and visualize how discounts would have an impact on total costs as the figures. By inputting different values into the table horizontally, volume discounts can be evaluated and marketed to increase future business.

CASE STUDY 21.1

In NASA's space explorations, it is critical to have high inventory accuracy for astronauts in space. In 2005, astronauts aboard the International Space Station (ISS) were informed from mission control that their outpost would result in abandonment if the space flight replenished consumables failed. The flight crews were given instruction to cut food intake by 5%–10% to maximize existing consumable food inventory. A space launch is estimated at $450 million, so emergency consumable inventory replenishments due to food supply shortage would be costly for NASA. Investigation of a "crew-free" inventory using automatic data capture (ADC) technology such as radio frequency identification (RFID) investigated the impact that ADC has in minimizing the situations of poor inventory control. Traditional inventory control planning methods; continuous and periodic review models such as EOQ, and (Q, r) models, where evaluated. These models described an integrated system to eliminate daily logging of inventory while maintaining high inventory accuracy. The results included a system that allowed astronauts minimum inventory counts and reduced excessive inventory weight for the space payloads. Investigating RFID in conjunction with EOQ, and Q, r models supported an effective system. This system effectively supported NASA's concern pertaining to inventory control issues in present space explorations and implementation is being considered for further space explorations. Image provided is in respect to ISS cargo transport bags (CTBs). PhD student Maurice D. Cavitt is conducting this research at the University of Nebraska-Lincoln Radio Frequency Supply Chain and Logistics (RfSCL) lab with the support of Dr. Erick Jones and other graduate students (Figure 21.2).

(continued)

CASE STUDY 21.1 (continued)

FIGURE 21.2 NASA ISS module with CTB. (From Fink, P.W. et al., *Unified Communications for Space Inventory Management*, p. 3.)

21.6 INVENTORY FLOWS

In the previous section, we utilize the figures from the EOQ and the safety stock calculations as an analysis tool. We cannot utilize these same calculations to determine inventory policy. We must first take the given information and use it to develop an *inventory flow* diagram. Assume that the EOQ in this instance has been determined to be 120 units, that the safety stock level is 60 units, that average demand is 30 units per day, and that the replenishment or order cycle is 2 days. On day 1 (in Figure 21.3), an EOQ of 120 units arrives.

We will consider a common inventory flow diagram suggested in other texts. Consider the following: total inventory (point *A*) is 180 units (one EOQ plus 60 units of safety stock). Demand is steady at 30 units per day. On day 3, total inventory has declined to 120 units (point *B*), which is the reorder point, because it takes 2 days to receive an order and during this time 60 units would be sold. If the inventory policy mandates that safety stock is not to be used under normal circumstances, reordering at 120 units means that 60 units (safety stock) will be on hand 2 days later when the EOQ arrives. The EOQ of 120 units arrives at point *C*, and then total inventory increases to 180 units at point *D*.

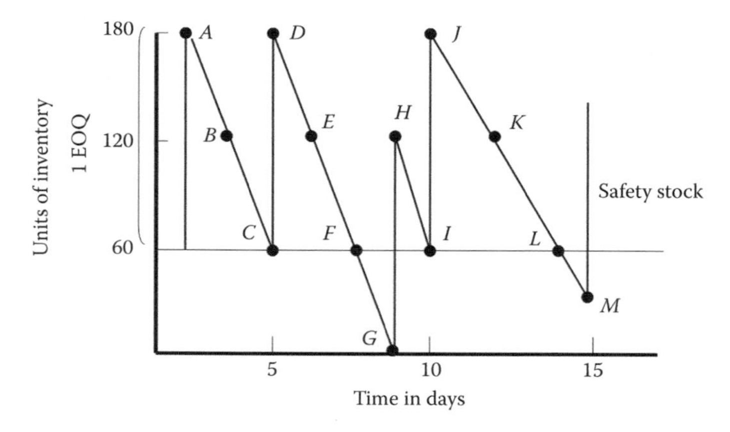

FIGURE 21.3 Inventory flow diagram.

If the rate of sales doubles to 60 units per day, the reorder point is hit at 120 units (point *E*), and an additional EOQ is ordered. However, it will not arrive for 2 days. A day after the reordering, the regular inventory is exhausted, and at point *F*, the safety stock is starting to be used. At point *G*, the EOQ arrives just as the safety stock is about to be exhausted. If the EOQ arrived later than day 8, a stock-out would have occurred. The new EOQ boosts the inventory to 120 units, which is also the reorder point. Therefore, at point *H*, another EOQ is ordered. Starting on day 8, the demand settles back to the old average of 30 units per day.

If it appeared that the demand rate of 60 units per day was going to become the average demand rate, the EOQ will need to be recalculated. Recall that a basic input into the EOQ formula is annual sales of the product. If this number changes, then the EOQ must be determined again.

Starting at point *H*, demand is again 30 units per day. The next EOQ arrives on schedule at point *I*, and total inventory increases to 180 units at point J. The reorder point is at 120 units, and an EOQ is ordered on day 12. Demand stays constant, but the transportation mode delivering the EOQ is delayed 1 day. Instead of arriving on day 14, it arrives on day 15. Safety stock is entered at point *L* on day 14. A stock-out is again prevented because the EOQ arrives at point *M*. Note that safety stock protects against two problem areas: increased rate of demand and an increased replenishment cycle.

When an EOQ is used, as illustrated in Figure 21.2, the time between orders varies. The normal time between orders was 4 days, but when sales doubled, the time between orders was only 2 days. One requirement for the effective utilization of an EOQ is that the level of inventory in the system must be monitored constantly. The ability of RFID to allow for this type of monitoring holds great promise for using EOQ theories more effective. Then, when the reorder point is hit, an EOQ is ordered. With the advent of computerization, many firms have the capability to constantly monitor their inventory and hence have the option of using an EOQ system. A reorder point for each item can be established in the computer's memory so it can indicate when the stock has been depleted to a point where a new order should be placed. The integration of RFID will allow for middle wear, decision support, and execution systems to transmit the purchase order to the vendor electronically.

A variation of the EOQ method is the *fixed-order quantity* method, used in repetitive purchases of the same commodity. This method can be initiated with RFID technologies triggering the reorder points. An example would be a materials retailer located in China buys product by the barge load (approximately 1000 tons per load). The retailer would wait until its product is out of stock before ordering another barge load. RFID would enable this type of activity by triggering the check out counter point of sell (POS) check out system to order the next lot of products from the China manufacturer directly. Tying the technical knowledge of the quantity to order with the automatic information capture of RFID will provide tremendous value in the future.

21.7 FIXED-ORDER-INTERVAL SYSTEM

An alternative inventory concept that is also commonly used is known as the fixed-order-interval system. In this system, EOQs are not used; instead, orders are placed at fixed intervals, such as every 3 days or twice a month. In the EOQ system, the time interval fluctuates, with the order size remaining the same. In fixed-interval systems, the opposite holds, and order sizes may vary.

Fixed-interval systems are used in many situations. One situation is when the firm does not maintain automatically updated stock levels. Such firms are manually checked to verify the levels of all items and determine which stocks are running low. This task is assigned on a regular basis and may be laborious. Another situation is when vendors offer the firm significant discounts if it will place its orders at certain fixed time intervals. Because the discounts are greater than the advantages of using the EOQ system, the fixed-interval ordering system is utilized. Further and additional condition is when the firm buys free on board (FOB) origin and tries to utilize its private trucking fleet whenever possible. If one of the firm's trucks travels empty in one direction without freight, commonly called deadheads, from a point near a supply source back to the firm's plant on a regular basis, the firm may decide to buy FOB origin and carry supplies in its own truck.

The fixed-order-interval system is commonly used with a safety stock inventory. It usually requires more safety stock than the EOQ system because the EOQ system requires constant monitoring of its inventory levels. In an EOQ system, if sales start to increase, the reorder point will be moved to an earlier time and a new order for an EOQ system need to be placed. Stock-outs can still take place, but only during the restocking cycle after the new order has been placed. With the fixed-order-interval system, the inventory levels are not monitored and a stock-out can occur during both the order cycle and the time before order placement.

Most fixed-order-interval systems do borrow one element from EOQ system next to each bin or slot in the warehouse is a bar code, card, or indicator that will allow for determination of the minimum quantity for that product. When the order pickers note that the stocks have been reduced to this level, they notify their supervisor, who decides whether the reorder should occur immediately or on the next scheduled date.

Cyclical buying is a very specialized form of fixed-interval ordering. This practice occurs in the women's fashion industry, in which retailers place their orders directly with the manufacturer for each season's fashions, and there is almost no possibility c reorder. Another example is a grocery retailer's purchase of Halloween pumpkins or Christmas trees.

21.8 JUST-IN-TIME INVENTORY SYSTEMS

An inventory system that has received widespread attention is the just-in-time (JIT) system. The concept is related to the fixed-order-interval system, and customers place orders with their suppliers on set schedules that frequently involve daily or hourly deliveries. In comparison to the EOQ system, the concept is based on the assumption that ordering costs are negligible; hence, firms order frequently to minimize inventory holding costs. In JIT systems, inventory is kept at a minimum because the processes create perpetual motion and continuous movement.

In addition to the JIT inventory systems, there are several other, more traditional systems for replenishing inventory stocks. Nearly all inventory systems require some formal stock-level monitoring capability. In practice today, the JIT systems may be incorporated into execution systems that are stand alone often termed best of breed or within a larger enterprise-wide system. Software applications that create perpetual motions and execution include warehouse management systems (WMS), transportation management systems (TMS), and order management systems (OMS).

21.9 RFID AND INVENTORY CONTROL

Some researchers suggest that operational labor can reduced in distribution operations by as much as 30%. A.T. Kearney, a notable supply chain consulting firm, suggests that labor savings of 7.5% are possible from reduction inventory cycle counting by using RFID. Distribution inbound receiving along with inventory cycle counting inventory reductions were recognized by Accenture consulting. Other researchers have report savings in stocking and retail checkout operations.

Opportunities in which RFID passive implementations can save money in operations in the future include

- Automatic replenishment from reserve stocking area
- Safety-stock reduction
- Automatic picking and stocking routing
- Automatic order generation from current inventory availability

21.9.1 Automatic Replenishment

Optimizing replenishments within warehouse or distribution centers (replenish primary picking locations, cross-docking, and kitting operations), within retail operations (replenish shelves from the back room to the retail floor), and within the supply chain (replenish or stock inventory between different nodes in the supply chain) are the practical applications for implementing RFID.

21.9.2 Safety-Stock Reduction

Researchers have investigated inventory control models and their impact on safety stock. Some research suggests transactional errors lead to variability in planning and inflate the need for safety stock. Transaction errors create excess inventory due to miscounting of inventory and buffer stock becomes necessary to meet service for these errors. The buffer stock for errors and the excess stock maintained due to the bullwhip effect created by inventory timing creates a large amount of excess inventory.

So, commonly, the optimal amount of inventory ordered traditionally derived by the EOQ or that includes a reorder component is represented by a continuous review (Q, R) system. Oftentimes, to account for the scheduling and lack of real-time information, an adjusted periodic review system is used, which includes safety stock inventory. Recently, researchers have addressed how RFID can influence the amount of inventory. Consider the following model:

1. Here, lead time is defined as placing an emergent order q ($q = \alpha Q$, $\alpha < 1$) at any time point b based on RFID real-time information as l and cost to place the emergent order is $k(l)$.
2. l is much less than the lead time of the regular order (Q). Additionally, the probability the emergent order will arrive before the regular order is $p(l)$ if the regular order is already on its way and an emergent order is released anyway.
3. Moreover, assume that expected total cost associated with inventory position and RFID reading point b without releasing emergent order is $C_0(IP, b)$ and the according total cost with emergent order release is $C_1(IP, b)$.
4. So, we can compare the two different costs under periodic review without RFID implementation and continuous review with given RFID real-time information in order to decide whether an emergent order should be placed:

$$C_1(IP,b) = K(l) + P(l) * C_0(IP,b) + (1 - P(l)) * C_0(IP + q, b)$$

Other researchers utilize similar continuous review application models to determine inventory levels in a real-time manner. In summary, current research models suggest that inventory can be reduced using RFID technologies due to the fact their real-time data capture abilities allow for common periodic models to move closer to the theoretical optimal continuous review models. The largest challenge is moving this theory to practice. Currently, in order for this model to work in practice, a fixed infrastructure of antennas and readers at the operational level, which would in reality cost operations more in process redesign than the theoretical inventory reduction savings.

21.9.3 Picking and Routing

Using RFID technology such as real-time locator systems, the promise of capitalizing on employee location information and inventory status to optimize employees order picking and stocking routes. This use of RFID builds on picking routes and stocking strategies commonly used in WMS.

An RFID system's ability to provide real-time information will further help optimize order picking schedules. The opportunities to reduce labor cost in these activities may be realized and the significance in labor savings may be large. Labor reduction of 20% for picking and stocking labor is commonly mentioned by users of WMS systems that employ these types of algorithms.

21.9.4 Order Batching of Waves

Using RFID to group orders commonly called waves automatically based on the latest inventory availability. The opportunity to use inventory as it is being received at the dock for immediate shipments can be utilized dramatically reducing labor for stocking, replenishment, and picking. This theory commonly called cross-docking requires a considerable amount of receiving labor to be realized. The promise of RFID will truly enable this one of many opportunities to effectively organize, group, and fulfill orders automatically.

Next, the importance of order wave batching is critical in high-speed operations, but inventory inaccuracies can reduce the quality of these batching of orders and reduce their effectiveness. Specifically, in order management modules of WMS or ERP systems, the order bid process within the software schema orders compete against one another for resources to meet their specific goals, described as the bid process in the negotiation schema. The real-time ability of RFID to provide inventory accuracy and identify secondary location of the inventory immediately can improve the accuracy of the wave batches.

21.10 SUMMARY

In summary, the contribution of RFID systems to closed-loop distribution logistics can produce significant contributions in the three areas mentioned above: (1) inventory reduction with respect to safety stock reduction, (2) optimization of order grouping and releases, and (3) and labor reduction with respect to picking and stocking labor.

The real benefit of RFID will be based on its ability to provide the inventory accuracy benefits of current technologies such as bar codes and translate the non-line-of-sight benefits into inventory savings. The areas to investigate in the future from a closed-loop perspective in distribution operations include the aforementioned areas that are listed as follows:

1. Physical and cycle inventory counts
2. Inventory replenishment
3. Order picking
4. Inventory stocking
5. Order cross-docking
6. Order kitting
7. Many other common distribution operations

The realization of these savings will more than likely result as RFID integration into common execution software such as WMS, LES, and TMS will quantify labor savings.

We previously discussed the benefits of RFID technology as a closed-loop systems framework. Traditional automatic identification (auto-id) systems are closed loop in the fact that they are static and traditionally are used within one tier of the supply chain. Consider a tote bar code or pallet tag that is used within a company's distribution center or between other distribution centers. Rarely is this bar code integrated between other unknown partners or even the customer to evaluate the history of that unit load. As opposed to a system in which information is passed from one intelligent auto-id technology to another.

Consider each technology acting as an intelligent agent. So, if a customer desires to know the history of a product, the UPC transfers information to a bar code that transfers information to an passive RFID tag, which transfers information to a active RFID tag that passes information to a global positioning system. This daisy chain type of interlinking provides an open-loop concept that provides all potential players into the supply chain effective visibility. This nesting of auto-id technologies may be an intermediate step to profitability for most intermediate RFID implementations. We next consider a framework that describes this type of technologies.

22 Transportation and RFID Applications

22.1 INTRODUCTION

A supply chain can be expressed as the parts that are involved, directly or indirectly, in fulfilling a customer request (Chopra and Meindl, 2007). By this definition, it can be seen that a supply chain consists of manufacturers, warehouses, retailers, transporters, and customers. The purpose of a supply chain is to maximize the value generated for the customer, namely, maximizing the difference between the final product worth and the total expended by the supply chain to provide the product to the customer.

In order to succeed, the supply chain must be conducted to minimize the costs incurred. Supply chain management (SCM) is responsible for the optimization of the flows within the stages and for minimizing the total cost at the supply chain. This term, SCM, is a unification of a series of concepts about integrated business planning joined together by recent advances in information technology (IT) (Shapiro, 2007). Despite all that, many companies have not completely taken advantage of this process.

In today's world, the competition between companies, more demanding customers, and reduced margins make the scenario more difficult for companies to succeed. In this context, SCM became a very important practice for companies that want not only to keep in business but also have their results optimized and meet the clients' expectations.

Responsiveness in the supply chain has gained importance and it is a trend that apparently will dictate future decisions regarding supply chain design. According to (Novack et al., 1995), the themes that will have an influence on logistics in the near future are the following:

- Strong corporate leadership will enhance logistics value through focusing on efficiency, effectiveness, and differentiation
- Value realization requires marketing of logistics capabilities within the company and to external customers
- Emphasis on the "scientific" aspect of logistics management in order to enhance the "art" of creating customer satisfaction
- Enhancing logistics value through integrating product, information, and cash flows for decision-making linking external and internal processes
- Logistics value enhanced by ownership of responsibility internally and externally to the firm
- Focus of successful companies is to create internal value for their organization and external value for their suppliers and customers

By those themes, it can be seen that SCM plays and will continue to play an active role in successful companies' routines. In order to achieve better results on the supply chain and better responsiveness to customers' necessities, new techniques such as real-time inventory and dynamic supply chain have been developed. The role of transportation is critical in the SCM. We provide an overview of transportation and nodes in the following sections. We refer to definitions from general descriptions from industry and *Contemporary Logistics* (Johnson, 2006).

22.2 TRANSPORTATION DEFINITIONS

Transportation is the movement of goods and people between two points. There are five different types or modes of transportation:

- Truck
- Rail
- Air
- Water
- Pipeline

Intermodal transportation occurs when two modes or more work closely together on a regular basis utilizing the advantages of each. Generally, intermodal transportation refers to transporting containers from the vehicle of one mode to a vehicle of another without the contents of said device being reloaded.

Transportation has the following impacts upon logistics system:

1. Determines the cost of the firm's plants, warehouses, vendors, and customers.
2. Inventory strategy is influenced by the mode of transport used. High-speed, high-priced transportation systems require smaller amounts of inventories near customer locations.
3. Determines the packaging required, and carrier classification rules often dictate package choice.
4. Dictates a manufacturing plant's materials handling equipment.
5. Customer service goals influence the type of carrier selected by the seller.

As supply chain managers attempt to integrate all facets of their systems, they often find it desirable to have the capability of tracking each and every shipment so that they can determine its geographic location. Hence, their own equipment or that provided by carriers is now frequently equipped with some tracking-type device.

Global tracking devices can be found by earth satellites. Dispatchers handling local delivery and collection fleets can use them to determine which truck should be assigned to make a "pickup" that has just been phoned in.

Parcel carriers have labels on each parcel, and at various stages in the parcel's journey, it is scanned and that information saved in case the shipper or consignee asks about its most recent recorded location. When some parcel companies deliver parcels, that fact is recorded on a handheld device the delivery person is carrying, and that information is transmitted almost immediately to the parcel carrier's main computers.

In late 1997, the U.S. postal service ordered 300,000 handheld scanners to be used for verifying delivery of packages, in a step that made it more competitive with UPS. One of the authors was at UPS in the strategic systems group when this happened. The foundations for using ADC including radio frequency identification (RFID), were tested at UPS during this period to make tracking more efficient.

22.2.1 Small-Volume Shippers

The smallest of businesses are probably operated out of people's homes. Mail and parcel post can reach any address in the United States and virtually any address elsewhere in the world. At the very beginning level, one can purchase supplies on a delivered basis, and the seller has to arrange for transportation. One can also sell free on board (FOB) source, in which case customers will have to come to pick up their purchases. One's own auto or light truck can be used for local carriage of goods. In some cities, taxis can be used to deliver packages; there are also local delivery services that will pick up and deliver packages within a certain geographic area. For purposes of discussion, we will assume that we are dealing with packages weighing up to 100 or 150 lb. These are often referred to as *parcels*, and firms that specialize in their carriage are called *parcel carriers*.

Probably the best-known parcel carrier is UPS, which now operates in nearly 200 countries, and financially dwarfs any other transportation company in the United States. The firm has over 300,000 employees in the United States plus 37,000 in other countries, and it operates nearly 160,000 trucks and 500 aircraft. This company has experienced growth because it has earned a reputation for very reliable service. UPS rates include both pickup and delivery. It offers a range of service relying on several modes of transport, and users of its air service can purchase next-day, second-day, or third-day deliveries. UPS also provides computer software to assist with documentation and to allow the customer access, via a modem, to those segments of the UPS computer system, which the customer needs to learn about the status of his or her shipment. UPS has several imitators. UPS's dominant role in the country's transportation system was evident during its employee strike that took place in 1997. Some firms that were completely dependent on UPS had to shut down as the strike took its toll.

FedEx runs on a similar concept, relying on a huge fleet of planes to carry parcels to and from several major hubs each night. Its specialty is overnight delivery.

22.2.2 LTL Shippers

As firms work with larger shipments, the next step in the progression is referred to as less-than-truckload, or LTL, traffic. Shipments in this category range from about 150 to, say, 5,000 or 10,000 lb. They are often too big to be handled manually, yet they do not fill a truck. Trucks that carry LTL freight have space for and plan to carry shipments of other customers at the same time.

The majority of the nation's trucking firms are LTL carriers. Since deregulation, a handful have developed high-quality, nationwide service for LTL amounts. Leaders are Crete Carriers and Werner Trucking along with others such as Roadway Express and ABF Freight Systems in this industry, which may change by the time of this. All operate in a similar way. They have multiple terminals spread throughout the nation. From each terminal, small trucks go out to customers, delivering and picking up shipments. These shipments are then taken to the terminal; goods are unloaded from the carrier, moved through the terminal, and loaded aboard a small truck for local delivery.

Consignees are receivers of freight. Some consignees consolidate their inbound freight by specifying that all shipments made to them be routed via a specific LTL or parcel carrier.

22.2.3 Freight Forwarders and Other Consolidators

Freight forwarders are not limited to just one mode of transportation, and generally consist of two types of domestic freight forwarders; ground and air. Freight forwarders operate as agents. They generally give volume discounts to customers shipping large quantities of freight at one time. Some forwarders also function as traffic departments for small firms, performing other traffic management functions. Oftentimes, they do not compete but work in partnership with LTL carriers and airfreight carriers.

Any forwarder–carrier partnership should include

1. Space and capacity commitments made by forwarder
2. Fixed rates
3. Rebates
4. Tender whole containers
5. Preferred access to capacity during peak periods

22.2.4 Shippers' Cooperatives

The cooperatives can act similar to air freight forwarders, except they do not operate as profit-making organizations. All profits achieved through their consolidation program are returned to members. This type of consolidation program has been well received by shippers.

The term *broker* is used frequently in transportation. A broker is a facilitator that brings together a buyer and seller. Some brokers handle LTL shipments. They consolidate shipments and then turn them over to truckers, forwarders, or shippers' associations.

22.2.5 Truckload and Carload Shippers

When shipments reach weights of 20,000–30,000 lb, one can start thinking of truckload (TL) or surface container load shipments. The exact weight depends on the product, and it is close to the amount that would physically fill a truck trailer. The shipper has experience in working with these types of weights and handles these shipments like a truck shipment; the trucker, however, may turn the container over to a railroad for a portion of the move.

TL shipments may be cheaper than LTL for these weights because the consignee can negotiate the following:

1. Who unloads the trailer
2. The load may go direct to the shipper
3. Back-office functions such as billing are the same

22.2.6 Large Bulk Shippers

Bulk materials are loose rather than in packaged form and are handled by pumps, scoops, conveyor belts, or the force of gravity. The decision must be made as to where in the distribution system the bulk materials should be placed into smaller containers for sale or shipment to the next party in the supply chain.

22.2.7 Bulk Cargo

An ideal equipment configuration for one bulk cargo may not be able to handle another. Another consideration is the size of particle of the cargo in question; there are costs involved in pulverizing to a uniform size so it can be handled by pneumatic or slurry devices.

Materials shipped in bulk move in TL, railcar load, vessel-size lots, or via pipeline. Lot sizes differ. A TL may be 20–30 ton, a rail carload run from 40 to 80 ton, a barge holds about 1000 ton, a Great Lakes vessel holds 25,000–50,000 ton, and the largest of ocean vessels can carry 500,000 ton.

22.2.8 Truckload Hauls

Trucking of bulk materials involves either for-hire or private trucks, with specialized bodies if necessary. For-hire trucks are retained for a specific haul or for a span of time or for a task. Brokers are often used by shippers to find and contract with independent truckers; the broker takes a certain percentage off the top of the rate.

22.2.9 Railroads

Rail rates and contracts encourage multiple-car shipments because the railroad can switch and haul a number of cars as easily as one. The largest of rail hauls are handled on *unit trains*. This is a train of permanently connected cars that carries only one product nonstop from origin to destination. It can be thought of as a conveyor belt. Once the product is delivered, the train returns empty to its origin and makes another nonstop run. Unit trains benefit both the railroads and their customers:

The trains achieve a very high percentage of car utilization and usually provide less expensive and more dependable service. Currently, over 90% of all coal movement is by unit trains.

22.2.10 Water Carriers

There are domestic movements of freight by water on the Great Lakes and on our inland waterways, or barge, system. There is also waterborne commerce via ocean-going vessels between the mainland ("lower 48") states and Alaska, Hawaii, and Puerto Rico. One of the largest domestic movements is oil from Alaska, which moves on large takers from Valdez to Panama. The tankers are nearly 1000 ft long, carry 1.5 million barrels, and are too large to transit the canal. At Panama, the oil is unloaded at a tank farm and pumped via pipeline across the isthmus to a tank farm on the Atlantic. There, it is reloaded aboard smaller tankers and taken to U.S. gulf and east coast ports. Inland waterway system, not counting the coastal routes, the Great Lakes, or the St. Lawrence Seaway system, is made up of about 16,000 mi that are dredged to a depth of 9 ft, which is the minimum required for most barges. Most of this system is concentrated in the southeastern region of the United States along the Mississippi River and its tributaries.

Domestic water carriers have specialized in transporting bulk products at very low prices at slow average speeds (6 mph). Petroleum and related products account for 36% of total barge commerce. Coal is second, with 28%. Other products that move extensively in the inland waterway system are grain and grain products, industrial chemicals, iron and steel products, forestry products, cement, sulfur, fertilizers, paper products, sand and gravel, and limestone.

22.2.11 Pipelines

There are two types of oil pipelines: crude oil and product. Crude oil lines transport petroleum from wells to refineries. There are approximately 150,000 mi of crude oil pipelines in the United States. There are two types of crude oil lines. Somewhat more than half of the crude oil line mileage is in the form of gathering lines, which are 6 in. or smaller in diameter and are frequently loaded on the ground. These lines start at each well and carry the product to concentration points. Trunk lines are larger-diameter pipelines that carry crude oil from gathering line concentration points to the oil refineries. Their diameter varies from 3 to 48 in.; 8 to 10 in. pipe is the most common size. A large pipeline's capacity is also impressive. The 48 in. Trans-Alaska pipeline, which is 789 mi long, has the discharge capacity of 2 million barrels (42 gal each) per day.

The other type of petroleum pipeline is called a product pipeline and carries products such as gasoline or aviation fuel to tank farms located nearer to customers. The products are stored at the tank farms and then delivered to customers by truck or by rail.

22.2.12 Slurry Systems

Slurry systems involve grinding the solid material to a certain particle size, mixing it with water to form a fluid, muddy substance, pumping that substance through a pipeline, and then decanting the water and removing it, leaving the solid material. Railcars can also carry slurry.

22.2.13 Comparison of Modes

Logistics manager must decide which transportation mix will best meet the company's objectives. Most common mode comparisons are

- Terms of speed
- Dependability
- Rates
- Fuel efficiency

The cost per *ton mile* or the cost of 1 ton of freight carried 1 mi has become popularized. In a benchmarking study of several modes performed for the Council of Logistics Management by Andersen Consulting, shippers were asked to rate the carriers by several criteria using a scale of 4 (outstanding) to 1 (below average). Other criteria included

- Level of communication
- Ability to handle volume peaks
- Geographic coverage
- Availability of carrier executives
- Responsiveness of operations: TL

22.3 TRANSPORTATION IN SCM

As a supply chain driver, transportation has a large impact on customer responsiveness and operational efficiency. Faster transportation allows a supply chain to be more responsive but reduces its efficiency. The type of transportation a company uses also affects the inventory and facility locations in the supply chain. The role of transportation in a company's competitive strategy is determined by the target customers. Customers who demand a high level of responsiveness, and are willing to pay for the responsiveness, allow a company to use transportation responsively. Conversely, if the customer base is price sensitive, then the company can use transportation to lower the cost of the product at the expense of responsiveness. Because a company may use transportation to increase responsiveness or efficiency, the optimal decision for the company means finding the right balance between the two.

The transportation design is the collection of transportation modes, locations, and routes for shipping. Decisions are made on whether transportation will go from a supply source directly to the customer, or will go through intermediate consolidation points. Design decisions also include whether multiple supply or demand points will be included in a single run or not. Also, companies must also decide on the set of transportation modes that will be used.

The mode of transportation describes how product is moved from one location in the supply chain network to another. Companies can chose between air, truck, rail, sea, and pipeline as modes of transport for products. Each mode has different characteristics with respect to the speed, size of shipments (parcels, cases, pallet, full trucks, railcar, and containers), cost of shipping, and flexibility that lead companies to choose one particular mode over the others. Typical measurement for transportation operations includes the following metrics:

- Average inbound transportation cost is the cost of bringing product into a facility as a percentage of sales or cost of goods sold (COGS). Cost can be measured per unit brought in, but typically included in COGS. It is useful to separate this cost by supplier.
- Average incoming shipment size measures the average number of units or dollars in each incoming shipment at a facility.
- Average inbound transportation cost per shipment measures the average transportation cost of each incoming delivery. Along with the incoming shipment size, the metric identifies opportunities for greater economies of scale in inbound transportation.
- Average outbound transportation cost measures the cost of sending product out of a facility to the customer. Cost should be measured per unit shipped, oftentimes measured as a percentage of sales. It is useful to separate this metric by customer.
- Average outbound shipment size measures the average number of units or dollars on each outbound shipment at a facility.
- Average outbound transportation cost per shipment measures the average transportation cost of each outgoing delivery.
- Fraction transported by mode measures the fraction of transportation (in units or dollars) using each mode of transportation. This metric can be sued to estimate if certain modes are overused or underutilized.

The fundamental trade-off for transportation is between the cost of transporting a given product (efficiency) and the speed with which that product is transported (responsiveness). Using fast modes of transport raises responsiveness and transportation cost but lowers the inventory holding cost.

22.4 INFORMATION TECHNOLOGY AND SCM

It is no surprise that IT played a big role in enabling many processes and ideas in SCM, which seemed impossible prior to its inception. The first advance was the decreasing of inventory level by managers abandoning rules of thumb and adopting the setting of inventories based on service level desired and historical demand (Shapiro, 2007). IT allowed the analysis of a great quantity of units and the process of recalculating the inventory level as the demands change. This ability made the companies more agile, while decreasing inventory levels and increasing service levels.

Another important fact that gave a great contribution to SCM was the electronic data interchange (EDI). This technology allows the direct data interchange between companies using computers. EDI changed the relationship between the company and its customers, with its suppliers, and also with the employees. The ability of trading data almost instantly across the supply chain gave companies the ability to manipulate more up-to-date information in a shorter period of time. This reduced the need for printing and transporting paper copies, enabled JIT practices, and helped to restructure logistics supply chain relationships. Together with EDI, we can also mention the importance of the Internet on global business by the electronic mail and the World Wide Web (Johnson et al., 1999).

Artificial intelligence systems are responsible for many advances achieved by society and by SCM as well. Computers can be programmed in order to execute routine functions and according to the rules imposed to the computer, it can be capable of behaving as an intelligent system that can execute complex activities in reduced time, and it is even capable of learning with its attempts during the time. This brought to logistics a much bigger capacity of processing information and executing tasks. Many activities including warehouse operators can operate without the human interference, their results in a more responsive and accurate supply chain (Johnson, 1999).

Some technologies, discussed later in this paper, can be used to make real-time adjustments to the supply chain. Those adjustments could be due to many events such as manpower shortages or equipment breakdowns. For example, if a problem occurs to a truck or the roads conditions change due to weather, the system, supplied with this updated information, should be able to make the necessary corrections to the transportation routes of other trucks to compensate for the truck failure.

This system would be very useful for natural disasters such as Hurricane Katrina. With real-time information (RTI), the system would reallocate transportation and production to a place that would be the optimum solution. This kind of modeling would reduce the response time for such events from months or weeks to days or hours. This system can also be expanded to urban transportation within a city or long distances between two cities.

22.5 REAL-TIME TECHNOLOGIES

RFID and global positioning systems (GPS) are emerging technologies that will allow for real-time data collection to assist with decision support in SCM. RFID has a wide variety of applications. Some examples of RFID usage are library checkout stations, automatic car toll tags, animal identification tags, and inventory systems. Real-time data collected using RFID allow a supply chain to synchronize reorder points and other data. RTI can also be used to design and operate logistical systems on a real-time basis. GPS is currently used solely as a means to locate equipment and derive navigation directions.

An RFID system consists of a reader, tags, and an air interface. The reader, also known as an interrogator, sends out a signal through an antenna. This signal is usually in the form of an electromagnetic (EM) wave. Because the signal is in the form of an EM wave, a direct line of sight is not needed to read the information on the tag. This is a major advantage of RFID. The signal is received by the tag and a response signal is sent back to the reader. This response signal contains a unique identifier associated with tag. The response signal can be powered in two ways corresponding to the type of tag. Passive tags utilize the energy of the original signal to send a response signal back to the reader. Passive tags have a limited amount of energy to power the response signal. Therefore, the amount of information transmitted by a passive tag is fairly small: quite similar to the information carried in a bar code. Active and semi-active tags use energy from an attached battery to power the response signal. The use of the embedded battery allows the response signal to contain more information and travel farther. The reader receives the response signal, decodes it, and sends that information to a database. Often, the information in the response signal is connected to additional information in the database.

RFID technology can be used throughout the supply chain in order to promote visibility. This visibility helps coordinate actions between entities in the supply chain. Figure 22.1 shows the relationships within the supply chain that can be affected by the implementation of the RFID technologies. An example of RFID implementation is the use of active tags for detecting, tampering, and monitoring security of maritime containers. Those types of tags also have the tracking advantages of RFID and can be used to improve operations management. Those tags can be seen in Figure 22.2.

GPS systems consist of a series of satellites orbiting the Earth and receivers. GPS works by calculating the distances from a receiver to a number of satellites. With each distance between a receiver and satellite, the number of possible locations is narrowed down, until there is only one possible location. A receiver must calculate its distance from at least three satellites to determine a location on the surface of the Earth. However, four satellites are usually used to increase the location accuracy (Dommety and Jain, 1996). This process of location would be

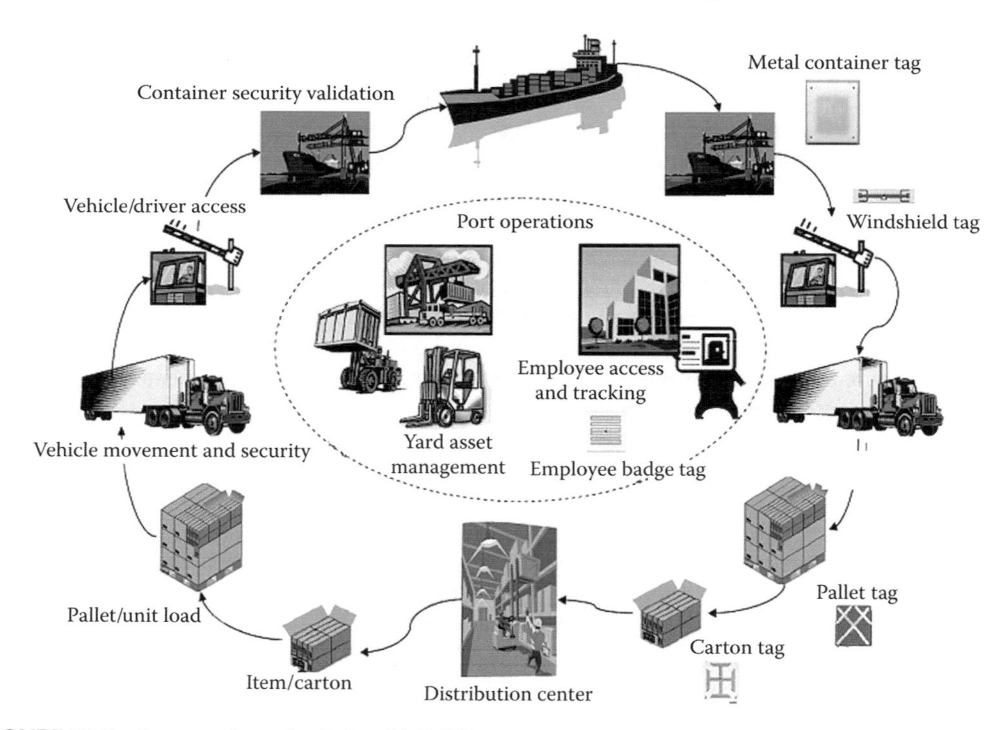

FIGURE 22.1 Integrated supply chain with RFID.

FIGURE 22.2 RFID container seal.

controlled by the positioning module of GPS system. An average GPS positioning and navigation system would also have the following modules:

- Digital map database
- Map matching
- Route planning and guidance
- Human–machine interface
- Wireless communication

There are three positioning technologies that can be used: radio wave-based positioning, dead reckoning, and signpost. The use of GPS for navigation can have direct and indirect impacts on intelligent transportation systems. GPS navigation systems can provide information about local surroundings. Also, emergency personnel can be provided with a precise location for situations, thus reducing response times. Asset tracking is one of the most popular uses of GPS. One of the limitations of GPS is that receivers cannot communicate with satellites when indoors (Feng and Law).

RFID and GPS are radio wave-based technologies that are currently used by many organizations. RFID is primarily used in inventory and material handling processes. Tags are placed on items. When these items pass by checkpoints where readers are located, the tag is read and the appropriate action can be taken. Real-time inventory can be kept by monitoring tag reads at strategic points like loading docks. RFID can also be useful in material handling. Items on a conveyor can be diverted at the appropriate times based on the information received from the RFID tag. GPS is primarily used to track assets such as vehicles and other expensive equipment. For example, if a truck breaks down, it is possible to locate the truck and get the shipment moving again in the fraction of the time it would take with a GPS receiver.

22.6 FUTURE TECHNOLOGIES

Current applications of RFID and GPS systems have allowed for more effective tracking of inventory and assets. These technologies can be used in conjunction, but, the data have to be captured and written to a database to be correlated to other tags or receivers. If these technologies can be

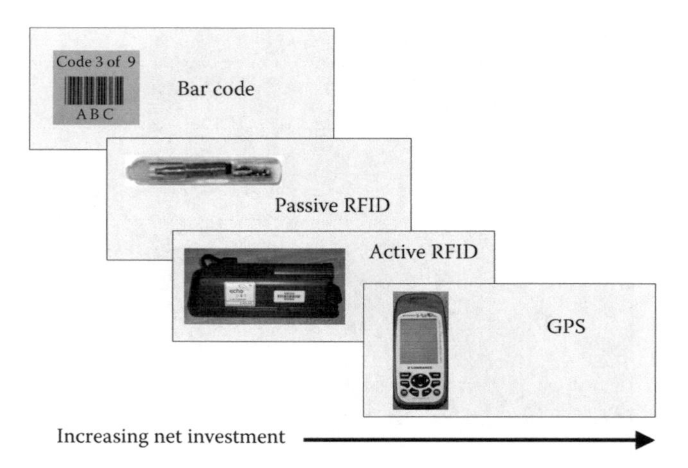

FIGURE 22.3 Nesting diagram.

combined to produce hybrid systems, greater gains can be achieved. One focus of research is the nesting of GPS receivers and various RFID tag types. If tags and receivers were able to communicate with one another, even more accurate real-time data collection could be achieved during transportation. This would also reduce equipment costs, because fewer readers would be required. The nesting would follow the form in Figure 22.3.

If these technologies can be nested, it would allow the information in a bar code or a passive RFID tag to be collected by an active tag. This information could then be combined with the information contained within the active tag and transferred to a GPS receiver. The GPS receiver could then send not only its location but all of the information about the cargo being shipped (Reade and Lindsay, 2007). A possible application of this nested technology approach would be in the railroad industry. Currently, there two passive RFID tags attached to the sides of all rail cars in the United States. In addition, most railroads are using GPS receivers to track locomotives. If nesting became possible, implementation would be easy in this case. Active tags could be used to capture the information correlated to the cargo in all of the rail cars and transmit it to the GPS receiver and thus to the inventory databases.

In addition to nesting technologies, more advanced tags can be developed to allow more detailed data collection. Tags that utilize sensors to capture and write data to the tag are being developed. Some tags have been developed, but are still very unreliable. These sensor tags could be used to monitor physical parameters like temperature and humidity, as well as security parameters. The main problem faced by these passive sensor tags is the limited power supply. The sensor cannot use any energy while outside the range of the reader. Also, the amount of energy available while in read range is very small. This limits possible measurement techniques (Want, 2007). With these sensor tags, perishable goods could be monitored to guard against possible safety issues. This could include salmonella outbreaks caused by frozen chicken reaching too high of temperatures for too long and medications being held at temperatures that reduce potency.

22.7 CONCLUSION

Technologies are being used to allow real-time data collection. This allows for more dynamic SCM systems that are able to adjust to varying market and environmental conditions. RFID and GPS facilitate this dynamic SCM. RFID allows for up-to-date inventory levels, and when combined with GPS, can provide a means of tracking inventory as it moves from supplier to customer through the supply chain. New technologies are being developed to further the amount of information to decision support systems for SCM.

23 Optimizing RFID Portal Locations in Distribution Using Systematic Layout Planning

23.1 INTRODUCTION

The manufacturing facility layout design (FLD) has been discussed by a number of researchers. Continuous improvement has been achieved through the use of simulation and computer-aided programs for designing facilities in actual manufacturing and warehouse environments. However, FLD still is a complex and broad area that cuts across several specialized disciplines. Basically, the facility layout problem is to determine the "most efficient" arrangement of cells or functional departments subject to flow and capital constraints imposed by the original layout, management, and site requirements.

The optimum solution for these facility layout problems is not only controlled by a numerical function, but more depends on the accepted baseline of the application of site and relevant requirements. Therefore, the solution for each single layout problem should not be a single solution with the optimum result based on the ratio of each functional department and its weight value. Most of the research on facility layout utilizes the classical concept about classification of layout problems by either the quadratic assignment problem or a large-scale mixed integer programming problem whereas nonlinear programming formulations have been solved by numerical methods by simulated annealing or by genetic algorithm approaches. Mix-integer programming (MIP) formulations have been solved by ad hoc interactive designer reasoning or by reducing the MIP to a linear programming optimization problem either by qualitative reasoning or, once again, by ad hoc interactive designer reasoning and by genetic approach. Although integer and non-integer problems have solved complicated layout problems, which are two-dimensional (2D) with flow and capital consideration, particular situations and single-case problems may have to be evaluated in other ways.

Radio frequency identification (RFID) facility layouts with warehouse applications introduce a new type of parameter to the traditional FLD problem. The following sections illustrate the differences.

SLP Muther developed a layout procedure known as systematic layout planning (SLP). It uses as its foundation the activity relationship chart that is described in the facility layout process. SLP is based on input data and an understanding of the roles and relationships between activities, a material flow analysis (from-to-chart) and an activity relationship analysis (activity relationship chart). This analysis results in a relationship diagram. The next two steps involve the determination of the amount of space to be assigned to each activity. Based on modifying considerations and practical limitations, a number of layout alternatives are developed and evaluated. The SLP procedure can be used sequentially to develop first a block layout and then a detailed layout for each planning department. Figure 23.1 discusses the application of SLP with an RFID warehouse design procedure.

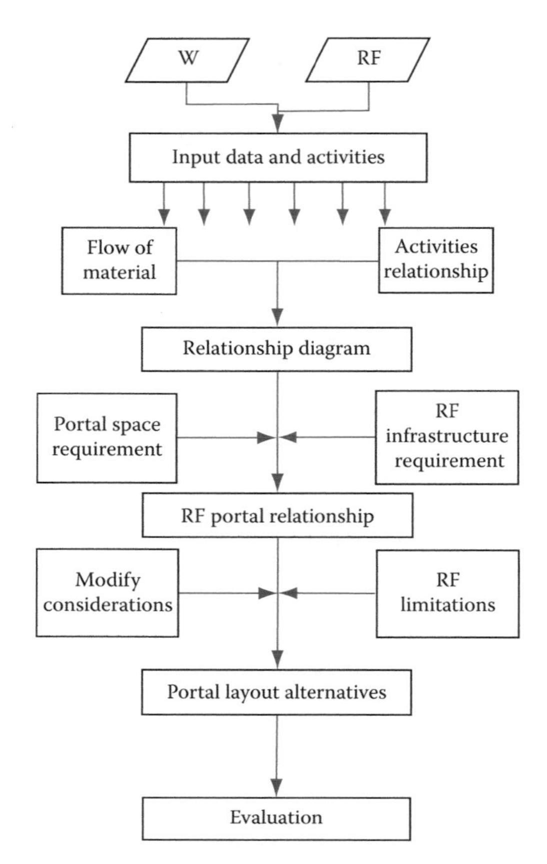

FIGURE 23.1 SLP procedure.

23.2 MODELING PROCEDURE

23.2.1 Phase 1: Multi-Objective RF Warehouse Architecture

The overall RFID warehouse implementing system includes three main parts: RFID edge layer, RFID physical layer, and enterprise integration network. The RFID physical layer is the connection of the other two layers. The RFID system is designed to process streams of tag or sensor data coming from one or more readers. The edge layer has the capability to filter and aggregate data prior to sending it to a requesting application. For example, an action (tag read) is triggered when the object moves or a new object comes into the reader's view. The RFID edge servers filter and collect the tag data at each individual site and send it over the Internet to the third layer—enterprise integration layer. The localized data are identified by moving actions and stationary actions separately, which divide the RFID reading type to portal door distribution process within limited range and mobile reader inventory checking. The fundamental tenet of warehouse portal distribution system is that they must be able to accommodate changes that may occur on a network. The portal devices provide real-time, positioning access capabilities to user communities, delivering and searching personal data. It allows external customer and partner access with data protection and secure access.

We can now divide the RFID warehouse system into three parts as we discussed before, the physical layer, the logic layer, and the system integration layer, as seen in Figure 23.2. Each layer has different components depending on what functions the RFID system needs. By understanding the flow in the warehouse, we can determine the types of tag and antennas needed in the warehouse.

Basically, the RFID implementation in any process has two to three layers. The physical layer produces log events for RF sensor during process executions. Logic layer records the log

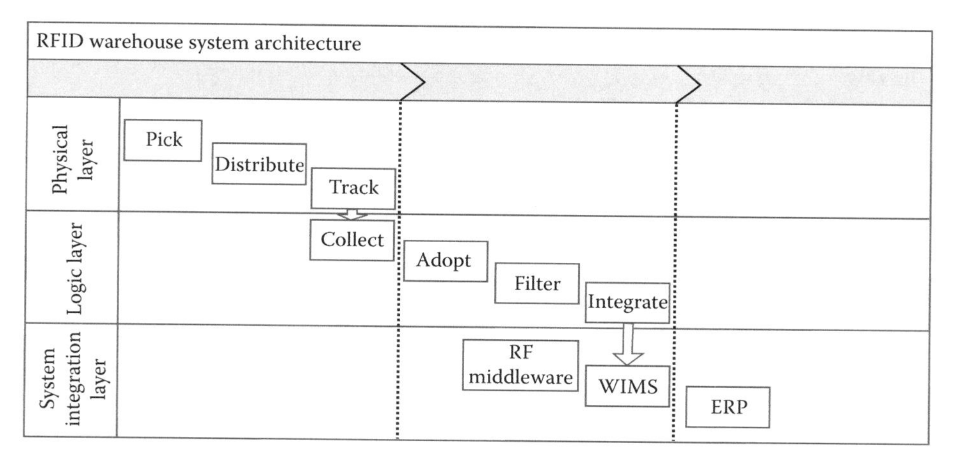

FIGURE 23.2 RFID operation in distribution process.

events-related data including filter and integrate functions. The analysis of the physical layer activity has been discussed in facility layout research. The difference between previous research and RFID facility layout is that the data flow should be added as a factor that influences the RFID warehouse efficiency and performance.

First, the production process has its own upstream and downstream flow. Each department and activity function has multiple interactions with the others whose layouts are defined by traditional facility layout algorithms that define the overall workflow for sites and how and where the functional part fits into it. However, for an RFID-enabled warehouse, the labeling function part is substituted by a 2D portal door with installed antenna.

23.2.2 Phase 2: Data Environment Analysis

The data flow through the distribution process in a warehouse is one of the design components of RFID warehouse layout. The goal of such an activity is to define the input and output data in order to confirm the efficiency of data flow and its physical flow. Data standards standardize data formats and data organization to ensure that the required data can be smoothly exchanged within the supply chain. The workflow and data flow are both generated by production flow from physical layer to logical layer. All the data through picking to distributing process are generated by RFID equipment including the tags on each pallet or antenna on the portal. Therefore, the location of RFID equipment has influential power on the accuracy of the distribution process, which will form the individual data flow according to the workflow. The location of the RFID antenna, which we call sensor in this section, will be discussed. "Sensor" is used to refer to a device that is connected, via a network or RF communication medium, to other sensor devices in the network. The location of sensors in the warehouse relates to either its environment or data traffic flow itself, which is detected by fixed antenna on portal door. Similarly, the data flow will be employed by sensors specifically in the picking entrance portal and distributing portal. Therefore, data traffic through two portal doors and its layout will be considered. The other communication between the nodes in the warehouse will be discussed in future work. In order to measure the accuracy and efficiency of RFID performance in a warehouse, we use ratios to evaluate the relationship of performance and efficiency of RFID readability, which is equal to the simple relationship between input and output data that can be related to regression analysis to show fair performance:

$$\sigma_r = \frac{\alpha_I}{\beta_o} \tag{23.1}$$

where

σ_r is the ratio
α_I is the input
β_o is the output

However, this ratio only gives an average performance for RFID readability. The components of input require precise data to evaluate the environment and performance. We measure the benchmark of the performance used to compare the different input data and data flow. For example, the different amounts of workflow reflect the different data flows in a warehouse, but the benchmark gives us reliable data against which to measure different warehouse environments and workflow.

The statistical power analysis estimates the power of the workflow to detect a meaningful effect, given product flow size, significance level, and standardized effect size. Product flow size analysis determines the product flow size required to get a significant result, given the statistical power, test size, and standardized effect size. These analyses examine the sensitivity of statistical power and product flow size to other components, enabling researchers to efficiently use the research resources. According to the power of the data analysis, we can know whether the workflow during the distribution process is too low or too high, which will influence the capital loss for the warehouse. If the sample size is too large, time and resources will be wasted, often for minimal gain. For the benchmark, as we discussed before, we used GPOWER, high-precision power analysis software, to determine the product flow size we needed so that we can draw a powerful conclusion. The inputs of GPOWER for determining the flow size in linear multiple regression model are effect size, the alpha level, power value, and the number of predictors (Figure 23.3). GPOWER uses "f" as a measure of effect size, which has a relationship with R (coefficient of determination: the total proportion of the dependent variable variability that is explained by predicted variables) as the following equation describes:

$$f^2 = \frac{R^2}{1-R^2} \tag{23.2}$$

In this experiment, we used $f = 1.5$ ($R = 0.6$), as well as alpha (0.05), power (0.90), and the number of predictors (4) as the other three inputs. Special considerations should be addressed when setting up an RFID system with multiple interrogators that have overlapping interrogation zones. For instance, a pair of readers in a portal door interrogation zone may interrogate multiple tags in a dynamic environment.

The concept of interrogation will be abstracted as "read zone" for both the practical and real environment in the following content. By considering the warehouse environment requirement for RFID application in the distribution process, the portal door RF read zone will be limited in some

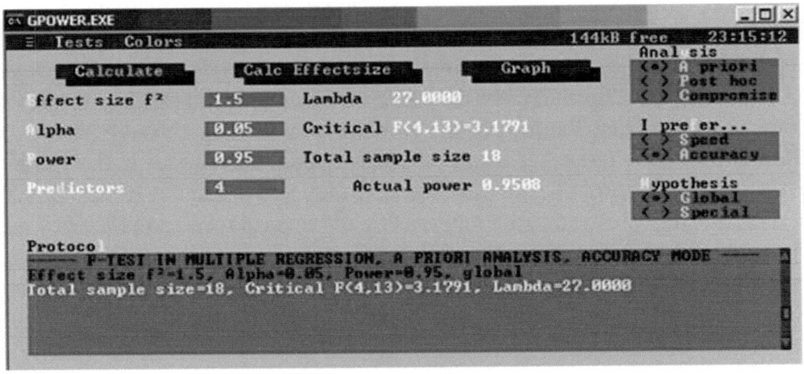

FIGURE 23.3 Results screen from GPOWER program.

ranges between dock equipment and RF interrogation range. The ranges of RF antenna (portal) and physical range of dock door layout can be described as n—vertex graph G ($V = \{1,\ldots,n\}$, E), and for each edge— its Euclidean "length." Denote a 2D layout of the graph where the coordinates of vertex i are $p_i = (x_i, y_i)$. Denote $d_{ij} = \backsim p_i - p_j \backsim = \sqrt{\{(x_i - x_j)^2 + (y_i - y_j)^2\}}$.

In the non-noisy version of the problem, we know that there exists a layout of the antennas that realizes the given edge lengths (i.e., $d_{ij} = l_{ij}$). Our goal is then to reproduce this layout. Fortunately, there is additional information that we may exploit to eliminate spurious solutions to the layout problem—we know that the graph is a complete description of the close antennas. Consequently, the distance between each two nonadjacent antennas should be greater than some constant, r, which is larger than the longest edge. This can further constrain the search space and eliminate most undesired solutions. Formally, we may pose our problem as follows:

Layout problem given a graph G ($V = \{1,\ldots,n\}$, E), and for each edge—its Euclidean "length," find an optimal layout ($p_i,\ldots,p_n$), where $p_i \in R^d$ is the location of the antenna i, which satisfies for all: $i \neq j$:

$$\left\|p_i - p_j\right\| = l_{ij} \quad \text{if } \langle i,j\rangle \in E \tag{23.3}$$

$$\left\|p_i - p_j\right\| > R \quad \text{if } \langle i,j\rangle \notin E \tag{23.4}$$

where $R = \max_{\langle i,j\rangle \in E} l_{ij}$.

An optimal layout is similar to that generated by common force-directed graph drawing algorithms that place adjacent nodes closely while separating nonadjacent nodes. Therefore, we may estimate the distances between nonadjacent antennas and then give constructive suggestions to minimize the blind spot within the reachable zone.

From the graph in Figures 23.4 and 23.5, the interrogation zone from a pair of antenna gives us a visual description for the range we calculated in formulas 23.3 and 23.4. The center red zone shows the high readability zone of 2 in. from each side of the portal door. The accuracy reduced with

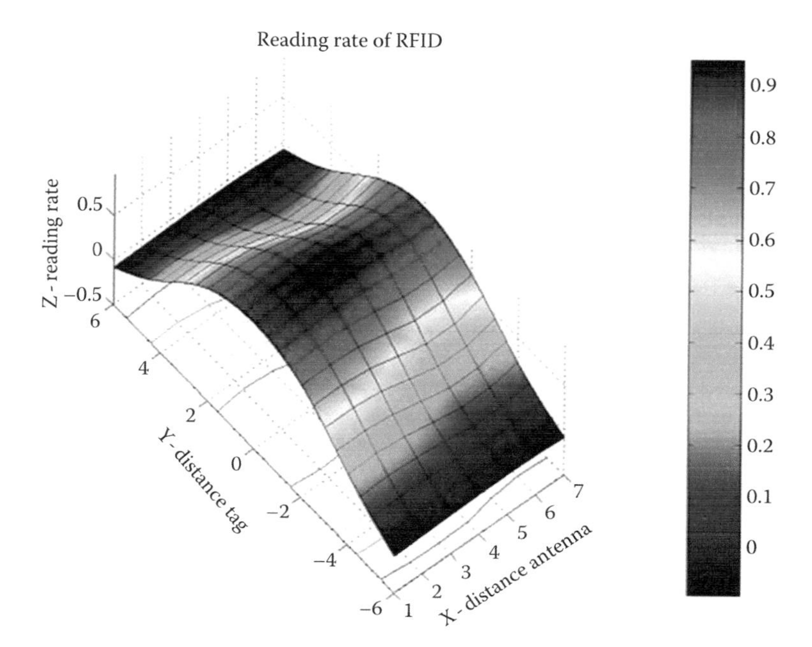

FIGURE 23.4 Dimension RF interrogation zone (effective range).

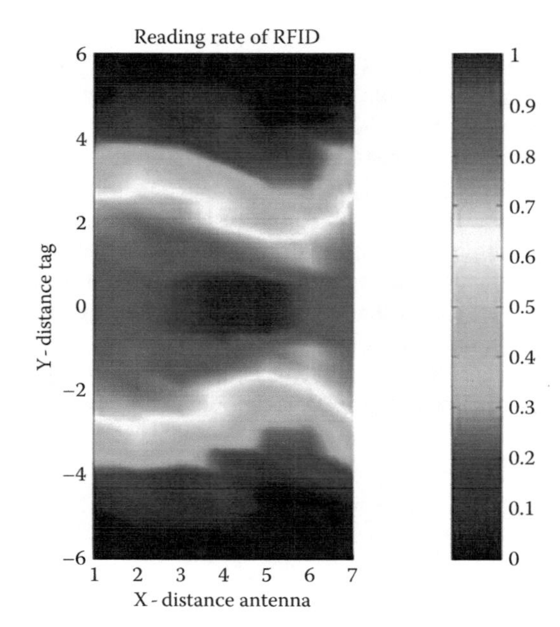

FIGURE 23.5 Dimension RF interrogation zone (effective range).

increasing distance from tag to each side of the portal. Therefore, the estimation of an antenna's physical coordinates should be done according to the features and requirements of the RFID and warehouse system. The data that are collected by the antenna should be accompanied by an indication of where in space that data were reported. The bandwidth and limitations of the antenna network make it necessary for data location coordinates of the physical location of the portal door in a warehouse to be established. In many cases, location itself gives the range of data that should be sensed—localization drives the need for an RFID antenna network in a warehouse and distribution process, which is able to locate items and tagged parts. In addition, the accuracy of geographic routing and graph algorithms requires validation of portal and other functional parts in the warehouse.

TABLE 23.1
Minimum Maneuvering Distance between the Back of the Dock Leveler and the Beginning of the Staging Area and Recommended Dock Staging Dimensions

Equipment Used	Distance (ft)	Item	Dimension (ft)
None (manual)	5	Served road width	
Hand truck		One-way traffic	12
Two wheel	6	Two-way	24
Four wheel	8	Gate openings, vehicles only	
Hand lift (jack)	8	One-way traffic	16
Narrow aisle truck	10	Two-way	28
Lift truck	12	Gate openings, vehicles + pedestrians	
Tow tractor	14	One-way	22
		Two-way	34

Source: Data from Tompkins, J.A. et al. *Facility Planning*, 2nd ed., John Wiley & Sons, New York, 1996.

The design of a portal door and the layout of RFID antennas combined both frequency interrogation and physical portal length so that the tagged pallets will be tracked and the employed frequency from the antenna can record the data from moving tags (Table 23.1).

23.3 LAYOUT IMPROVEMENT ALTERNATIVES AND NUMERICAL RESULTS

Relying on the basis of implementation and RF facility layout principle, the facility layout algorithm will follow the baseline model we discussed above. Therefore, the qualitative algorithm is deployed to analyze the overall function parts. The layout algorithm continues to develop the relationship between function parts, including warehouse layout and RFID distributing zone. Because of the limitations of the RFID interrogation zone, we consider its correlation to the RF facility layout.

23.4 COMPUTER-AIDED PROGRAM ALGORITHM APPROACH (BLOCPLAN)

The program generates and evaluates block-type layouts in response to user-supplied data. It is used for single story layouts. BLOCPLAN uses a "banding" procedure to develop layouts. This permits a large range of possible layouts for a problem. For a nine-department problem, the number of possible layouts is close to 20 million, and for a 15-department layout, there are more than 2.6 × 1013 possibilities. Each department will also be rectangular in shape. The structure that holds the departments will also be rectangular in shape, and the user may select the length/width ratio of the structure.

23.4.1 Relationship Data

BLOCPLAN uses the relationship codes described by Muther in *Systematic Layout Planning*. Each sub-procedure we discussed in the SLP flowchart shows that the functional departments are defined by the material flow. We take one of the typical warehouses as an example. BLOCPLAN uses adjacencies for one type of layout analysis. We define the departments as picking/receiving, storing, inspecting, forward picking, sorting, shipping, and dock to dock. The differences between classic warehouse layout algorithm using BLOCPLAN and RFID applied warehouse is that the consideration of adjacent function zones is separated for the reason of interfaces between sensors. For instance, the picking, forward picking, shipping, and dock-to-dock zones are considerably separated according to the amount of product flow.

For this example, Figure 23.6 that shows layout data illustrates the departments and the square footage required for each department.

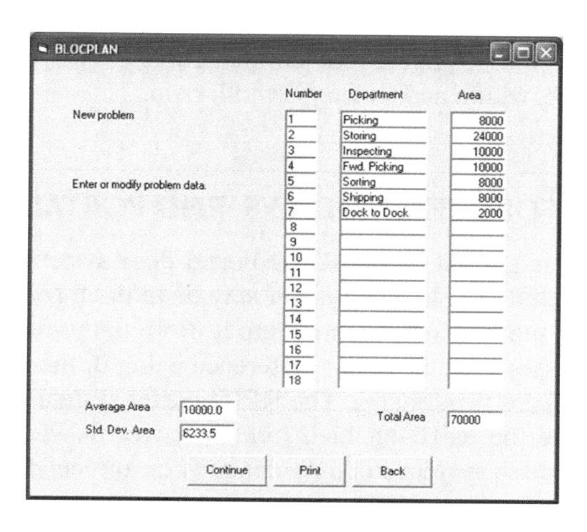

FIGURE 23.6 Layout data.

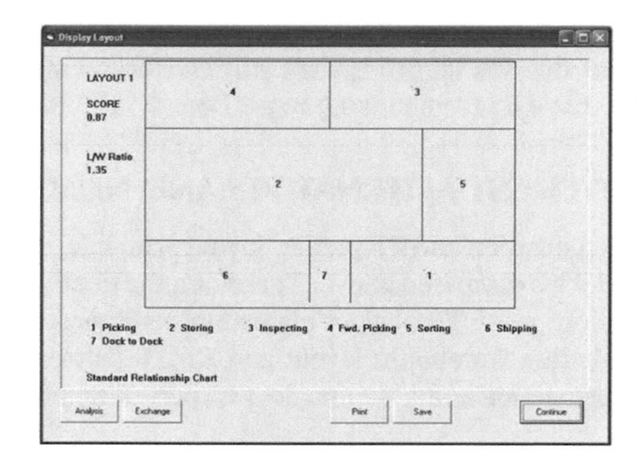

FIGURE 23.7 Layout.

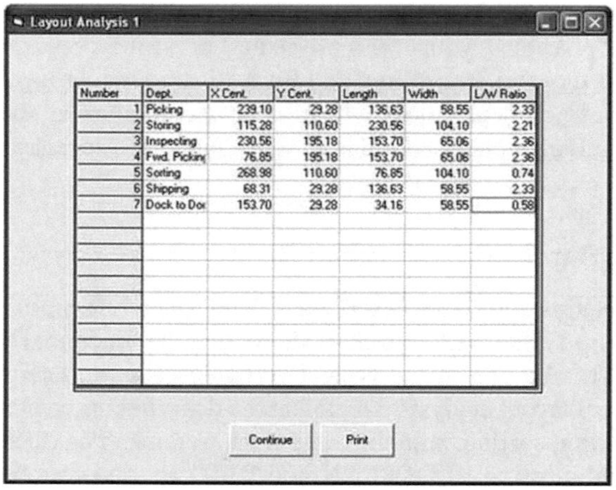

Layout Analysis 1

Number	Dept.	X Cent.	Y Cent.	Length	Width	L/W Ratio
1	Picking	239.10	29.28	136.63	58.55	2.33
2	Storing	115.28	110.60	230.56	104.10	2.21
3	Inspecting	230.56	195.18	153.70	65.06	2.36
4	Fwd. Pickin	76.85	195.18	153.70	65.06	2.36
5	Sorting	268.98	110.60	76.85	104.10	0.74
6	Shipping	68.31	29.28	136.63	58.55	2.33
7	Dock to Doc	153.70	29.28	34.16	58.55	0.58

Continue Print

FIGURE 23.8 Layout analysis.

Figure 23.7 that shows layout illustrates one possible solution to the given problem. This solution has a layout score of 0.87. Layout scores may range from 0 to 1.00. Higher scores indicate greater satisfaction of the adjacency relationships specified in the problem parameters.

Figure 23.8 that shows layout analysis proves additional analysis results in the solution provided in Figure 23.7 that shows layout. This figure provides the X and Y centroid of each department along with the department's length, width, and its length/width ratio.

23.5 RFID-ENABLED FORKLIFT SYSTEM VS. RFID PORTAL SYSTEM

An RFID-enabled forklift is similar to the RFID portal door system but consists of a forklift with an RFID system mounted onto it. The system may be in the form of a separate antenna and reader or a single unit with the two incorporated into it. Both the portal door and forklift system have strengths and weaknesses with the major difference being in the fact that one is mobile and flexible while the other is fixed (stationary). The RFID portal system waits for the tags to move through its read zone while the RFID-enabled forklift moves to where the tags are and reads them. The decision as to which system to apply depends on the needs and requirements of the warehouse. For given operations, an RFID-enabled forklift system may provide more capabilities with the ability to maneuver through the warehouse even in confined spaces as long as there

FIGURE 23.9 An RFID reader with an acoustic sensor. (Courtesy of Rush Tracking Systems, Lenexa, KS.)

FIGURE 23.10 An RFID-enabled forklift. (Courtesy of Rush Tracking Systems, Lenexa, KS.)

is enough room for the forklift to go through. In addition, depending on the type of tags being used, longer read ranges and greater capacities can be achieved. Solving the facility layout problem for a warehouse utilizing the RFID-enabled forklift is simpler than the former. With a few adjustments, even a traditional warehouse could easily be adapted to use RFID-enabled forklifts, making its implementation less complex than that for an RFID portal. Better still; the use of both systems in various warehouse functions can lead to greater efficiencies. Figures 23.9 and 23.10 show an RFID reader with an acoustic sensor and a typical RFID-enabled forklift used by Rush Tracking Systems.

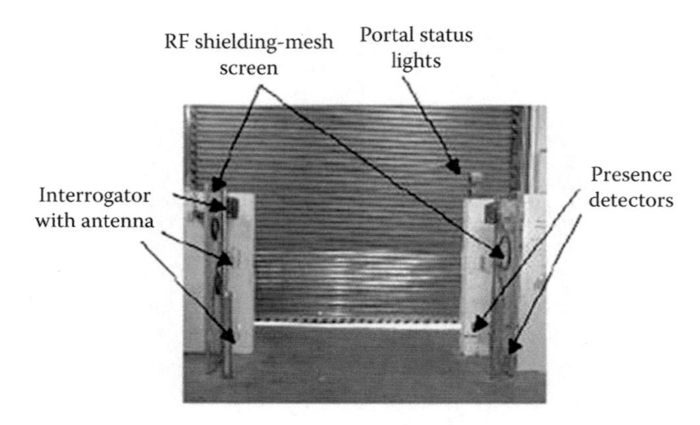

FIGURE 23.11 An RFID portal.

23.6 DISCUSSION AND CONCLUSION

The use of RFID systems in both existing and new facilities requires rethinking traditional layout approaches (Figure 23.11). This is necessitated by the need to take into consideration the department relationship requirements added by RFID system components. What may have previously been an optimal facility layout may no longer be optimal.

This chapter describes a layout methodology that takes an integrating multi-objective architectural approach involving data environment analysis, and RFID interrogation zone optimization. The effectiveness of the resulting layouts can be evaluated using facility layout software such as BLOCPLAN for Windows.

24 RFID Uses in Warehousing

24.1 INTRODUCTION

It is important to discuss software when we describe radio frequency identification (RFID) and operations such as warehousing. Since the mid-1990s, warehousing and other operations have become computerized. To realize any benefit from technologies such as RFID, operations must be computerized. In this section, we describe the different types of systems that allow for efficient operations. Because software and middleware are the most important pieces of an RFID solution, these packages are needed to make use of information collected by RFID technology with all the other systems operating in the warehouse: warehouse management systems (WMS), transportation management systems (TMS), event management systems, order management systems (OMS), and enterprise resource planning (ERP) systems.

The ability to capture, store, rationalize, and integrate information captured by RFID technology, including product information, location, volume, and transactional data, allows organizations to more efficiently pick/pack, ship, route, track, and distribute materials. This operational improvement can result in lower inventory levels and improved labor and equipment productivity. Integrating the information from RFID tags into an ERP system allows alerts to be sent that pre-set conditions have occurred such as inventory max–min levels have been realized. System standards and compatibility problems can result in expensive software implementation process. Standards are currently being developed at EPC global.

24.2 WAREHOUSE APPLICATIONS

Manhattan Associates, the largest WMS vendor, has built their business by implementing software that allows for warehousing best practices. Bobby Collins, SVP of national accounts, suggests that the warehousing problems drive efficiencies and costs in most large and small companies. He describes that WMS implementations seek to drive value by solving the warehousing problems. The top 10 warehousing problems are as follows:

- Inventory accuracy
- Space utilization
- Picking information
- Slotting
- Order picking
- Order accuracy
- Returns
- Vendor coordination
- Performance reporting
- Strategic planning
- Warehousing is a requirement of a successful business
- Warehousing delivers customer satisfaction

When implementing WMS systems, a standard implementation process includes the following master planning methodology:

- Document current warehouse operations
- Determine future requirements over the planning horizon
- Identify and document deficiencies in the existing warehouse
- Identify and document alternative warehouse plans
- Qualitative and quantitative evaluation of alternatives
- Select and specify a plan
- Detail planning
- Implementation

In the following sections, we provide a brief overview of relevant warehousing operations and how RFID may support improvements in these areas. Also in the text, there is an overall presentation of warehousing and WMS donated by Global Concepts on best practices for warehousing and WMS to further describe usage improving warehousing operations.

24.3 RECEIVING

RFID technology eliminates the need to physically check the bill of lading and/or the packing slip during the receiving process in a warehouse. This represents a significant labor reduction and inventory accuracy improvement in most operations. RFID can alert most WMS systems to indicate if a product needs a cross-dock movement. Cross-docking is the process in which product received can be identified as an immediate need to fulfill an order and is immediately loaded into outbound trailers to fulfill the order. This cross-docking process reduces the labor and time to store, replenish, pick, pack, and ship a product. The system requirement consists of a WMS interfacing with an OMS to determine if this product is needed so that a task can be created to ship the product "across the dock" to the outbound dock so the order can be completed and placed on the waiting vehicle. RFID makes the identification of these types of immediate need orders easily identifiable and possible more reliably than traditional bar code scanning.

If using a conveyor receiving process or conveyor in general, RFID provides greater efficiencies by eliminating the need to ensure that cases/items are placed properly on the conveyor so that the bar code can be read accurately. RFID allows for accurate reads regardless of product position, resulting in fewer reading errors.

24.4 STORAGE

RFID system can eliminate the need to scan the bar code on the pallet and at the storage, replenishment, and picking locations for the different types of storage racks. RFID scanners can continuously scan locations using WMS specification, and create task from identification of inventory inaccuracies. Since the RFID tags can be read from anywhere, products and pallets do not have to be placed in specific or assigned locations such as the golden zone illustrated in Figure 24.1. Material handling principles such as using random storage locations system, minimizing honeycombing, and replenishing to fast picking zones can be realized.

24.5 PICK/PACK

RFID readers can integrate with the WMS to validate that the correct items and amounts are picked and measure productivity in the warehouse.

FIGURE 24.1 Golden zone concept.

24.6 SHIPPING

An RFID reader can confirm that each item is placed onto the correct outbound vehicle, which can improve the accuracy of the shipping process. This verification can be made as the product moves through the portal to the outbound dock door. RFID allows for an automatic check of the items loaded into the trailer against the bill of lading. By using RFID readers or portals at exits of the facility and employee areas ensures that all items leaving the building are accounted for.

24.7 RELIABILITY

The reliability of the RFID tags is problem with many pilot implementations. Currently, RFID accuracy for Walmart implementations has averaged between 70% and 75%. General problems including problems with accuracy are related to multiple reads and no reads because of readers' inadvertently scanning adjacent products and/or double scanning the same product. Others include products containing metal or liquids that will reflect respectively, the signal from the RFID scanner. Metal racking systems could also pose a problem of reflected signals. Additional problems occur with data overload from the high-speed movements of products. "No reads" create a unique problem or RFID technology at the present. With bar code technology, the reader can detect if it did not read a bar code. With RFID technology, a no read goes undetected.

24.8 IT INFRASTRUCTURE ISSUES

One of the major concerns is the potential bandwidth requirement or an RFID system capturing all the available data from every RFID tag in a given warehouse. The potential volume of information from real-time scans moving between multiple applications or every single case or pallet in a warehouse can easily overwhelm even the most robust information systems. Hewlett Packard uses RFID in its facilities in Memphis, Tennessee; Chester, Virginia; and Sao Paulo, Brazil. These sites generate 1–5 TB of data a day. Therefore, organizations must analyze the potential data from an RFID tag and determine what information needs to be captured in real time and what information can wait for a batch update. The information systems also need to be robust enough to handle the speed increases associated with a successful RFID implementation. Shorter scanning intervals, faster product movements, and shorter order cycle times must be handled without sacrificing system integrity.

Other problems include the differences between storing UPC bar codes, which are 11 digits, and storing RFID serial numbers, or EPCs, which are 13 digits. The Uniform Code Council, a standards body for the retail and manufacturing industries, states that their Sunrise 2005 initiative requires all U.S. and Canadian companies to be capable of scanning and processing up to 14-digit bar codes by January 2005.

Slap and ship RFID implementation approach in which requires a minimal amount of investment to slap tags onto a subset of outgoing shipments to comply with the current mandates. The second approach relies on larger investments to develop an internal capacity that impacts the supply chain upstream, in an effort to both comply with mandates and capture operational efficiencies from RFID.

The slap and ship approach is driven by the mentality that RFID is a cost of doing business due to the mandates set forth by both Walmart and the DoD. Organizations employing this strategy are not looking for a short- or long-term ROF on their investment; they are only concerned with being compliant with their customers so that they are able to continue doing business. This approach oftentimes is just as costly as a well-thought-out long-term strategy to use the technology. Most pundits suggest that the second approach should be utilized in order to increase business efficiency. The next section provides some implementation examples in which the companies sought to integrate RFID into operations.

24.9 RFID WAREHOUSE IMPLEMENTATION EXAMPLES

The following are examples of RFID system implementations in various companies in different business sectors.

24.9.1 Gillette

In January 2003, Gillette bought 500 million Class I EPC tags from Alien Technology. Gillette has been using the order to tag all pallets and cases of women razors. Gillette worked with its WMS and TMS provider; Provia, to ensure that the RFID information can be integrated into the appropriate systems. Below is an explanation of how Gillette has incorporated the tags into their processes.

24.9.2 International Paper

International Paper, the world's largest paper and forest products company, went live with their first fully automated RFID warehouse tracking system in August 2003. The use of truck-mounted RFID readers and proprietary tracking technology provides forklift operators with execution task information. The elimination RFID portals in tracking inventory movement for the use of mobile forklifts truly integrating RFID in operations has provided more efficiency.

24.9.3 Proctor & Gamble

Proctor & Gamble performed a pilot project in which they used RFID at an international manufacturing plant in Spain to send pallets to domestic operations. Results indicated this was a cost-effective way to implement RFID tagging.

CASE STUDY

RELATED BACKGROUND TO ADC, RFID, MILITARY, OR LOGISTICS

Military logisticians can use RFID technology to track troop movement as they use different modes of transportation. For example, they may use a bus to get to the military base, a carrier to get to the next port, a train in an international country, and possibly a tank as they are in theater. We look at a simple example of how one public transportation agency uses RFID technology to support data analysis of passenger demand.

In 2008, gas prices peaked and Metropolitan Transit saw a record increase in the number of passengers it served. To avoid spending a small fortune on gas, local residents began to take the train for their daily commute to work. Over the past year, passenger use of the trains has fluctuated with gas prices and economic conditions. Methods for tracking passenger counts have been manual, leaving room for error. Variability in passenger usage has made it more difficult to determine if the variability in count was due to an error or due to actual changes in usage. In addition, Metropolitan Transit is facing its own budget cuts and is looking at potentially reducing the number of trains it runs.

To more accurately provide trains based on passenger demand, Metropolitan Transit has launched a new fare collection system that uses smart card technology to collect data as passengers enter and exit the train station. With the new fare collection system up and running, Metropolitan Transit is ready to begin analyzing passenger demand.

Prior to boarding a train, passengers can load a one-way or round-trip fare onto a plastic card that looks similar to a credit card but that has an embedded RFID microchip. To enter and exit the train station, passengers tap the card to a card reader that records the station location, time of day, and type of fare loaded on the card.

METHODS AND RELATED MODELS

The data set for the first month of operations since the launch of the new smart card system has been collected and is ready for analysis. The passenger demand, D, followed a uniform distribution $D_{\min} = 200$ and $D_{\max} = 400$. The supply of trains can now be calculated with the Newsvendor model by using the number of passenger seats to determine the number of trains needed. Each train has 50 seats. The optimal quantity, q, using the newsvendor model is

$$q = F^{-1}\left(\frac{p-c}{p}\right)$$

where F^{-1} is the inverse cumulative distribution function of D, with price, p, and cost, c.

SAMPLE PROBLEM

Metropolitan Transit assumes that there is a cost, c, and a price, p for each seat on the train. Assume that the price is the passenger fare for one trip: $p = 2$ [$/passenger seat] and the cost of each seat on a train running along the route is $c = 1$ [$/passenger seat]. The optimal quantity of passenger seats is calculated as

$$q_{\text{opt}} = F^{-1}\left(\frac{2-1}{2}\right) = F^{-1}(0.5) = D_{\min} + (D_{\max} - D_{\min}) * 0.5 = 300$$

Therefore, the optimal number of trains that Metropolitan Transit should send out is six trains.

24.10 CONCLUSION

RFID technology provides a method to more accurately gather data on how passengers interact with a system. As passengers register their smart cards with identifying information, Metropolitan Transit can also adjust other components of its transportation system as passengers move from train to bus to commuter bus.

Part VI

Military RFID Initiatives and Applications

25 U.S. Army RFID Initiatives

In this chapter, we present past, current, and future U.S. Army RFID initiatives. This includes both active and passive radio frequency identification (RFID) efforts.

25.1 INTRODUCTION

The U.S. Army has been utilizing active RFID technology since as early as 1993. These early efforts were a direct result of the logistical problems encountered in the Middle East operations, Desert Shield and Desert Storm between 1990 and 1991. Early efforts included tagging cargo for North Atlantic Treaty Organization operations in Haiti and Macedonia, with approximately 35% of the cargo tagged. Over the next few years, the Army increased the percentage of cargo being tagged until 2001 where approximately 85% of the cargo shipped to Afghanistan for Operation Enduring Freedom was tagged. The next most significant event was the 2004 Undersecretary of Defense for Acquisition, Technology and Logistics directive that all the services use high data capacity, active RFID technology to support the in-transit visibility (ITV) of supplies. According to DOD policy, all containers being shipped from the Continental United States (CONUS) to Outside the Continental United States (OCONUS), OCONUS to CONUS, or between combatant commands must have data-rich active RFID tags.

25.2 TOTAL ASSET VISIBILITY AND IN-TRANSIT VISIBILITY

In 1990 and 1991, during Operations Desert Shield and Desert Storm, significant weaknesses were identified in the then current logistical supply deployment system. Most critical of these was the realization that shipping containers received from OCONUS had to be opened, inventoried, closed prior to continuing through the supply chain. This was because there was little or no ability of the system to determine the exact contents of the shipping containers and where the contents needed to be forwarded. At times, significant numbers of shipping containers were awaiting inspection while the combat units needing the contents went unfulfilled. As a result of this logistical deficiency, the Army began experimenting with active RFID tags. This effort has now developed into the current worldwide Radio Frequency—In-Transit Visibility (RF-ITV) system. To date, the ITV system is the most successful application of active RFID in the U.S. Armed Forces.

The RF-ITV system consists of four regional computer servers that collect, process, and report the location of tagged assets in 40 different countries across the globe. This is performed by a series of active RFID tag interrogator readers at more than 4000 nodes along the worldwide supply chain network of military and commercial marine ports, airports, and supply depots. As vessels, airplanes, and trucks are processed through each of these facilities, the tag readers interrogate the RFID tags on each container. The tags return their identification as well as any other data that may have been written to the tag. The local computer network then uploads the container tags identification and the reader's location to one of the four regional servers. This enables the system servers to maintain a continuous record of each container's last known position along the supply chain. Thus, users may query the ITV system to find out the current status of their supplies. Equally importantly, the system can also alert the monitor personnel in the event that a container is accidentally mis-shipped or stolen.

In addition to establishing the position of the containers, the system can also utilize tags with sensors. These tags such as the Savi ST-675 tag can determine when the container has been unsealed, exposed to light, or subjected to shock. This additional capability has proven useful in storage depots where a container might wait until transportation has been arranged for movement further along the supply chain. When an intrusion or damage has been sensed, the tag can send an alert to the closest interrogator–reader. This will allow the storage facility to investigate the incident.

Temperature sensors have also been integrated with Savi ST-674 tags for medical shipments to Iraq. In this case, the medical shipments had specific temperature and shelf life times. In the event that these parameters are violated, the tags can send out an alert similar to an intrusion alarm. The end event is similar, the system is notified as to the fact that an unacceptable situation has occurred.

25.3 TRACKING TRAINEES

RFID has also been used by the Army to track soldiers during training exercises. Since 2005, the Army has been using a combination of RFID technology to determine the physical position of soldiers training in the Ft. Benning, GA Mission on Urbanized Terrain (MOUT) training site. This is a mock-up of a small village complete with multistory buildings for training soldiers on the intricacies of fighting in built-up urban environments. In this application, an ultrawide band RFID tag developed by the Ubisense Corporation is used for the trainee tracking.

The system consists of a 900 MHz transmitter with a GPS locator. This system provides continuous position data as long as the trainee is outside of a building. Once inside, the GPS cannot acquire the position signal. To overcome this limitation, the system also uses an RFID wristband. As the trainee enters a building and loses the GPS signal, the wristband is interrogated by a set of fixed-position RFID readers. The strength of the RFID signals enables the system to determine the soldier's location within 1 ft. The transmitter is then fed the position data along with the soldier's identification.

By being able to continuously track the trainee's positions both indoors and outdoors, the training staff can recreate the scenario in three dimensions. This allows the training staff to assess the effectiveness of the trainees in the MOUT center.

25.4 GUN MONITORING

Another area in which the U.S. Army is developing RFID applications is in the area of ground vehicle gun monitoring. The concept behind this effort is that RFID technology can be used to track the number of rounds that have been fired in a tanks cannon for maintenance purposes. Currently, the tank's crew must manually record the number and types of rounds fired. This information is transferred to the maintenance component to determine their maintenance requirements for the barrel. The difficulty with the current system lies in the fact that it is unlikely that this information will be accurately recorded under battle conditions.

The initial RFID system was developed by August Systems and uses American Science and Technology RFID technology. With the initial RFID when the tank's gun is fired a microelectromechanical (MEM) sensor determines that the round has been fired and keeps a running tally on a tablet PC. Subsequent development efforts have replaced the MEM sensors with piezoelectric sensors. These generate an electrical impulse that is then used to record the firing of the gun. The gun crew can then view the history of the barrel on the tablet PC.

In the latest generation, the tablet PC has also been eliminated in favor of just an RFID tag that contains the same information as the previous tablet PC. However, in this iteration, the information cannot be viewed by the gun crew, only by the maintenance facility. When the tank returns to the maintenance facility, the RFID tag is read by a fixed-position Intelleflex I-beam reader. Ongoing test includes the possibility of replacing this reader with a handheld version in order to improve system flexibility.

25.5 USING RFID TO TRACK ARMY MEDICAL RECORDS

25.5.1 INTRODUCTION

In 2006, the U.S. Army awarded 3 M a $3.76 million dollar 3 year contract to develop and implement a passive RFID file tracking system for the 150,000 military personnel and dependent medical records at Ft. Hood, Texas. This marked the first use of RFID technology to track medical records on U.S. Military base. The particular RFID technology that was used in the project is the 3 M Smart Shelf System with Sirit Infinity 510 ultrahigh frequency Generation 2 tag readers.

25.5.2 Application

The medical records at the Ft. Hood facility are spread among 300 file cabinets at six clinics on the post. Since as many as 70,000 records could be accessed in a given month, the retrieval, filing, and merging of these files at each and between each of these clinics created significant records management issues for the base personnel. Previously, the medical records utilized bar codes and handheld readers. However, bar codes had to be deliberately read and records could only be manual read by a reader one at a time. More importantly, if a file was removed, but not scanned, there was no way of determining the location or the user of the file.

With the Smart Shelves systems installed in the file cabinets, the medical administrators can perform a number of file management operations automatically. This is possible with a unique numbered Generation 2 passive RFID tag affixed to each individual's medical record. Information, associating each tag to an individual and the file management record, is stored in an external database. For example, when a file is removed, the system records the time, date, identity of the user, and the location from which the file was removed. This makes it impossible for a record to be removed without its removal being recorded.

The system also allows for automatic inventory checks and inventory error warnings. For example, the hardware/software system can identify which medical records should be present in each file cabinet. Similarly, if a record is misfiled, the system can automatically warn the administrators of the problem. As a Generation 2 RFID-based system, the system is capable of faster and more reliable reads than Generation 1-based systems. In some cases, Generation 2 readers can conduct 1000 reads per second. This capability enables systems such as the Smart Shelf to continuously rather than intermittently poll for the presence of passive RFID tags. Generation 2 tags are also more reliable in terms of tag reads than Generation 1 tags.

What this effectively means is that the system should be able to determine where any medical file is at any time. If the needed medical record is not in its normal file cabinet location, the system can identify the current user. The user can be contacted and the file can be returned in a timely manner, rather than waiting perhaps indefinitely until the file is eventually returned on its own. This enables the military medical community to provide more effective medical treatment because the correct and needed medical records should be available when needed.

25.5.3 Summary

The Ft. Hood system was adopted in 2007. Between 2007 and 2008, the implementation team investigated with a variety of tags, frequencies, and standards. Installation was completed in 2008. It is currently under evaluation for deployment at other military installations.

25.6 CHAPTER SUMMARY

In this chapter, we have described a few of the past, current, and future Army applications for RFID technology. These include the ITV System, the use of RFID to track soldiers during training, and the use of RFID and sensors to automate gun barrel maintenance procedures. As the technology becomes more sophisticated, it is expected that the use of RFID, particularly passive RFID technology, will become more widespread in the Army logistical system.

26 Marine Corps and Navy RFID Initiatives

In this chapter, we present past, current, and future Marine Corps and Navy radio frequency identification (RFID) initiatives. This includes both active and passive RFID efforts.

26.1 MARINE CORPS ACTIVE RFID INITIATIVES

26.1.1 Introduction

In 2003, the Undersecretary of Defense for Acquisition, Technology, and Logistics directed the services to use high-data-capacity active RFID technology to support the in-transit visibility (ITV) of supplies. The U.S. Marine Corps initially responded by purchasing nearly 200,000 active RFID tags and acquiring over 1200 RFID reader interrogators for positioning throughout the Marine Corps operating forces. The Marine Corps has also acquired nearly 100 RFID portable deployment kits to support these efforts.

26.1.2 Active Tagging Requirements

According to Department of Defense (DOD) policy, all containers being shipped from the Continental United States (CONUS) to Outside the Continental United States (OCONUS), OCONUS to CONUS, or between combatant commands must have data-rich active RFID tags. When necessary, combatant commands may also specify additional tagging requirements to support operations or exercises for purposes of ITV. These tagging requirements are in addition to those specified by the global transportation network. Marine Corps units that fill a container or build a pallet are required to attach the tags, write content-level information to the tags, and send the data to the ITV servers.

Common containers and pallets include the following:

- SEAVANS (20 and 40 ft military containers moved by sea)
- MILVANS (military-owned, demountable containers)
- Quadcons (57.5 × 96 × 96 in. container with a metal frame and pallet base that can be strapped together to form 20 ft ISO containers)
- Sixcons, 8 × 8 × 4 ft fuel/water containers
- Palcons, 40 in. L × 48 in. W × 41 in. H pallet or palletized container
- 463 L pallets, standardized pallet for military air cargo made of aluminum with a balsa cored and measuring 88 in. (224 cm) wide, 108 in. (274 cm) long, and 2–1/4 in. (5.7 cm) high. 463 L pallets can hold up to 10,000 lb.

The contents for the RFID tag information is written by the unit though one of several automated information systems (AIS) used by the Marine Corps including the

- Automated Manifest System-Tactical (AMS-TAC)
- Cargo Movement Operations System (CMOS)
- MAGTF Deployment Support System II (MDSS II)

AMS-TAC is an Army and Marine Corps jointly developed automatic system for generating and managing shipping manifests and database records using a variety of bar code readers and RFID tags. CMOS is an air force developed AIS for installation cargo movement during peace time and the deployment of both cargo and passengers during wartime or contingency operations. Lastly, MDSS II is a stand-alone system used by the Marine Corps alone to read and write military shipping labels and RFID tags. It prepares data in a format for the ITV server and can exchange information with other services logistics systems.

Using one of these systems, the tag data are then sent to one of the four ITV servers positioned around the world. At this point, the container is active in the defense transportation system (DTS).

26.1.3 Containers inside the Defense Transportation System

The DTS consists of distribution nodes with ports of embarkation, ports of debarkation, and a variety of different checkpoints and container storage lots such as theater distribution centers. Each of these areas contains both fixed and portable RFID reader interrogators. The fixed interrogators are positioned at key locations so as to easily capture the arrival and departure of containers and pallets. The identity and location of each fixed reader is registered in the ITV server database.

As cargo containers reach each of the mode areas in the DTS, the interrogator records the interrogator number, the tag identification, and the date time group. These data are sent to one of the four worldwide ITV servers. The ITV servers continuously build a transit record of where the container currently is and where it has been. This goes on until the container reaches its ultimate destination.

26.1.4 Active RFID Tags and Placement

The active tags used for this application are required to have at least 128 K of memory and operate at a frequency of 433.92 MHz. Because they utilize active RFID technology, the tags require the use of internal batteries. These tags have an effective range of 300 ft. The average cost of these tags is between $75 and $100. Currently, the majority of these tags are being provided by the Savi Corporation, currently owned by Lockheed.

Because of the limited power and range of the active RFID system, the placement of the tags on containers and pallets is of importance. For example, the signal has difficulty transmitting through metal and liquids. For this reason, the placement of the tag must be in a position where the transmission through either metal or liquid is minimized. The best locations are on some sort of central point rather than concealed in side or off to one side. In addition, the tag must also be securely attached to the container or pallet from at least two points.

Due to the cost and limited availability of the tags, the tags must be conserved and recycled. When a container no longer has the need to be tracked, the tag is to be removed by the receiving organization. The memory specifying the containers contents is to be erased. To conserve battery power, when the tags are removed, the batteries are reversed, which essentially turns the tag off. It will now not respond to interrogation transmissions by the reader interrogators. The used tags are then recycled for use in retrograde shipments or sent back for reuse at either the distribution depots at Susquehanna, Pennsylvania or San Joaquin, California.

26.1.5 Summary of Marine Corps Active RFID Initiatives

The U.S. Marine Corps active RFID initiatives are well under way. Part of this success is due to the overall DOD effort to ensure ITV of assets. By utilizing the ITV server resources and combining their front-end resources including the AMS-TAC, CMOS, and MDSS II, the Marine Corps can determine where assets are along the DTS.

26.2 MARINE CORPS PASSIVE RFID INITIATIVES

26.2.1 Introduction

In 2004, the Deputy Undersecretary of Defense for Logistics and Material Readiness issued a policy directive requiring each service to develop an implementation plan for the use of both passive and active RFID technologies. In response to this directive, the Marine Corp developed a series of documents for implementation guidance starting in 2005. This included the use of pilot passive RFID programs within the Marine Corp system with continuously increased funding from $684,000 in FY 2008 to $9,996,000 in FY 2013. The pilots included initial testing and evaluating the cost of the passive RFID technology; testing the technology with future logistical requirements; receiving and passing RFID information; and testing the technology in austere operating environments, particularly with the use of handheld devices.

26.2.2 Passive RFID Pilot Project: 2004

An initial passive RFID pilot project was performed at Camp Lejeune, North Carolina with the supported activities supply system (SASSY) Management Unit (SMU) of the second supply battalion. The purpose of this pilot project was to test the tracking of shipments from the Defense Depot Susquehanna, Pennsylvania to the supply battalion. This project consisted of a Susquehanna server encoding the RFID and generating an advanced shipping notice (ASN) file with content data similar to DD form 1348. The tags were affixed to the shipment. The ASN was then transferred electronically to the server at the SASSY management unit of the second supply battalion. When the tagged pallet was eventually received in Camp Lejeune, a passive reader scanned the RFID tag on the pallet and matched it with the ASN file on the SASSY server. This pilot project although small in scope demonstrated the potential of using passive RFID technology in this capacity.

26.2.3 Implementation Plan Pilot Project 1: 2008

The first pilot project under the 2005 implementation plan was scheduled to take place in 2008 at the Marine Corps Logistics Command. There are two major objectives in this pilot project. The first objective is to assess the level of different forms of RFID technology with respect to reads, read rates, and the transfer of RFID data at different levels of the Marine Corps logistical system.

The pilot was conducted in a receiving warehouse with six fixed interrogators and eight handheld receivers. As the pallets are received at the warehouse, both the pallet and case-level tags were read. Any discrepancies between the pallets and the case-level tags are resolved at this time. The pallets are then moved to transshipment points where the individual cases on the pallets are read. The costs associated with this pilot project will be recorded for both the equipment acquisition and integration and the personnel requirements for receiving, processing, and accepting the shipments.

26.2.4 Implementation Plan Pilot Project 2: 2009

Pilot Project 2 is to be conducted at a Marine Corps Supply Management Unit level facility. The focus of this pilot is to assess the ability of RFID technology to be used with future Marine Corp logistics needs. In particular, this process focused on the ability of the technology to be integrated with the Global Combat Support System—Marine Corps (GCSS-MC) and the National Management Inventory System (NMIS). In this pilot, inventory planning, receiving, receipting, and acceptance activities will be performed with Generation-2 passive RFID tags. This pilot is similar to Marine Corps Logistics Command test as pallets will be received through warehouse doors with both fixed and handheld RFID readers. Similarly, both the pallet and the individual item case tags will be read and checked for discrepancies. Pallets will then be broken down and the individual components inserted into the distribution system. In addition, this pilot project will include the use of printers to process additional tags for initiating new orders for orders with previously identified discrepancies including missing items.

The metrics for this pilot project will be somewhat more encompassing than the Marine Corps Logistics Command pilot test as it will include receiving, receipting, acceptance of pallets and items, as well as the effect on inventory levels and the associated personnel costs. The pilot will also include a metric on the ability of the system to effectively support the end user, the war fighter.

26.2.5 Implementation Plan Pilot Project 3: 2010

The third implementation plan pilot project is scheduled for execution in 2010. The purpose of this project is to assess the feasibility of receiving ASNs through the Marine Corps AMS-TAC. ASNs will then be distributed to the SASSY Management Unit so that the location of the inventory can be determined. This pilot project is expected to utilize up to six fixed readers and 10 hand readers. Four printers will also be used for sending and receiving inventory. The metrics for this pilot project include inbound inventory read rates and an assessment of the benefit of determining inventory location.

26.2.6 Implementation Plan Pilot Project 4: 2011

The fourth implementation plan project is expected to be conducted during 2011. The focus of this project will be on the use of handheld portable devices in difficult environments. This includes high heat, low temperatures, and high humidity. Similar RFID portal tests will be performed as in a number of the previous pilot projects. However, these tests will be performed exclusively with the handheld portable devices. Similar to the other tests, shipments will be scanned for both pallet and case RFID tags. Tag discrepancies will be rectified during this process. Printers will also be used to generate distribution and return documents. The metrics associated with this pilot project will also include personnel effects of the system, as well as the impact on receiving, receipting, and acceptance. A separate metric will also be utilized to determine the value of the system to effectively support the end war fighter user.

26.2.7 The Future of Marine Corp Passive Applications

The following section is a direct excerpt from the USMC RFID implementation plan. It illustrates the potential of a completely integrated RFID system within the USMC supply system.

> In an ideal scenario, an advance shipping notice (ASN) for a box containing 10 HMMWV brake shoes is sent through Defense Automated Addressing System (DAAS), and received by a retail-level warehouse (SMU) via GCSS-MC. This information is passed to the receipting area at the warehouse and as the case enters the warehouse bay doors, interrogators located on the sides and top of the door read the tag on the case (a 96-bit passive RFID tag). The tag is linked to a database that ties to the warehouse management, inventory management, order management, distribution management, and financial management systems.
>
> The database then tells the Warehouseman receiving the shipment to open the case and place 8 brake shoes in the HMMWV brake-shoe storage bins. The system "knows" the location in the warehouse and that information is passed to the warehouseman (warehouse management). The system automatically updates the inventory by eight brake shoes (inventory management) as they are placed on the warehouse shelves through the use of "smart shelving".
>
> Simultaneously, the database informs the warehouseman that of the two remaining brake shoes, one is to be delivered to 2nd Battalion 11th Marines and the other is going to 7th Engineer Support Battalion (order management). The Warehouseman places them in appropriate bins for pick-up and delivery.

Another signal is sent to Transportation Support Battalion (or other distribution unit/organization) informing them that they must stop and deliver repair parts to 2/11 and 7th ESB (distribution management). Finally, information is sent to the financial system charging 1st FSSG (Supply Bn for eight brake shoes and 7th ESB for one brake shoe) and 1st MarDiv for (2/11 for one brake shoe) the expense associated with the delivery of the brake shoes (financial management).

26.2.8 Summary of Marine Corps Passive Initiatives

The Marine Corps RFID initiative plan described in this chapter was a response to the 2004 DOD directive to develop an implementation plan for the use of both active and passive RFID technologies. The Marine Corps has already begun this process with the passive RFID tests performed at Camp Lejeune, North Carolina. The additional four scheduled pilot projects include consideration for evaluating both the technology and the ability to integrate the technology into the existing and future Marine Corps logistical systems. The metrics associated with the pilot projects involve both personnel impacts on receiving, receipting, and acceptance of supplies and the overall benefit to the war fighter.

26.3 PERSONNEL RECORDS APPLICATIONS

The U.S. Marine Corp Installation Personnel Administration Center (IPAC) in Hawaii maintains over 8000 personnel records including Marine service record books (SRB) and officer qualification records (OQR). These records include the past training, assignments, and evaluations of the individual marines and officers assigned to the base. These records are frequently moved within the base for various purposes, as well as being moved with Marines as they are deployed or transferred. The location of the records at any given time is important in the event that a record updating process is delayed or if a record is needed for multiple purposes. To ensure the integrity of the record storage system, the individual records are subject to monthly inventories. Due to the large number of records and limited staff, typical personnel record inventories can take up to 1 week to complete.

To improve the monitoring of record locations and to reduce inventory times, the center implemented a new RFID system on the Marine SRB and OQR file system. The Hawaii-based Marine Corp system utilizes the recently developed 3M passive RFID tracking tag system for filing.

26.3.1 Application

The 3M RFID system includes several components. These are a tracking tag for each individual record, a tracking pad, a handheld tracker, and tracking software.

26.3.2 Tracking Tag

The passive RFID tracking tag is 2.25 × 1.89 in. with an adhesive backing. It operates on a frequency of 13.56 MHz with a 1.8 × 1.8 in. antenna. As a passive RFID tag, it is powered by the radio frequency interrogation field generated by a variety of the systems readers. Like any passive tag, barring damage, the tag has the capability to be read an unlimited number of times. The tag may also be written to a total of 100,000 times. Each tag has a permanent identification number and 256 bits of writable memory. The writable memory capability allows the tag data to be updated or changed as necessary. This is an enhanced feature not available in earlier passive tags. Normal use of the RFID tag includes physically adhering the tag to a file or file box.

26.3.3 Tracking Pad

The tracking pad is an RFID antenna reader, which is connected to a computer. The pad is normally used for recording the removal and replacement of tagged records from a designated area. Although an individual file may still be removed without being properly recorded, the ease of use of the tracking pad increases the probability that the removal and replacement of an individual file will be properly recorded. The read range of the tracking pad is 11 in. The tracking pad provides both audible and visual feedback of a successful tag read. This means that movement of multiple files can be simultaneously recorded by placing the stack of files on top of the tracking pad.

26.3.4 Handheld Tracker

The 3 M handheld tracker is a portable device with an antenna and a rechargeable battery. The handheld tracker can be used in conjunction with the 3 M tags to perform file inventory, identify shelf ordering errors, and search for specific files. File inventory functions include determining the ID of the file and the file location. The shelf ordering error function notifies the user of incorrectly placed files and their correct locations. Lastly, lists of file can be uploaded into the unit for location. This capability alone has great potential as a single missing file could have been easily misplaced anywhere within the entire filing system. Rather than have to do manual search through individual files, the handheld scanner can be passed over open file drawers. The handheld tracker unit is fairly robust with a battery life of approximately 4 h of active use. The handheld tracker has a read range of approximately 8 in.

26.3.5 Summary

The use of the 3 M RFID system at IPAC is considered as a success by the staff personnel. Monthly inventories that used to take 1 week have been reduced to a little as 2 or 3 h. This has resulting in huge savings in manpower and man-hours for the IPAC. The use of the system worldwide at Marine Corp bases is under consideration.

27 Marine Terminal RFID Applications

27.1 INTRODUCTION

According to U.S. Customs and Border Protection, approximately 95% of the cargo tonnage that enters the United States comes by sea. In FY 2005, this amounted to 11 million containers (USCBPa). The container security initiative (CSI) was created in 2002 to identify and examine high-risk containers prior to being loaded on vessels intended for the United States. Currently, 50 ports in foreign countries are participating in the CSI program. This covers approximately 82% of all cargo containers headed to the United States.

The four foundations of the CSI program are

- Identify high-risk containers
- Prescreen and evaluate containers before they are shipped
- Use technology to prescreen containers
- Use smarter more secure containers

27.2 TRACKING CONTAINERS

The fourth foundation is where radio frequency identification (RFID) technology is being applied with respect to shipping containers. The use of RFID technology is directly covered in the Customs Trade Partnership against Terrorism program. This program is an industry-driven effort to develop a global security network from the point of origin to the point of delivery. This includes tracking across countries and different modes of transportation. Tracking and security is achieved through the mounting of RFID tags on each shipping container. The tags are monitored at each RFID-enabled facility that is linked to the global security network.

27.3 CONTAINER TAGS

The container tags, CSI and C-TPAT programs were developed by the Savi Corporation. The designation for these tags is ST-676. These tags are similar to the Savi ST-656, but incorporate both RFID circuitry and sensor technology. The tags include sensor technology for both security tampering and environmental monitoring. The security sensors protect the container from possible theft or terrorist-based intrusion. The security sensors monitor both light and whether or not the container doors are open or closed. The environmental sensors monitor temperature, humidity, and shock. The user can set acceptable environmental standards to the tag's memory.

These environmental sensors can help prevent spoilage or other damage to the contents of the container.

27.3.1 Tag Operation

Under normal operation, the tag sensors monitor the container. In the event that an event occurs, the type of event, time, and date is recorded in the memory. A signal is then sent to the facility RFID

FIGURE 27.1 Savi ST-676 container tag.

reader. The reader can then transmit an alert message via the Internet or telephone. The facility personnel can then investigate the cause of the alarm. In the event that the alarm was intrusion based, security personnel can be dispatched to the exact location of the container.

27.3.2 Tag Mounting

The tags are designed to be fixed to the left hand door of ISO shipping containers without any special procedures or tools. The mounting process can be completed in seconds. It is weatherproofed to function under adverse environmental conditions. The tag itself is positioned inside the container. The antenna remains outside. The tag operates on 433 MHz. It has a range of approximately 100 m. The tag has a replaceable 3.6 V lithium battery. The battery is expected to power the tag for up to 4 years based on two data collection operations per day. Figure 27.1 illustrates the Savi ST-676 container tag.

28 Other Foreign Countries Armed Forces Adoption of RFID Technology

In this chapter, we explore the use of radio frequency identification (RFID) technology by foreign countries. The RFID applications presented include those from Spain and France.

28.1 NEW ZEALAND MILITARY CLOTHING RFID APPLICATIONS

The benefits of RFID technology have been extended to the process of outfitting new military recruits in the New Zealand defense forces. Traditionally, during uniform issuing process, new recruits try on different sizes of shirts, trousers, shoes, underwear, and other articles of clothing. When the recruit has determined the correct sizes, they report to a clothing clerk. The appropriate sizes each of these items for each recruit is hand recorded and later hand entered into a computer along with the recruit's identification information. This could include up to 32 lines of information. As can be imagined, the traditional way of doing this is a laborious and time-consuming process. Of course, with any process in which data are manually entered, the data can be easily entered inaccurately. Sometimes errors would not be detected until the individual recruit was issued what was thought to be the appropriate-sized clothing later in the intake process.

The New Zealand system for issuing uniforms to recruits was superior to many other military forces uniform issuing processes in that it used the sizing stock approach with subsequent issuing of the actual uniforms. The more common approach was to determine the size using the actual issue uniform components. By using the sizing stock approach, the New Zealand defense forces split this process into two parts. This reduced the logistical burdens of maintaining a huge warehouse of all the different uniform parts at the sizing facility.

To further increase the speed and accuracy of the uniform sizing and issuing process, the New Zealand defense forces investigated using various forms of automated information systems. Initially, consideration was given to using traditional bar codes. However, since bar codes are optically based, they cannot be read through several layers of clothing. Thus, direct visual access was required to record the size of each individual piece of the uniform. Another problem that was encountered was that the printed bar codes that were used on the clothing could not hold up well to the industrial laundry processes that were used to regularly clean the sizing clothing stock. These limitations led the New Zealand defense forces to design and implement a passive RFID system for the uniform issuing process in 2005.

28.1.1 Application

The New Zealand clothing company Yakka and IT provider Integral Technology Group implemented the system by imbedding passive EPC Generation 1 Class 1 RFID tags into each type and size of uniform clothing sizing stock. When the recruit arrives at the fitting location, they are also tagged with a Gen 1 Class 1 tag. The recruit then finds a complete set of clothing that correctly fits from the RFID tagged uniform stock. When this process is complete, a handheld RFID scanner collects the size for each type of clothing stock along with the recruit's identification number while the recruit is still wearing the clothing.

The capability of the scanner to read through several layers of clothing at the same time makes the RFID system far more efficient than having to individually remove and scan items as would be necessary with a bar code system. This information is then electronically downloaded on what is known as a scale of entitlement. In addition, the scale of entitlement contains fields for all of the authorized items of clothing, so it will immediately know if any of the recruit's clothing items are incomplete. Thus, data entry errors are eliminated. The electronic scale of entitlement system is also more flexible than a manual paper form. Deletions, additions, or changes in the uniform requirements can be easily executed and in real time. Different scales of entitlement can be used depending on if the recruits are in the New Zealand Army or Navy. If necessary, the scale of entitlement could be also varied on an individual or unit level for military operations in different theaters of war.

With the data recorded in the scale of entitlement, the RFID tagged sizing clothing stock is taken from the sizing process and laundered for future sizing with other recruits. As the passive RFID tags are more environmentally robust, they can survive the cleaning process, whereas bar codes could not. Once laundered, the RFID-tagged uniform components are returned to the sizing stock for the next set of recruits.

The record of clothing sizes is processed by another New Zealand defense force computer that generates purchase orders for the clothing manufacturer. As the process is now computerized, what previously took days to complete, now takes only hours. Once the purchase order is approved by the defense forces, it is sent to Yakka, the clothing manufacturer. When the order is received by Yakka, stock is drawn from the clothing manufacturer's warehouse and delivered to the defense forces logistical system. The recruit is then issued his correctly sized uniform set. With this new process, the recruit now typically receives the uniform items within 5 days of fitting at the sizing facility.

Because the entire inventory maintenance process is handled by the clothing manufacturer, the New Zealand defense forces are able to relieve themselves of stockpiling uniform components for recruit issue. From the perspective of the New Zealand defense forces, this is similar to a just-in-time system. The New Zealand defense forces do not need to concern themselves with inventory theft, damage, spoilage, or obsolescence. Back at the clothing manufacturer's site, removed stock is replenished as necessary in order to fulfill the next order of recruits in a timely manner.

28.1.2 Summary

The New Zealand defense force approach to issuing uniforms to recruits is an effective use of RFID technology where other forms of automatic identification have failed. The use of RFID tags provides both a non-line of sight and an environmentally robust solution to the sizing stock problem. The electronic entry of data with the RFID tags is leveraged throughout the process, greatly improving the effectiveness of the system.

The New Zealand defense force users of the Yakka system report that the fundamental concept is sound. A few difficulties were initially encountered. However, these had to do with electronic record issues beyond the scope of the actual RFID application. Currently, the New Zealand defense forces and the Yakka clothing manufacturer utilize the system to fit up to 200 army and 1000 navy and air force personnel a day.

28.2 SPANISH ARMED FORCES RFID EFFORTS

Beginning in 2006, the Spanish Armed Forces began installing Savi Technologies RFID interrogators at Spanish military and logistics posts. This includes posts in Afghanistan, Haiti, the Balkans, Kosovo, and Bosnia–Herzegovina. An additional four posts are being created within Spain. The purpose of this system is to track containers including items such as food and clothing, but not ammunition.

28.2.1 Application

A total of between 2000 and 4000 of the Savi 433 MHz active RFID tags were to be purchased for use within the system. The intent is to attach these tags to the cargo containers that are to be tracked. The program was scheduled to be completed in three stages. In the first 6 month stage, fixed readers were to be installed at specific locations. The purpose of these readers is to detect the presence of incoming shipments. In the subsequent phases, additional readers were to be installed at other installations as well as more remote outposts. It is likely that these outposts would also utilize the Savi portable deployment kit. The Savi SmartChain Site Manager software was to be utilized to keep global track of the tagged containers. This allows the Spanish Armed Forces to be compatible with existing NATO systems.

28.2.2 Summary of Spanish Armed Forces RFID Efforts

The Spanish Armed Forces RFID project differs from many other new RFID applications in that this implementation was being conducted initially at a full scale rather than as an initial pilot project. Savi officials are quick to note that this is a result of other successful deployments.

28.3 FRENCH ARMED FORCES RFID EFFORTS

The French Army has adopted RFID tags for tracking of emergency equipment such as life rafts and life jackets. The tag chosen for this application is the Texas Instruments 13.56 MHz laundry RFID tag. This tag was chosen for its resistance to exposure including water, mud, and high temperatures. The tag has the ability to record 2000 bits of information. In this application, the memory can be used to record the date of manufacture and other important information such as the performance of scheduled maintenance activities. This is expected to improve the French Army's ability to have emergency equipment where it is needed, when it is needed, and in operational condition. The tags have been approved by NATO for use by other countries in addition to France.

Part VII

Other Potential Military RFID Applications and Radio Theory

29 Military Physical Fitness RFID Applications

29.1 INTRODUCTION

A fundamental military activity is individual and unit readiness. Aside from military occupational specialty proficiency, this also includes physical fitness. Each year the military invests millions of man hours ensuring that its soldiers and sailors are capable of performing their wartime mission. To this end, the military branches have various military and physical competitions to promote individual and unit fitness. Some examples include the following:

- International Military Sports Council Games
- Armed Forces Sports Championships
- Marine Corp Marathon
- Best Ranger Competitions

In the civilian world, radio frequency identification (RFID) technology has been widely used in similar competitive sports. RFID technology has actually been used in some individual competitive sports such as running as early as 1994. In that year, the passive RFID ChampionChip was utilized in the Berlin Marathon. Since then, RFID chips have been used in cycling, skating, and even triathlon competitions.

The use of RFID technology in these types of events is particularly advantageous as competition race committees need to process large volumes of competitors in relatively short periods of time. For example, in a marathon race, there may be hundreds or even thousands of runners at the start of the race. Though the numbers of competitors passing though the checkpoints at the same time will diminish over the course, each individual competitor still needs an accurate time record.

Unlike bar codes or digital video records, RFID technology has the capability of being able to record the large volume of competitors nearly simultaneously. As competitors enter the antenna portal, the powering and transmission of the RFID chip and subsequent receipt of the signal is measured in milliseconds. Depending on the number of RFID readers in position, the time records for large numbers of competitors can be read at the same time.

For the purposes of illustration, in this chapter, we will initially focus our discussion of RFID technology to individual running-type sports competitions. These include normal distancing running, marathons, and walking events. However, the same technology is similar for other sports competitions such as cycling and skating.

29.2 APPLICATION CONSIDERATIONS

The use of RFID tags in individual sports competition is dependent on the organizing committee. Some organizing committees may accept compatible competitor-owned RFID chips, other committees may require competitors to use race-specific chips.

The use of compatible competitor-owned RFID chips reduces the logistical costs to the organizing committee. However, other problems may be introduced. Unfortunately, a common problem is that before the race, competitors using their own chips must ensure that their tag identification number (TIN) is properly associated with their entry. If the competitor fails to do so, there is no way that the committee can properly record their start, splits, or finishes.

Conversely, the race committee can require competitors to utilize the organization's RFID chips. This has the advantage of insuring that each competitor's TIN is properly recorded. However, the purchase of the anticipated number of RFID tags plus a reserve for a particular competition can significantly add to the operating costs of the event.

The cost of the event also depends on how the organizing committee intends to set up the RFID technology. RFID tags can be used to record any combination of individual

- Starts
- Splits and checkpoints
- Finishes

The exact choice of RFID tag reading may also be a function of the funding available to the race committee. Obviously, the most expensive system will be one that uses RFID tags to record the start, a number of splits or checkpoints, and the finish. There must be reader antenna systems positioned at each of these locations. In addition, if the data are to be centrally recorded, additional networking hardware will be required. This type of system will actually produce the most accurate type of performance data. The net time resulting from the difference between the finish and the start can be precisely calculated for each competitor. Competitors can also utilize the split times for race analysis and future training. Similarly, race organizers can utilize the checkpoint times to ensure that competitors have negotiated the entire course in realistic times. This approach virtually eliminates the possibility of competitors cheating as has happened with many competitions in the past.

On the other end of the spectrum, to minimize costs, some race committees will limit themselves to a conventional start, no splits or checkpoints, and only an RFID-enabled finish. In these types of applications, the level of technology is greatly reduced. It is possible to use a single reader and a local computer system.

29.3 TAGS

The dominant tag in the individual sporting event arena is the ChampionChip. This is a 134 kHz passive RFID tag and specialized housing. The specialized housing is designed to enable the chip to be securely fastened to the laces of a shoe or an arm or ankle band.

As a passive tag, the ChampionChip is dependent on the reader antenna to energize the chip's circuitry. When energized, the tag transmits its unique identification code. The ChampionChip identification code consists of seven digits. Some characters are not used. However, generally speaking, the first two digits are alphabetical characters, the third digit is numeric, and the fourth through seventh digits can be either numeric or alphabetical.

Figure 29.1 illustrates both the ChampionChip's tag and housing.

As can be seen from the illustration, the actual passive RFID tag is the small cylindrical capsule. The antenna and the circuitry can be observed. The RFID tag is by necessity waterproof. It is normally held inside the mounting unit, which in turn is attached to shoe laces or an arm or ankle band. Different styles and colors are also available; however, the RFID tag is universal.

29.3.1 Mounting Considerations

Despite increasingly robust components, some consideration must be made to protect the RFID chip from damage. RFID chips are sufficiently robust to absorb the shock received from being mounted on the top of footwear. The chip housing is typically secured with the shoe laces. In this position,

FIGURE 29.1 ChampionChip RFID tag.

the shock and possible damage to the RFID chip is mitigated by the shoes padding. Figure 29.2 illustrates a ChampionChip affixed to a runner's shoe.

In other competitions such as triathlons, competitors will wear a variety of footwear. It would be disadvantageous for the competitor to repeatedly dismount and remount the unique RFID tag. In this case, competitors normally affix the RFID tag to some portion of clothing or a wrist or ankle band. This is illustrated in Figure 29.3.

If an area other than near the competitors foot is selected for the chip placement, this must be taken into consideration when designing the antenna reader system. Similarly, competitors are also

FIGURE 29.2 RFID tag affixed to shoe.

FIGURE 29.3 Ankle band mounting.

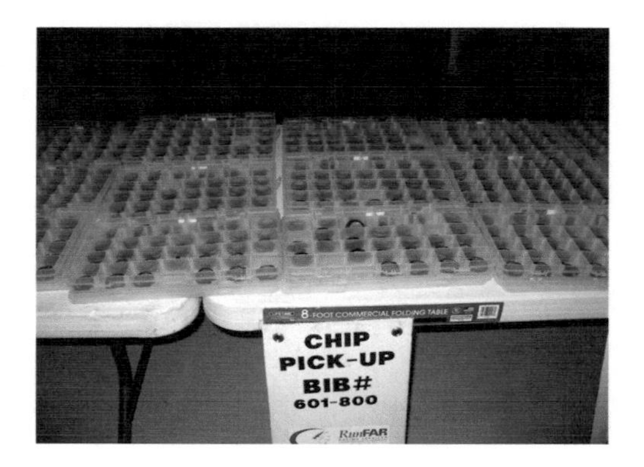

FIGURE 29.4 Chip check-in table.

advised not to finish in unusual body postures such as walking on their hands, as this would be outside of the finish line's antenna's normal interrogation zone.

Competitors may use either race committee-provided chips or individual-owned chips. Figure 29.4 illustrates the chip checking table at a race competition.

29.4 ANTENNA SYSTEMS

Due to the volume of reads required at potentially the same time, specially designed antenna portals may be required for individual sporting events. The most obvious requirement is that the portal must be wide enough to allow a substantial number of competitors to pass through at the same time. The other issue that drives the antenna design is the fact that short-range passive RFID chips are utilized. This means that the antenna must be relatively close to the position of the tags as they pass by.

These two design considerations have led to the development of antenna mats. The antenna structure is housed inside of the mat itself. If properly set up, competitors who step on the mat will be inside the antenna's interrogation zone. Up to four of the antenna mats illustrated in Figure 29.5 may be attached to a single reader. The use of four mats at one time effectively quadruples the width of the antenna portal.

29.5 READER SYSTEMS

In contrast to the antenna systems, the sports application reader systems are relatively conventional. The only significant difference between these and other RFID readers is the possibly increased need for portability and the ability to operate under adverse environmental conditions.

FIGURE 29.5 Antenna mat.

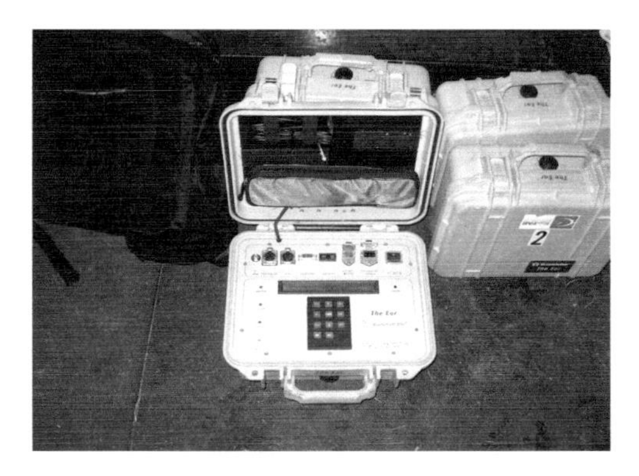

FIGURE 29.6 RFID reader.

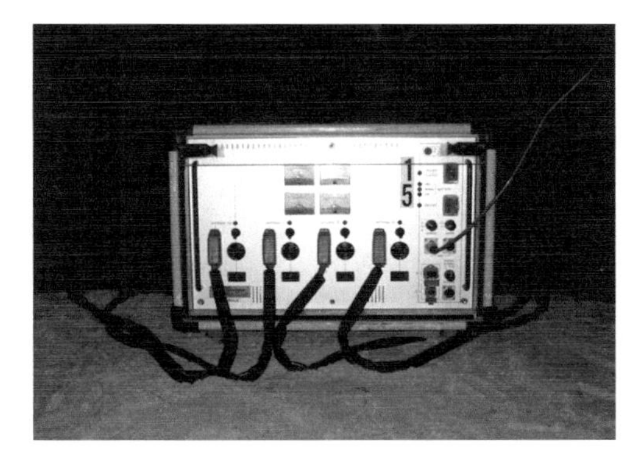

FIGURE 29.7 RFID antenna control unit.

The portability requirement means that the RFID reader antenna system cannot necessarily rely on AC power. In order to function in power inaccessible areas, the system must be capable of operating off of DC power. This means that the system must be powered off of a battery. The battery must be capable of powering the system long enough between the first and last finishers.

The environmental requirements mean that the system must be able to operate under various temperature and humidity conditions. Thousands of dollars are expended on even the most modest events. So, competitions are not normally cancelled unless really extraordinary conditions exist. Thus, for example, the RFID reader antenna system must be able to operate in the rain. Both the portability and the environmental condition requirements have been addressed through housing the system in waterproof boxes and powering the system with a battery.

Figure 29.6 illustrates a portable RFID reader and storage unit.

Figure 29.7 illustrates the antenna mat control unit.

29.6 PERFORMANCE ISSUES

A significant issue with the use of any type of RFID technology is reliability. This is not necessarily an issue with the technology itself, but rather in recognition that anytime that technology is used, it may also fail. The type of technology used in individual sports competition is relatively mature passive RFID technology in which only the TIN is transmitted. However, at any given time, an individual antenna reader system may experience difficulty reading an individual tag. This is a real issue with the short-range passive RFID tags typically used in sporting applications. This concern is definitely an issue since in many cases the positioning of the tag with respect to the antenna is not at all consistent.

For this reason, many organizing committees utilize both a primary and a secondary antenna reader system. This has the potential of reducing the probability of a bad read caused by any type of tag interference. It also provides a backup in the event that either of the two antenna reader system fails. What it does not protect against is RFID tag malfunctions. The only way to reduce this source of problem is by assigning two or more tags to an individual competitor. Although this may seem to be an extreme measure, in most cases, individual sporting performances cannot be repeated in the same manner as improperly shipped goods can be reordered.

29.7 EXTENSIONS TO OTHER SPORTS COMPETITIONS

The basic RFID technology discussed in this chapter has so far focused on applications to individual running-type sports competitions. This same RFID technology has been adopted for other individual sports competitions such as cycling and skating.

29.7.1 Special Cycling RFID Considerations

The same passive RFID tag that is used in running sports may also be utilized for competitive cycling. However, some modifications must be made as to how the RFID tag is positioned with respect to the competitor. In cycling, the tag cannot be attached to the competitors shoe as with running sports. The reason for this is that the cyclist's shoe travels in a circle subscribed by the pedal. At the top of the circle, the cyclist's shoe will be the maximum distance away from the antenna mat. With the relatively short range of passive RFID tags, this position may be beyond the range of the interrogation zone of the antenna mat. If this occurs, the competitor's RFID tag will not be recorded.

To prevent no-reads of this type, the current practice is to mount the RFID tag in a holder specifically designed to be attached to some position on the actual cycle itself. The exact position does not specifically matter as long as it fulfills two requirements. First, the metal in the bicycle frame cannot block the tag's signal. Second, the position of the tag must be as close as possible to the antenna mat. One practice is to affix the RFID tag housing to the front tire axle. This provides a consistent position with a direct line of sight to the mat.

29.7.2 Special Skating RFID Considerations

Special considerations must also be made when this RFID technology is utilized in skating type events. The first consideration is that the RFID tag must be positioned so that the blade of the skate does not interfere with the RFID tag's signal. The second consideration is that with ice skating in particular, for obvious reasons, it is not possible for the skaters to skate directly over the antenna mat. Because of the short range reads for the passive tags, reader antennas cannot be positioned on either side of the skating track. This means that an antenna mat must still be used. However, in this case, there is no alternative but to position the antenna below the ice. Testing will be required to ensure that this method of antenna positioning provides reliable reads.

29.8 SUMMARY

The use of RFID technology in individual competition sporting events began in 1994. Since then, RFID technology has been used in running events as large as city marathons. The popularity of using RFID tags continues to increase as race organizers appreciate the increased timing accuracy available only through the use of this technology. Although not a primary consideration, RFID technology can also help ensure that competitors properly pass through a number of checkpoints. This eliminates the possibility of competitors improperly completing a competition route. These advantages will ensure that RFID technology continues to become increasingly widespread in the coming years. The technology is currently sufficiently widespread that many competitors have a permanent RFID chip registered to them.

QUESTIONS

29.1 Why is RFID technology so advantageous over bar codes and digital video for recording accurate start and finish times in sports competitions?

29.2 Why can RFID chips not be affixed to competitor's shoes in triathlon competitions?

29.3 How can the use of RFID technology help ensure that the competitors complete the course properly?

29.4 Why is it necessary to provide both a primary and a secondary RFID system in an important sporting event?

30 Marine RFID Security Applications

30.1 INTRODUCTION

While radio frequency identification (RFID) systems are beginning to be used widely in land-based logistics systems, their use is much less developed in marine applications. Marine applications that are currently in existence focus primarily on the tracking of shipping containers on cargo freighters. In these types of applications, active RFID tags can be used for both customs and storage location purposes in the same manner as more conventional land-based applications.

However, RFID systems, both now and in the future, have great potential for application with respect to marine security-related systems. These include among others, registration tag and illegal vessel identification applications. Registration tag applications can be used to track the movement of both recreational and commercial vessels. Illegal vessel identification applications include the tagging of suspect vessels at sea for subsequent apprehension by law enforcement authorities at coastal ports. Prior to examining these two potential applications, it will be first necessary to address special hardware and infrastructure considerations that are necessary in order for an RFID system to operate reliably in the marine environment.

30.2 SPECIAL RFID HARDWARE CONSIDERATIONS

The marine environment, in general, presents more demanding requirements on RFID systems than those typically found in warehouse environments. These include the obvious increase in exposure to moisture and corrosive agents such as salt. Additional demands are placed on the equipment due to the constant motion encountered at sea. Sophisticated electronics such as those found in RFID tags are notorious for failing in the marine environment. RFID tags, antennas, and readers can expect to suffer the same fate, if not properly hardened for these types of applications.

Hardening the RFID tag is specifically required in the following areas:

- Housing water resistance
- Circuit board hardening
- Shock resistance

30.2.1 Water Resistance and Water Resistance Ratings

A special hardware modification that is necessary in the marine environment involves the sealing of the instrument housing from moisture. All electronics specifically designed for the marine environment are specifically protected with rubber gaskets. In this respect, passive tags that can be permanently sealed have a distinct advantage over active tags that require a replaceable or rechargeable battery. The tag's battery compartment presents an additional weak point with respect to maintaining any level of water resistance. Not only will the battery compartment need to be gasketed, but provisions must also need to be made to secure the cover to the main housing. This will normally be performed with captive bolts. Compare these requirements with a non-marine grade electronics that will have simple ungasketed, snap-on covers.

The use of a properly secured gasketed design will help maintain the integrity of not only the battery compartment, but also all of the other electronics as well. The complete ability of electronics to function in the marine environment with respect to water is typically rated according to the Japanese Industrial Standard for Water Resistance. The following table illustrates the different levels of ratings for electronics.

JIS Rating	Description
0	No special protection
1	Vertically dripping water shall have no harmful effect (drip resistant 1)
2	Dripping water at an angle up to 15° from vertical shall have no harmful effect (drip resistant 2)
3	Falling rain at an angle up to 60° from vertical shall have no harmful effect (rain resistant)
4	Splashing water from any direction shall have no harmful effect (splash resistant)
5	Direct jetting water from any direction shall have no harmful effect (jet resistant)
6	Direct jetting water from any direction shall not enter the enclosure (water tight)
7	Water shall not enter the enclosure when it is immersed in water under defined conditions (immersion resistant)
8	The equipment is usable for continuous submersion in water under specified pressure (submersible)

Source: http://www.hy-com.com/jis.htm

Higher-end marine electronics are currently rated under the Japanese Industrial Standard JIS-7 for immersion of 30 min at a depth of 1 m. Whether or not this level of water resistance will be needed for RFID tags depends on the exact nature of the application. Applications where the tag is sheltered such as in the vessel's engine compartment need not have a JIS-7 rating. However, applications where the tag is either on the deck or attached to the side of the hull will definitely require JIS-7, if the tag is to survive more than minimal use. As with other JIS-7 electronics, RFID tags with JIS-7 rating will obviously be more expensive to design and manufacture than those with lower ratings. However, as has been observed with other RFID tags, volume use will be expected to lower per unit pricing.

30.2.2 Circuit Board Hardening

In the event that the outer housing of either a passive or an active RFID tag is compromised, it is important that the circuitry is properly hardened. This will not only increase the length of operating time after compromise, but will also allow the possibility of refurbishing a potentially expensive tag.

Proper hardening for the circuit board includes, but is not limited to, the use of special waterproof coatings. These will help reduce the possibility of corrosion on both the circuit board itself and on the components mounted on the circuit board. The composition of the circuit board is also an issue. Less stable circuit board materials may swell with the absorption of water. If this occurs, circuit reliability may be compromised. Thus, the use of more water resistant phenolic-based circuit boards is preferred over less expensive paper-based circuit boards.

30.2.3 Shock Resistance

Another special requirement that marine applications may place on RFID equipment involves shock resistance. In situations where the RFID tag is affixed to the hull of the vessel, the tags must be able to withstand water intrusion, but also the constant pounding of waves on the vessels hull. In other situations where the RFID tag may be more protected, it will still have to be able withstand the acceleration and deceleration effects of the vessel while it is underway.

30.2.4 UV Protection

A final consideration when selecting or designing marine security RFID systems is the issue of ultraviolet (UV) radiation. Constant, direct exposure to the sun is a major cause of many types of marine systems failures. The UV rays received directly from the sun are also augmented by UV rays that reflect off of the water's surface. The end result is that many components that do not specifically incorporate UV stabilizers are much more likely to suffer premature deaths. In particular, the outer housings of RFID tags will become less structurally sound. The housings may then crack or become brittle. Water intrusion or some other sort of failure is soon to follow. This consideration makes it imperative that not only the tags, but also the rest of the exposed system be protected from UV damage.

30.2.5 RFID Tag Positioning

Just as with inventory-type applications, the positioning of the RFID tag circuitry is of significant consequence for the application. In the automotive industry, some discussion has revolved around the use of RFID tags embedded into the engine system. The primary purpose of this tagging was to assist with stolen vehicles. The same concept could be applied to marine engines. One question that may arise is how to prevent the end user from disconnecting the RFID tag. While this question may present considerable debate, at least one solution is to hard wire the RFID tag into the electrical system of the engine. If the RFID tag is deliberately disabled, so may be the engine system. The topic of tag positioning is further discussed in the applications section of this chapter.

30.3 INFRASTRUCTURE CONSIDERATIONS

30.3.1 Marine Portals

Just as inventory passes through a series of specific portals, most commercial and recreational vessels pass through specific narrow navigational areas. These include jetty entrances, channels, the mouths of rivers and bays, bridge underpasses, and even marina breakwaters as illustrated in Figures 30.1 and 30.2.

These areas are typically marked with navigational aids. Navigational aids include both fixed channel markers on metal stakes or wooden pilings and floating buoys of various sizes. Many of these navigational aids are powered. Power is required to either illuminate the navigational aids at night or allow the navigational aid to emit a sound signal such as a fog horn. A common approach

FIGURE 30.1 Channel entrance.

FIGURE 30.2 Bridge underpass.

FIGURE 30.3 Typical navigation channel marker.

to providing this power is to have a bank of batteries, which is recharged during the hours of daylight by an array of solar panels. Figure 30.3 illustrates typical navigational channel markers. Note the battery box mounted to the left on top of the platform. The recharging solar panel array can also be observed on the back side of the platform.

Figure 30.4 illustrates a typical navigational buoy. Larger versions of these types of buoys have the same power and recharging capabilities as with the navigational channel marker.

These navigational aids offer an attractive previously existing infrastructure from which the system of RFID portals can be initially based. The navigational aids are already positioned in the ideal locations to construct a system of portals and the RFID systems can tap into the existing navigational aid power sources. This approach could easily minimize the cost of initially deploying the RFID system.

However, the existing system need not be constrained only to previously existing navigational aids. The RFID portals may be enlarged or entirely new portals can be easily constructed through the placement of new dedicated floating RFID buoys. Dedicated RFID buoys have the distinct advantage of being positioned as necessary for maximum effectiveness without disturbing the existing navigational aid system. They also allow the system operators to more easily remove and replace buoys for maintenance and repair.

FIGURE 30.4 Typical navigational buoy.

FIGURE 30.5 Modified navigational aid RFID portal.

Both modified navigational aids and RFID specific buoys will ultimately have to transmit the RFID tag identification to a reader on land. If the transmission is non-satellite based, the nearest law enforcement agency will obviously have to possess an antenna system to receive this information. This type of technology is currently in use today with the NOAA weather buoys operated offshore. Given the necessary components for operating an RFID portal, Figures 30.5 and 30.6 illustrate what modified navigational aid markers and platforms might look like. In both cases, the RFID antenna is visible.

30.3.2 Antenna Mounting

One significant performance consideration concerning the implementation of an RFID system pertains to the mounting of the RFID antennas. In land-based operations, the platform on which the antenna is mounted is level and static. In this case, static refers to the stationary positioning of the antenna with respect to the horizon. For marine RFID applications where a fixed navigational aid is used as the portal antenna mount, the application is similar. This is the case when, for example, a piling or tower is used as the mount. Unless the antenna experiences a collision with a vessel, the antenna will maintain a level and static orientation. While this is the more desirable approach, it is only possible when the depth of the water is relatively shallow.

FIGURE 30.6 Modified navigational aid RFID marker portal.

In marine applications where the depth of the water requires that the RFID portal antenna be mounted on a floating navigational buoy, antenna mounting considerations become somewhat more complicated. The same is true for an RFID antenna mounted on a vessel. In both cases, the antenna platform is subject to wave action. This means that the orientation of the interrogation zone of the RFID antenna will be constantly changing. Since many RFID systems require a certain period to successfully acquire a signal, this could present a significant performance issue.

In similar situations with other electronic systems such as radar and GPS, their antennas have been positioned on gimbaling mounts. These mounts typically pivot on one or two axis. This allows the antenna to maintain a more horizontal position with respect to the horizon. To prevent uncontrolled movement with respect to the horizon, these systems typically contain some sort of dampening mechanism. Successful approaches to date include hydraulic and friction adjustable systems. In any event, the use of similar gimbaling mounts for the RFID system can be expected to help address this difference between land- and sea-based antenna systems. Figure 30.7 illustrates an example of a hydraulically dampening antenna mount.

FIGURE 30.7 Questus hydraulically dampened antenna mount.

30.4 OTHER CONSIDERATIONS

The list of items previously presented is by no means comprehensive. There are many other considerations that must be taken into account when implementing a marine RFID system. Some of these have yet to be discovered. However, one that is obvious is associated with the movement of vessel-mounted tags relative to antenna portals. In land-based systems, this is usually not an issue since the tags move on a relatively stable two-dimensional (2D) grid. In contrast with a marine application, the vessel and the portal may be moving in as many as six different axes. These include the usual x and y planes. It also includes the z plane, as well as, pitch, roll, and yaw. Pitch is rotational movement up and down. Roll is rotational movement around the direction of movement. Lastly, yaw is rotational movement around the z-axis. Where this may come into play is that the vessel-mounted tag is constantly changing orientation with respect to the antenna portal. The antenna portal, if mounted on a floating platform, may also be experiencing the same type of movement at the same time. The combination of the vessel-mounted tag and the antenna portal may make acquisition of a good read more difficult than that of an equivalent land-based system.

While not all of the RFID technology necessary for successful marine applications may currently exist, the reader can rest assured that the need for such applications can drive the development of both modified and new technology.

30.5 MARINE RFID SECURITY APPLICATIONS

As previously identified, future marine RFID applications include

- Vessel registration tag applications
- Hostile vessel identification

In the following section, we will discuss the major issues involved in each of these applications.

30.5.1 Vessel Registration Tag Applications

One of the most promising marine security applications involves the use of RFID tags on both recreational and commercial vessels for registration and identification purposes. All recreational vessels are required to be registered. For example, in the state of Texas, any engine-powered vessel must be reregistered every 2 years. Currently, the tag consists of a waterproof preprinted vinyl square with the registration information printed on the face of the tag. Figure 30.8 illustrates a current recreational vessel registration tag from the state of Texas.

FIGURE 30.8 Recreational vessel registration tag.

A recreational vessel RFID tag could either be a passive or an active tag. Obviously, the passive tag would have the advantage of not needing an internal power source. However, the current level of RFID passive tag range may be unsuitable. The use of active tags is not as limiting as it might first appear. Since the registration must be renewed on a periodic basis, the tag's power source would only need to last that long.

As with conventional visual tags, the RFID recreational vessel tags could be mounted on the relatively narrow forward section of the hull of the vessel. This would provide a relatively unimpeded position for the RFID tag to transmit to the portal sensors. Placement of the tag elsewhere might result in signal degradation due to other components on the vessel.

30.5.2 Scanning Recreational Vessel RFID Tags

The use of RFID integrated registration tags would allow marine law enforcement officers to rapidly scan tags without having to visually observe or record the registration tag. This is particularly important, because the auditing of registered vessels at the dock could easily be performed from a vessel on the water. Since the RFID tag would not necessarily need to be visually acquired, it would not matter whether the vessel was docked bow or stern to. All tag reads could be achieved by simply passing between sets of docks. In the event that a vessel is not available for tag reads, the vessels could also be scanned by law enforcement officers by walking down marina docks. Again, it would not matter if the vessel were docked bow or stern to.

Recreational vessels underway could also be easily scanned by personnel positioned on a law enforcement vessel. The law enforcement vessel could be positioned alongside any commonly traversed channel. Tag reads could also be made while patrolling navigable bodies of water. As the recreational vessels come into the interrogation zone, the antenna will acquire the tag signal. The tag identification can then be matched in the vessel registration database.

In this application, RFID systems display their nonvisual-dependent acquisition advantages. The law enforcement vessel and the recreational vessel could meet bow to bow, bow to stern, port to starboard, or starboard to port. Some of these meeting situations could make it difficult if not impossible for a visual-based tag to be successfully acquired. For example, in a pursuit scenario, the law enforcement vessel will be coming up from astern the evading vessel. Since most registration tags are positioned near the bow of the vessel, there would be no way for the law enforcement personnel to be able to visually acquire the registration tag. The only other information that may be available is the name of the vessel on the stern of the hull. There is no legal requirement to display the same name on the stern of the vessel as that entered in the state registration or the USCG documentation database. In contrast, in any of these situations, as long as the RFID signal can be acquired, it does not matter what type of meeting situation is present.

Reliable recreational vessel tag reads can also be achieved when difficult sea or weather conditions are present. Salt spray or waves can easily obscure a visual-based tag. Similarly, driving rain can prevent law enforcement personnel from visually acquiring a tag even if there is no spray or waves present. Under these conditions, identification is easily possible when visual observation is impossible.

Vessels without readable RFID signals could be stopped in the same manner that land-based law enforcement officers stop vehicles for out-of-date registration tags. During the stop, the law enforcement officials could examine the cause of the failure of the RFID tag. If the tag was deliberately deactivated, the law enforcement officials could take appropriate action. On the other hand, if the tag failure were simply an electronic issue, a summons could be issued to have the tag replaced.

Although the previous discussion has focused on recreational vessels, the same concepts are applicable to commercial vessels. All commercial vessels are normally examined on an annual basis as part of the U.S. Coast Guard certification inspection. The primary purpose of this inspection is to ensure the safety and seaworthiness of the vessel. As part of this inspection, the USCG can examine the RFID tag for serviceability.

30.6 STOLEN VESSEL IDENTIFICATION

Another potential application for RFID in the marine environment is the identification and subsequent recovery of stolen vessels. In this situation, the report of stolen vessels could be maintained in a central law enforcement database. As each vessel proceeds through existing marine RFID portals, the tag identification is automatically captured and compared to the centralized list of stolen vessels. A match triggers an alert to the local law enforcement agency, which can then pursue the stolen vessel according to their operating policy.

In a stolen vessel RFID application, the most effective type of tag would be one linked to the propulsion engine. In order to navigate among through the RFID portals, the vessel's engine would most likely be running. The tag can tap into the engine's electrical system to ensure that electrical power is present in order for the active tag to operate. Whether or not the tag can be designed so that it must also be operational in order for the engine to operate is a more serious question. This might present objectionable safety concerns in the event that the tag becomes inoperative. If this is not a concern, the approach could be taken one step further. In this case, an integrated tag could also receive a signal to deliberately disable the engine.

An RFID system of this type holds special interest to antiterrorism concerns. Land vehicles used in illegal operations are usually stolen to prevent traceability back to the terrorist's infrastructure. It is equally likely that a vessel used in waterborne terrorist activity also be stolen. If this is the case, then the stolen vessel RFID system offers an additional layer or opportunity to help prevent terrorist attacks like that executed on the USS Cole in 2000.

30.7 HOSTILE VESSEL IDENTIFICATION

A third potential application for RFID in the marine environment is the apprehension of potentially hostile vessels. In this case, we are talking about vessels suspected of being engaged in terrorist activities, drug smuggling, or illegal alien trafficking. The apprehension of any of these three types of vessels can easily lead to a lethal situation. Even in cases where less-than-lethal force has been applied, death has resulted from an accidental chain of events. RFID approaches are inherently far less than lethal than even less-than-lethal approaches.

In a nonlethal type situation, law enforcement authorities will want to identify a suspect vessel for eventual search and/or seizure. However in some cases, the law enforcement personnel may be able to initially identify a suspect vessel at sea, but will not be able to maintain contact with the vessel. This situation occurs, in particular, whenever a suspect vessel is encountered by aircraft. With limited fuel capacity or poor weather, the aircraft cannot stay on station until a surface resource is able to reach the location of the suspect vessel. The end result is that the initial contact with the suspect vessel is broken. The suspect vessel may then change course after the initial contact. With the change in course, the suspect vessel may then proceed to any one of a number of ports. Unless some mechanism is present to detect the suspect vessel, it may proceed unimpeded at the port of entry. Once actually in the port, the suspect vessel may then become lost among the other legitimate commercial and recreational vessels.

In hostile vessel scenarios of this type, weather conditions may make it impossible to visually identify the vessel while it is underway. Attempts to visually identify the vessel would be further complicated by any contact made during the hours of darkness. If the vessel did have a registration RFID tag system aboard as described in the previous section, the contact aircraft still might not be able to get within the interrogation zone. Any RFID system that the vessel has may also have been deliberately disabled.

In these situations, what is needed is a system that can first be easily deployed by the aircraft making contact. Second, it must be able to alert the authorities to the presence of the vessel in any one of many ports at a later date. Third, the system must not be able to be easily removed by the hostile vessel. These three requirements are easily fulfilled through the use of specifically designed RFID systems.

30.7.1 RFID Chaff

These types of RFID systems operate on the same principle as radar chaff. In this case, a large number of small RFID tags are utilized in place of radar-reflective strips of metal. As the law enforcement plane or helicopter passes over the suspected hostile vessel, a payload of RFID tags is deployed. The tags rain down on the vessel in the same manner as radar chaff.

Vessels that have been marked with the RFID tags in this manner will obviously attempt to remove the tags. However, like chaff, completely sanitizing the vessel of the RFID tags will prove to be a difficult if not impossible task. As long at least one tag remains on the vessel, it will be at risk of being detected as it passes through the interrogation zone of the marine portal at the port of entry.

Since the tag identification can be linked back to their deployment, the shore-based law enforcement agencies can be alerted to not only the presence of the suspect vessel, but also the circumstances under which the vessel was marked by the tags. Once the local law enforcement personnel are altered, the search for the vessel can be reactivated.

30.8 SUMMARY

In this chapter, we have investigated some of the issues associated with implementing marine security RFID applications. Many of the advantages that RFID systems possess in land-based applications are transferable to these types of marine applications. The use of RFID systems for marine security applications introduces issues unique to the marine environment. While normal RFID tags and antennas are satisfactory for their intended purpose, the electronics must be hardened for marine use. The nonstationary nature of marine applications also requires that special considerations be given to antenna mounting in certain circumstances.

Some of the technological aspects necessary for the successful implementation of RFID systems for these applications may need further development. For example, can sufficiently powerful RFID tags be produced at a low enough cost to be utilized in place of conventional registration tags? Similarly, can the range of RFID reader antennas be increased sufficiently for use between any existing navigational aid devices? These are issues that may or may not present barriers to the use of RFID systems in these marine security applications. However, as normal land-based RFID technology becomes more powerful and sophisticated, these enhancements will also benefit marine security applications.

QUESTIONS

30.1 What special considerations need to be considered when adapting or designing RFID tags for the marine environment?

30.2 How can the cost of deploying a marine portal system be reduced?

30.3 Why do RFID systems have a distinct advantage over existing recreational vessel registration tags?

30.4 What special communications equipment may be necessary for mobile RFID systems operating far offshore?

30.5 Why might the technology to implement marine registration systems need to be further developed?

31 Military Mortuary RFID Implantation Applications

31.1 INTRODUCTION

In 2004, the U.S. Food and Drug Administration approved the use of the passive RFID technology VeriChip for use in medical applications, specifically in human implantation. The implantation of an RFID chip in human beings is or should be of great interest for military morticians. Currently, military personnel are positively identified through the use of ID tags, dental records, and DNA. Some military casualties may leave little evidence of the identification. This is prevalent in casualties resulting from the detonation of mines or improvised explosive devices or heavy munitions such as artillery or aircraft bombs. In these cases, dog tags may be missing or only partial remains may be recoverable making positive identification of the casualty difficult if not impossible. While the use of DNA identification is possible, this requires lengthy and possibly inconclusive laboratory analysis, which may not be feasible during conflict. The use of human-implantable RFID chips such as the VeriChip offers another alternative to identifying military personnel. So, even if as little a limb, foot, or hand is available, rapid identification of the casualty will be possible.

31.2 VERICHIP

The VeriChip is a small passive RFID tag approximately 12 mm long and 2.1 mm in diameter. The electron chip and the antenna are contained in a glass capsule. When energized by an RFID reader, the VeriChip broadcasts on a frequency of 121–132 MHz. The tag broadcasts a unique 16-digit identification number. Like other passive tags with limited capabilities, the identification number must be matched to an external database that contains additional information. The range of the tag is typical of small passive RFID chips at a few feet. Since the chip utilizes passive RFID technology, there is no internal battery. This allows the chip to last almost indefinitely provided that it is not damaged.

The VeriChip Corporation also offers VeriTrace software for use with the VeriChip. This system has been in use with the Florida Emergency Mortuary Operations Response System and the State of Hawaii Department of Health since 2006. The purpose of this system is for identifying, tracking, and accounting for the remains of victims. The system includes the VeriChip implantable RFID microchip, a VeriTrace Bluetooth handheld reader, a customized Ricoh Caplio Pro G3 Digital Camera that can receive both RFID data and GPS data, and a web-enabled database. With these components, the system can be used for gathering and storing information and capturing images during emergency response operations.

31.3 IMPLANTATION PROCESS

The implantation of the VeriChip is a relatively simple process. A special type of syringe is used. The tag can be implanted in a number of different areas where it will be unnoticeable to the host. A typical example is in the arm between the elbow and shoulder. However, standardization of a single or multiple implantation areas for military purposes would facilitate identification in situations involving dismemberment. In the case of mine or aircraft bomb casualty, only a limb might be available for identification. Implantation in each of the soldier's feet and arms, for example, might increase the probability of the RFID chip surviving a death involving the detonation of a significant munition.

The implantation process is fairly simple and can be conducted in a doctor's office. One reference citing the use of implantable chips for hurricane victims indicates that the medical insertion process can be performed in 20 min at a cost of approximately $200. On a whole scale military level, it would be expected that this cost would be reduced dramatically. For reference purposes, companion pets have been implanted for years with similar Trovan passive RFID tags for $40 or less.

31.4 POST MORTEM OPERATIONS

In similar fashion, a separate series of RFID chips such as the VeriChip could also be implanted in deceased military personnel if one did not previously exist. In this manner, the chip would be able to retain its identity whereas an alternate means of identification such as a toe tag could become separated, damaged in transit, or degraded from bodily fluids. Again, a specific location on the body could be used for this particular type of chip. A separate database for the 16 digit identifier could be maintained with the individual's cause and time of death. Alternatively, the main database could also contain the same information.

31.5 MEDICAL OTHER THAN MORTUARY USES

The same advantages of using the VeriChip for mortuary uses are also beneficial for military medical uses. In the event that the casualty cannot speak or otherwise identify themselves, the chip can reliably provide needed information. Ideally, the chip would contain the soldier's entire medical record; however, the limited capability of these types of RFID chips precludes that possibility. However, the 16-digit identification number can be readily linked to the same external database, which not only provides the soldier's name, but also their blood type and previous medical or casualty history. This could be a critical benefit under difficult military medical operating conditions. The ability to positively identify the soldier would greatly reduce the possibility of mistakes in transfusions or medication allergies. So long as the medical unit has access to the centrally maintained database, all of this information can be readily accessed. If necessary, major medical units could periodically download medical records in the event that communications were temporarily disrupted.

The database could also contain the rest of the soldier's personnel record. This would include the soldier's current unit information. This would enable the unit to be easily contacted and informed of the soldier's status. At the same time, the medical facility or the soldier's unit could also rapidly inform the soldier's family of their status.

31.6 SUMMARY

The insertion of medically approved RFID tags in military personnel does not face the same philosophical and security issues as those present in the civilian world. In the civilian world, concerns over implantation focus on privacy issues and individual rights. The military benefits of tagging soldiers outweigh any residual protests concerning these issues. By tagging military personnel, casualties from large explosive ordnance have a greater potential for being positively identified for both mortuary and medical treatment purposes.

32 Railroad Car Tracking by an RFID System to Organize Traffic Flow*

32.1 INTRODUCTION

Almost every driver who has come across a railroad crossing and has had to wait for the train to pass has probably wondered to themselves: what if they had been warned of the oncoming train? If they had known about it blocking their path, would they have gone another way? Since not every railroad crossing has an overpass to allow traffic to freely pass, traffic can get backed up during busy times causing potential delays. With a warning system somewhere along their route, drivers will be able to take alternate routes and potentially save time. Several issues can result from a traffic buildup with safety being the largest. The Southwest Research Institute recently presented the Texas Department of Transportation TransGuide with a system design document that looked right into the heart of the subject matter. The system is called **A**dvance **W**arning to **A**void **R**ailroad **D**elays (AWARD). This system would notify oncoming traffic of an oncoming train and that they would need to take an alternate route to avoid this train or expect a delay waiting for the train to pass.

The system, however, does not use RFID as a tool.

RFID has been around for a long time, but until recently, it has not been really utilized as well as it could potentially be. The *RFID Certification Handbook* (Harold et al., 2006) not only talks about RFID in general, but it also gives insight as to where it could and already is used. An RFID system consists of a tag (either passive or active), a reader, an antenna, edgeware, middleware, and some sort of IT system. The RFID system can be set up to desired settings if the appropriate components are being used. In this particular case, the RFID system can be used not only to set up a warning system for oncoming traffic, but also to tell what the train is carrying, where it has been, even how fast the train is currently traveling. This is why it would be interesting to see if an RFID system would potentially work in this particular situation. Information is invaluable to every component of a supply chain and RFID could potentially provide it readily and in real time.

Another aspect that needs some attention in this project is the traffic signalization and specifics regarding what can be done at an intersection or section of railroad. The book, *Traffic Engineering* (Roess et al., 2004), goes into these specifics. Although the specifics are not in great detail, they certainly are very helpful when tackling this particular task. The signalization times, design standards and regulations, and basic ideas to help create an ideal warning system are listed in this book.

When it comes to ideas on how to communicate with the drivers of the vehicles, the book, *Intelligent Transportation Primer* (The Institute of Transportation Engineers, 2000) has a lot of good ideas. There is everything from a simply LCD display sign strategically located along the route, to simply having the driver tune his or her radio to a particular radio station, which updates them with the information they need to know. Although there are many options, RFID may limit ones that can be used and ones that cannot be used.

* Dr. Erick C. Jones, Dr. Mehmet Eren, and Dr. James R. Gubbels have contributed equally to this chapter.

In the article, "Safety warning based on highway sensor networks," the authors go on to talk about a proposed system in which various sensors would be used to record and send information to inform drivers of what is happening ahead (Xing et al., 2005). They go on to say how and when a driver should be warned. Their proposed system is not limited to only railroad crossings, but can also be utilized for any traffic problem. The whole concept behind this system is not only safety, but to also eliminate any unnecessary delays.

32.2 CURRENT PROBLEM

Railroad crossings cause delays in traffic. In most places, an approaching train cannot be seen from a far enough distance that drivers are able choose an alternative route. Instead they head to the crossing and are forced to wait. In many situations, cars get trapped with no way out.

Freight trains generally have more cars and are much longer than passenger trains. Their speed tends to also be relatively low. They are the biggest part of the train network in the United States. Most freight trains have 100 cars on average, and each car is 51.51 ft (15.70 m). For safety purposes and certain laws, trains lower their speed when they get close to a crossing, which creates longer times for the cars to wait at the crossing. The average passing time through a crossing for freight trains is about 4.5 min. In some crossings, this time can be up to 12 min due to weather and conditions of the trains.

Many railroad crossing accidents are caused by the cars and pedestrians who get to a closed railroad gate and do not obey the traffic signalization for various reasons. Nearly half of the total crossing accidents are caused at the railroad crossings that have warning devices like the stop signs, advance warning signs, and pavement markings. Many of these accidents occur when the lighting is poor and the people cannot visually see the train coming or they simply misjudge its location. Even with many new devices and methods being tested, accidents are still regularly occurring and a need for a new system is evident.

32.3 REASON FOR IMPROVEMENT

The objective of the proposed system is to provide a warning system to traffic and prevent unnecessary delays. RFID can be used in place of other proposed systems, such as using radar sensors and underground sensors. These systems do not allow for flexibility and sometimes do not provide accurate information to the control center. Radar systems tend to be sensitive to the environment. Very cold weather or heavy rain can alter its functionality.

RFID eliminates this issue and provides extended information to the control center. This allows them to organize the traffic flow more efficiently. In doing this, not only does it organize traffic, but it also eliminates the unwanted backups at the crossings.

Another possibility, besides being a tool in a warning system, RFID may also be used for nationwide railcar tracking network to follow the movement of any train equipped with RFID system. There is also a possibility to use RFID for tracking passenger trains and for providing their expected arrivals/departures. This information can be updated on Internet, so the people who are going to use the train will be informed by using the Web site.

Being able to track railcars in real time opens the door to better supply chain management (SCM). Information about the location of supplies or finished goods could potentially speed up the entire production process and provide better customer service. This concept is not limited to only railcar tracking. It also can be applied to trucks, airplanes, and any other means of transporting goods.

The RFID tracking systems like every system has pluses and minuses. When all these are taken into consideration, RFID comes out on top as a great tool to solve the problem of traffic backups considering its other potential uses.

32.4 STRATEGY

The proposed RFID system would use active tags to send and record any necessary information. The rest of the RFID system would need to be task oriented and specific.

The reader, which reads the signal of the active tag mounted on the train, needs to send the tag data to traffic signal control center. This task could be accomplished a few different ways. First, a copper twisted pair cable line can be used to connect the reader to the control center. By using the cable, the cost would be more reasonable in the long term, in comparison to other methods. Also, communications cut offs would be minimized due to the cable's reliability. However, this method requires setting the cable underground, which would take a lot of time and it does not provide any flexibility for any kind of system changes if some conflict would arise.

Another possible method of communicating with the control center to consider is using a wireless technology. Bluetooth 802.11 technology is a wireless technology that connects the electronic devices wirelessly. Low required power can allow using sun light as energy source for this method, but because of current distance limitations of the Bluetooth 802.11 technology, it would not be suitable for this application.

This leads to another possibility, a global positioning system (GPS). The current space-based telecommunications environment is characterized by satellites orbiting the earth at various distances. The signals can be transmitted and received with GPS devices by using the satellites. This kind of communication is expensive and it requires the user to be registered in the network all the time. This potentially rules GPS out as an option for the final system design.

The final option is broadcasting the signal from the reader to the control center. A radio station uses small portion of the radio spectrum for broadcasting audible data. Radio data systems (RDS) use a portion of the unused spectrum called a subcarrier to transmit information. An RDS receiver receives the signal and then decodes it, or translates the information to text or audio information. This makes broadcasting a feasible option for the proposed system.

32.5 METHODOLOGY

32.5.1 Testing the Active Tags

The experiment consisted of an RF code mobile reader installed in a HP Palm Pilot, RF code active tags with a frequency of 303 MHz, and a car. The idea was to test the reader's ability to read while the tags were in motion much like a train would be. If the tags read consistently, then the system would be feasible. The test consisted of five active tags placed on the car at select locations. Then the car would drive past the stationary reader and the data could be recorded for various speeds of the car.

32.5.2 Test Results

The car was tested at speeds of 0, 10, 20, 30, 40, and 50 mph. Each time, the reader would capture certain tags based upon their location and also the speed of the car. The tag placed on the nose of the car proved to be the only tag that read every single time. The tags placed on other locations read based on the speed. The tags were numbered 1 through 5 with Tag 1 being the tag on the nose of the car. Tags 2, 3, and 4 were placed on the car's windshield and Tag 5 was located inside the car. Since there was always one tag reading ever single test, it can be concluded that speed is not a factor and that using active tags in the system is feasible.

32.5.3 Finding a Practical Way to Send Data to the Control Center

As mentioned before, each RFID reader placed around the crossings needs to communicate with the traffic control center. This can be established by wire or wireless communication. In the case of placing the readers around the crossing and assuming the traffic control center is to be no more than 4 mi away from the reader location, a radio transmitter or radio modem would be a good solution. The fact that the readers are placed outside and they communicate with the control center in open environment with no line of sight requirement makes the process easier. The frequency of 458 MHz allows a communication range up to 20 km in free space. A transmitter and receiver with that frequency are recommended to connect the RFID to the control center.

32.5.4 Determining the Appropriate Layout to Set the Devices

Since the connection from the reader to the control center can be made wirelessly, a certain tolerance for the distance of placing the readers must be considered. This tolerance allows the control center to organize the routes properly for specific cases. For the proposed system, RFID readers are to be placed about 1.5 mi away from the crossing.

32.6 SIMULATION RESULTS

To demonstrate a case regarding the wait time of the cars, SIMUL8 software is used. In the simulation

4000 unit = 1 min
66.667 unit = 1 s
Average passing time for a train = 5 min* (default)
Train speed = 30 mph (default)
Average wait time for a car due to closed crossing = 3.8 min
Average number of cars waiting = 73.6

From this result, each car looses 3.8 min depending on the time of the day, and there are approximately 74 cars involved this loss. Therefore, each time the crossing is closed, the average total time lost is 4.29 h. This might not represent any variable that can be used in the calculations, but it makes sense if some of the drivers are in duty and using their working time.

There are also some accidents that occur concerning railroad crossings. These accidents can be caused by signalization errors or the cars that do not obey the signalizations. An early warning system might be able to reduce these accidents. The idea is to divert traffic so that there are fewer backups. If traffic is diverted, there is less of a chance of an accident caused by a driver not being alert or impatient.

32.7 COST ANALYSIS

The essential equipment for an RFID system is the reader and a tag. As mentioned earlier, the recommended tags need to be active, which are more expensive than passive tags. The range for active tags is $35–$142. The tag that has the highest read range of 100 ft is about $140. Each train requires one of these active tags costing $140.

Because of the environment, the reader has to be durable, but at the same time it should not be costly. There are many kinds of active readers in the market that cost anywhere between $1200 and $7000. The reader chosen for the proposed system costs $1480.

If we look at the average total cost for a crossing including the other equipment required, the following are found:

The reader: $1,200–$7,000
Active tags: $35–$142
Transmitter–Receiver: $300–$800

* Based on the interview with Bhaven Naik, PhD std.—Mid America Transportation Department, Lincoln, Nebraska.

Electronic sign: $80–$600
Labor: $700
Average total cost for the proposed system: $4100 (two readers— tags not included)
Periodic cost: $350 each year (maintenance)
Total cost for 5 years: $5850
If the system were to be implemented at every intersection, not including tag cost:
280,000 crossings * $5,850 = $1,638,000,000

32.8 RECOMMENDATIONS

RFID readers can be connected up to four antennas. Taking advantage of this fact, instead of using a reader at each crossing, a reader antenna can be placed by itself at certain crossings. If two or three crossings are close enough to each other, only one reader would be necessary with the use of antennas. In this case, the communication can be established via special antenna cables that vary depending on the kind of the reader. This would in turn affect the cost analysis depending on what gets implemented.

An alternative solution could be to connect the reader directly to the traffic signs. This system would bypass the control center. Each warning sign in this case needs to have a radio transmitter to get the information from the reader directly. Although a cheaper alternative, this solution would take away from some of the benefits that RFID could create.

32.9 CONCLUSION

An RFID-based system can replace existing sensor systems with affordable cost and more effectiveness. The proposed RFID system has capability of sending more information about the train other than basic data such as the trains speed. The main question before the experiment had been, can the readers read tags in motion? The test results show that active tags are able to be read when they are at twice the expected speed of a passing train at an intersection. This makes the usage of RFID in a system a reality. A radio transmitter for communication between the reader and control center or directly to the electronic signs is the ideal recommendation for the proposed RFID system.

The primary concern of this study is saving time and avoiding backups. However, there is the definite possibility that the proposed system may help to prevent accidents. The total cost for the proposed system per crossing is $5850. The system could be paid for in several ways depending on who is using it. In terms of a warning system, if it can prevent accidents from happening or even save the right person time, the potential savings begin to add up. In terms of the other possibilities, if a train can be tracked in real time, then SCM becomes faster. Also, RFID may be able to help increase passenger traffic on passenger trains if a person could check the train's status in real time. The possibilities are endless and a warning system using RFID is only the beginning. The initial cost may be large, but the potential savings and usage help tip the scale back into the proposed systems favor.

33 Six Sigma

33.1 SIX SIGMA

Six Sigma (SS) methods are recognized by many companies as a means for reducing defects, increasing company productivity, and improving company profitability. The label "Six Sigma" originates from statistical terminology. In statistics, the sigma (σ) represents standard deviation. For a normal distribution curve, the probability of falling within a plus or minus SS from the mean is 0.9999966. This is more commonly expressed in production processes as a defective rate for a process, which will be 3.4 defects per million units. Thus, SS promotes high degree of consistency by designing operations with extremely low variability. Statistically, the goals for SS methodologies are to reduce operational variation to achieve small process standard deviations.

The term "Six Sigma" is credited by a Motorola engineer named Bill Smith. SS is a trademark of Motorola. In 1980s, Motorola developed the SS standard and the methodology associated with it. The company has documented over $16 billion in savings through SS. Companies such as GE have utilized the methods to reduce costs and improve business profitability.

33.2 SIX SIGMA METHODOLOGIES

As seen before, there are a few different methods within SS, which can be utilized depending on the process that needs to be improved. Operational SS and Design for Six Sigma (DFSS) are two of the most common methodologies. Operational SS is commonly attributed to the steps in the process DMAIC. DMAIC is the acronym used to describe the define, measure, analyze, improve, and control process; it includes the following elements:

- Define the project goals and customer (internal and external) deliverables
- Measure the process to determine current performance
- Analyze and determine the root cause(s) of the defects
- Improve the process by eliminating defects
- Control future process performance

33.2.1 Design for Six Sigma

DFSS has been publicized as a new product or process, development approach that follows operational SS integrations. Some suggest that DFSS cannot be utilized without some maturity with DMAIC processes, tools, and techniques. Others see this methodology as a stand-alone process that does not require prior knowledge of DMAIC. These alternating views have led to multitude of DFSS methodologies.

The major objective of DFSS is to "design it right the first time," to design or restructure the process in order for the process to intrinsically achieve maximum customer satisfaction and consistently deliver its functions, to avoid painful downstream experiences. Therefore, it is an upstream activity. The term "Six Sigma" in the context of DFSS can be defined as the level at which design vulnerabilities are not effective or minimal.

The best way to differentiate between DMAIC and DFSS is to use the former for improving existing products and services, and the latter for new products and services. DFSS contrasts with

TABLE 33.1
Process Comparison of Six Sigma Methodologies

Six Sigma Methodology	DMAIC	DFSS			
		DMADV	IIDOV	CDOV	DFSSR
Scope of application	Process improvement	New process and product development	Technology development	Product design	Environmental testing
Plan	Define, measure	Define, measure	Invention, innovation	Concept	Define, measure
Predict	Analyze	Analyze, design	Develop	Design	Analyze, design, identify
Perform	Improve, control	Verify	Optimize, verify	Optimize, verify	Optimize, verify

DMAIC in that the phases or steps of DFSS are not universally recognized or defined. It is more an approach than a methodology and typically used to design or redesign a product or service from the beginning.

Many organizations, enthusiastic to build on SS momentum, generated their own DFSS processes before a standard template emerged; therefore, organizations have adopted a variety of approaches that resulted in acronyms such as DMADV, IIDOV, CDOV, IDOV, DCOV, and IDEAS. Some of the major practices of DFSS are showed in Table 33.1.

DMADV includes the following elements:

- Define the project goals and customer (internal and external) deliverables
- Measure and determine customer needs and specifications
- Analyze the process options to meet the customer needs
- Design (detailed) the process to meet the customer needs
- Verify the design performance and ability to meet customer needs

Pyzdek's *Six Sigma Handbook* (2003) describes the differences between the two commonly implemented forms of SS known as DMAIC and DMADV. DMAIC is the acronym for the process improvement methodology that follows the define, measure, analyze, improve, control steps. DMADV is the acronym for the new product (or process) development methodology that follows the define, measure, analyze, design, verify methodology. The major differences are that DMADV focuses on measuring customer requirements in order to design a product that meets their demand whereas DMAIC focuses on determining what the root causes of defects are. DMAIC also focuses on controlling the process changes implemented whereas the final step in the DMADV is the verification that customer requirements are met.

33.2.2 Lean Six Sigma

Lean production has been around much longer than SS. This quality approach was heavily used by Henry Ford and has roots in the Toyota Production System. Lean concepts focus on eliminating waste and have shown significant success in manufacturing. The Lean philosophy consists of many tools such as 5-S, just-in-time (JIT), etc. As SS has been evolving, many companies have added a Lean aspect to this management strategy in order to develop Lean Six Sigma (LSS). The combination of Lean and SS leads to a common principle: "The activities that cause the customer's critical-to-quality issues and create the longest time delays in any process offer the greatest opportunity for improvement in cost, quality, capital, and lead time" (George 4). The methodology of LSS is the same as Operational SS, but it incorporates Lean tools as well.

33.3 SIX SIGMA AS AN INDUSTRIAL ENGINEER

The courses offered in the industrial engineering discipline enable the engineers to handle all aspects of the Six Sigma methodology; therefore, to fully receive all the benefits of SS, one might want to employ an industrial engineer (IE) to handle the heavy statistical calculations required by the methodology. There are several indications that an IE is the best person to aid in a SS project. The data collections can be handled by the education in time study and work management. The procedure to set up the experiments is helped by the study of design of experiments (DOE). Engineers are educated on the statistical evaluation tools in courses such as multivariate statistics and probability.

33.3.1 Roles in Six Sigma

The traditional roles of an IE within the SS hierarchy are black belt, master black belt, or champions. The academic education of an IE enables them to perform these roles. Through additional education, these IEs are to become champions, statistical evaluators, economic justifiers, and variety of the roles as required by the project. Figure 33.1 displays the typical SS roles inhabited by IEs.

33.3.2 Industrial Roles

IEs who are trained in engineering and additional courses are enabled to act as a change agent. The various tools used in LSS are part of the IE curriculum. Many IEs suggest that LSS is nothing more than using IE tools to focus on financial performance instead of efficiency improvement. The versatile nature of the IEs helps them to perform in any part of the LSS projects. Also, many tools used in the SS like flowchart, cause and effect, cause tree effect diagram, etc. are considered as non-IE tool, but actually they are part of the IE academic discipline.

33.3.3 Six Sigma/Industrial Engineering Interface Framework

A Six Sigma/industrial engineering interface Framework (SS/IE IF) can be introduced to describe a new discipline. The framework describes three components that relate SS initiatives with industrial engineering. The components include

1. Six Sigma industry interface
2. Six Sigma industrial engineering academic interface
3. Six Sigma industrial engineering specialized knowledge interface

We describe those components and demonstrate how the framework shows the value of using IEs in all of the SS roles. *The Six Sigma interface* shows the DMAIC process and the specific tools that are

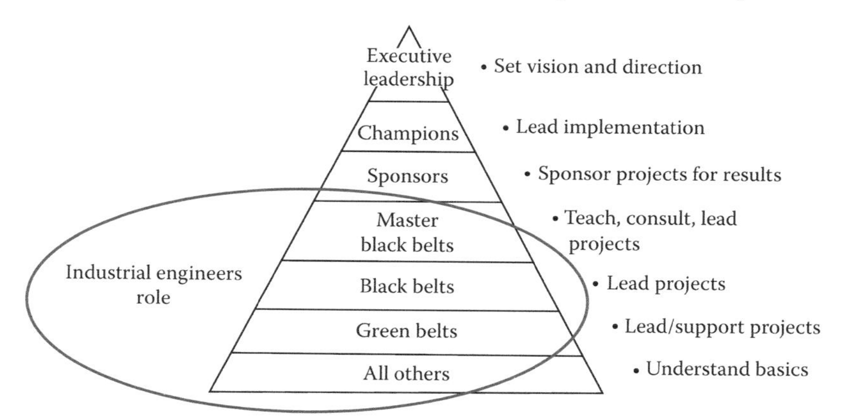

FIGURE 33.1 Role of industrial engineers in SS projects.

TABLE 33.2
Six Sigma Industry Interface

Define	Measure	Analyze	Improve	Control
Interview process	Hypothesis testing	Process capability analysis	Factorial design of experiments	Control plans
Language processing	Analysis of variance	x–y map	Fractional factorials	Visual systems
Prioritization matrix	Quality function deployment	FMEA	Data mining	5-S
System map	Flow down	Multi-vari chart	Blocking	TPM
Stakeholder analysis	Measurement system analysis	Chi-square	Response surface methodology	Mistake-proofing
Though process map	Graphical methods	Regression	Multiple response methodology	SPC/APC
Value stream mapping	Process behavior charts	Buffered tolerance limits	Theory of constraints	

utilized in those steps. These tools have been envisioned to support the successful implementation of SS projects. Some IEs suggest that most of the tools described are generally taught in IE undergraduate and most IE and engineering management curriculums. Table 33.2 shows tools that allow industry to execute SS. Typically, the Lean aspects of SS or the Lean in LSS are demonstrated in the control phase. Specifically, the Lean tools include 5-S, TPM, mistake-proofing, visual systems, and value stream mapping.

The Six Sigma industrial engineering academic interface describes the general academic classes that cover the topics and theory that support the correct use of the SS tools. The premise is that to correctly use tools such as regression, multivariable analysis, and statistical process control (SPC), it is important to have foundational knowledge of calculus-based statistics. Classes such as engineering statistics, applied statistics, and statistical quality control allow for mastery of these tools. Oftentimes, many experts who criticize SS identify that most statistical components are not correctly used or understood, which in turn minimizes the effectiveness of these tools. Utilizing IEs in the execution of these tools would provide a higher probability of using these tools correctly and better results for SS projects and initiatives. Table 33.3 shows interface between the industrial engineering academic curricula and the SS process. The course names used in this study are typical course names and are used to represent courses that are typically available in IE undergraduate and graduate programs. Additionally, some courses such as project management, and advanced classes in engineering economics may be provided as a graduate class in IE departments. These courses would teach the student how to perform a stakeholder analysis commonly used in the define phase.

TABLE 33.3
Six Sigma Industrial Engineering Academic Interface

Define/Academic Discipline	Measure/Academic Discipline	Analyze/Academic Discipline	Improve/Academic Discipline	Control/Academic Discipline
Project management, engineering management, industry quality control, engineering economy	Applied statistics and quality control, engineering statistics and data analysis	Industry quality control, engineering statistics and data analysis, applied statistics and quality control	Quality engineering: use of experimental design and other techniques, engineering statistics and data analysis	Industry quality control, applied statistics and quality control

TABLE 33.4
Six Sigma Industrial Engineering Specialized Knowledge

Special Knowledge	Industrial Tools	Academic Courses
Statistics	Multi-vari chart, chi-square, regression, factorial design of experiments, measurement system analysis	Applied statistics and quality control, engineering statistics and data analysis
Lean	5-S, TPM, mistake-proofing	Industry quality control, applied statistics and quality control
Leadership	Value stream mapping, though process map, stakeholder analysis, system map, interview process, prioritization matrix	Project management, engineering management, industry quality control, engineering economy

The Six Sigma industrial engineering specialized knowledge interface as shown Table 33.4 highlights the special knowledge of the IE. IEs can also participate in SS in specialized roles that can support SS initiatives. Some SS organizations create specialized positions that assist them in successfully implementing SS. Some of the positions include statistics specialist, Lean specialist, and leadership liaison (or change agent). Separating these roles from traditional SS roles provides the organization the ability to supplement their SS process. An additional role not shown is economist, which is a role in which the specialist evaluates and verifies the earnings or savings for the project. This role usually works in close partnership with accounting.

Many courses in the IE discipline provide the foundation for various roles in SS initiatives that are implemented in many companies. The foundational knowledge that is important in SS is supported by the coursework provided in most IE curriculums. The typical roles that are associated with SS can be better fulfilled by one who has been through an IE curricula or IE body of knowledge. The industrial engineering/Six Sigma interface framework shows how the SS roles utilize the IE body of knowledge in order to use the standard SS Tools. Through the use of IEs, the tools of SS are more likely to be utilized correctly.

33.4 SIX SIGMA IN LOGISTICS: A CASE STUDY

Throughout the book, we have discussed certain aspects of logistics that are important in the understanding of how the supply chain works. It is important to understand how SS fits within the framework of the supply chain. Through this case study, one can identify the critical areas of the supply that are impacted by the SS methodology.

33.4.1 Typical Issue within the Supply Chain

Technical organizations often face the challenge of aligning their supply chain. The engineering manager faces challenges in coordinating data collection and analysis efforts to evaluate the supply chain in a cost-effective manner. In some organizations, it may be prudent to utilize current technical personnel to perform this analysis. Oftentimes, companies consider utilizing costly software and consultants prior to using their in-house resources. Allowing the engineering manager to utilize an internal team to provide an analysis is more cost effective for several reasons:

1. Data collected will be utilized again if consultants are deemed necessary
2. The in-house team will understand the implications of solutions that the model may provide and can make adjustments for reality
3. Simplified assumptions can be agreed upon by internal stakeholders
4. The project will prepare personnel for change
5. The project provides a cost-effective solution

Also, this study will reveal if your supply chain network may be too complex to model using simply the Excel solver prior to investing in an extensive study. Though it is very important to perform supply chain analysis, many companies cannot justify the use of expensive software and consultant to perform these analyses continually. The engineering manager can provide good solutions by creating this type of study.

Previously, a project team of students and faculty from the University of Nebraska-Lincoln and material management personnel from a city located in the Southwest, United States began a SS project to reduce obsolete inventory. The supply chain consisted of a network of warehouses, storerooms, suppliers, and the internal end user that represented the customer. During the SS process improvement study, the team determined that customer service needs were not being met, obsolete inventory was being driven by purchasing behavior, and that facility costs could be reduced with facility consolidations. The team analyzed the supply chain network of the city's public works department using modeling techniques to recommend which warehouses could be consolidated. Based on recommendations, 96,000 ft^2 could be reduced and gross of $3.5 M would be saved over 5 years not including taxes and depreciation. This represents a cost reduction of 25%.

33.4.2 Background

An engineering manager's goal when locating facilities and allocating inventory should maximize the overall profitability of the resulting supply chain network while providing customers with adequate service. Traditionally, revenues come from the sale of product and costs arise from facilities, labor, transportation, material, and inventory holding. Ideally, profits after tariffs and taxes should be maximized when designing a supply chain network. In this scenario, the city government does not pay taxes or collect revenues, so their goals were to minimize overall operating cost and still be responsive to the customer.

Trade-offs must be made by the engineering managers during network design. For example, building many facilities to serve local markets reduces transportation cost and provides fast response time, but it increases the facility and inventory costs incurred by the firm. Engineering managers can use network design models in two different situations. First, those models are used to decide on locations where facilities will be established and the capacity assigned to each facility. Second, these models are used to assign current demand to the available facilities and identify lanes along which product will be transported. Managers must consider this decision at least on an annual demand basis, prices, and tariff charge. In both cases, the goal is to maximize the profit while satisfying customer needs. The following information must be available before the design decisions can be made:

1. Location of supply sources and markets
2. Location of potential sites
3. Demand forecast by market
4. Facility, labor, and material cost by site
5. Transportation costs between each pair of sites
6. Inventory costs by site as well as a function of quantity
7. Sales price of product in different regions
8. Taxes and tariffs as product is moved between locations
9. Desired response time and other service factors (Chopra, 2004)

Given this information, a choice of model type can be made. The capacitated plant location model seeks to minimize the total cost of the current supply chain network; the problem is formulated into an integer program. The gravity location model's goal is to locate an optimal location based on cost inputs. Beyond optimization models, the engineering manager could build a simulation of their supply chain (Chang, 2004).

In this study, the capacitated plant location model (Chopra, 2004) in order to determine the minimal number of facilities that could hold inventory and meet customer demand. In our study, the city has chosen to consolidate warehouse facilities. Management is questioning whether all 12 facilities are necessary. The goal is to formulate the model to minimize total costs taking into account costs, taxes, and duties by location. Given the taxes and duties do not vary between various locations, and that the city does not pay taxes, the team decided to use the existing facility locations and allocate demand to the open warehouses to minimize the total cost of facilities, transportation, and inventory.

33.4.2.1 Capacitated Plant Location Model

A linear programming model, similar to that seen in the aggregate planning chapter, allows for a nice way to solve the network optimization problem. The capacitated plant location network optimization model requires the following inputs:

N is the number of potential locations
M is the number of demand points
D_i is the annual demand from market i
K_i is the potential capacity of plant i
F_i is the annualized fixed cost of keeping factory i open
C_{ij} is the cost of producing and shipping one unit from factory i to marker j (cost includes production inventory, transportation, and duties)

and the following decision variables:

Y_i is the 1 if plant is open, 0 otherwise
X_{ij} is the quantity shipped from factory i to market j

The problem is formulated as the following integer program:

$$\text{Min}\left(\sum_{i=1}^{n} F_i Y_i + \sum_{i=1}^{n}\sum_{j=1}^{m} C_{ij} X_{ij}\right)$$

Subject to

$$\sum_{i=1}^{m} X_{ij} D_{ij} \quad \text{for } j = 1\ldots m \tag{33.1}$$

$$\sum_{j=1}^{n} X_{ij} \le K_i Y_i \quad \text{for } i = 1\ldots n \tag{33.2}$$

$$Y_i \in (0,\ 1) \quad \text{for } i = 1\ldots n \tag{33.3}$$

The objective function minimizes the total cost (fixed + variable) of setting up and operating the network. The constraint in Equation 33.1 requires that the demand at each facility market be satisfied. The constraint in Equation 33.2 states that no plant can supply more than its capacity. (Capacity is 0 if closed and K_i if it is open. The product of the terms $K_i Y_i$ captures this effect). The constraint in Equation 33.3 enforces that each plant is either open ($Y_i = 1$) or closed ($Y_i = 0$). The solution will identify the plants that are to be kept open, their capacity, and the allocation of regional demand to these plants. The model is solved using the Solver tool in Excel (Chopra, 2004).

33.4.3 Network Modeling Steps Incorporated into a Six Sigma Service Project

The typical SS DMAIC approach was used with the addition of a network model within the analyze phase. DMAIC stands for define, measure, analyze, improve, and control. These are the steps in a standard improvement model for a SS directed project.

33.4.3.1 Define

The main work in the define phase is for the team to complete an analysis of what the project should accomplish and to confirm their understanding with their sponsor(s). They should agree on the problem, understand the project's link to corporate strategy and its expected contribution to ROIC, agree on the project boundaries, and know what indicators or metrics will be used to evaluate success. The last two issues often prove particularly important in service environments (George, 2003). The problem defined for this project was to reduce obsolete inventory.

33.4.3.2 Measure

One of the major advances of SS is its *demand* for data-driven management. Most other improvement methodologies tended to dive from identifying a project into improve without sufficient data to really understand the underlying causes of the problem. The measure phase is Six Sigma's stage for data collection and "measuring" the problem. This phase is generally broken into several steps, including establishing baselines, observing the process, and collecting data (George, 2003). The measure of success was reducing the percent of obsolete inventory in the supply chain.

33.4.3.3 Analyze

The purpose of the analyze phase is to make sense of all the information and data collected in measure. A challenge to all teams is *sticking to the data*, and not just using their own experience and opinions to make conclusions about the root causes (George, 2003). There are many tools available in the analyze phase, including network modeling. Network models provide a rich and robust framework for combining data, relationships, and forecast from descriptive models. They provide managers with broad and deep insights into effective plans, which are based on the company's decision options, goals, commitments, and resource constraints (Shapiro, 2001). After using regression analysis and design of experiment analysis, the team chose to use supply chain optimization for a more robust solution.

The network model used within this project followed several steps, including

1. Collect input data and establish baseline
2. Set optimization constraints
3. Run alternatives with the capacitated plant location model (Chopra, 2004)
4. Show alternatives in revenue, savings, and customer service
5. Select an alternative

These steps led to an alternative that minimized the cost of the supply chain. This alternative then directs the tasks within the improve stage.

33.4.3.4 Improve

The sole purpose of the improve phase is to make changes in a process that will eliminate the defects, waste, costs, etc., that are linked to the customer needs identified in the define stage (George, 2003). The improve stage differs for every SS project. The common underlier is that the improvements should be centered on the largest issues found in the analyze phase. The recommendations for consolidating facilities (the supply chain model recommended) and using a more robust criterion for eliminating outdated inventory were recommended for the improvement.

33.4.3.5 Control

The purpose of control phase is to make sure that any gains made will be preserved, until and unless new knowledge and data show that there is an even better way to operate the process. The team must address how to hand off what they learned to the process owner, and ensure that everyone working on the process is trained in using any new, documented procedures. Six areas of control phase are critical: document the improved process, turn the results into dollars, verify maintenance of gains continually, install an automatic monitoring system, pilot the implementation, and develop a control plan. Key performance indicators were identified to be tracked with SPC charts for the following year. This is further elaborated in the results section.

The DMAIC process with the capacitated plant location model in the analyze phase was utilized to study the city's public works warehousing operations.

33.4.4 Case Description

33.4.4.1 Organizational Description

The organization used for this case study is a city in Southwest United States, public works, materials management branch (MMB). The MMB is responsible for the processing and coordination of all procurement and contract-related activities as well as warehousing and distribution of all general inventory items for the department.

The branch facilitates purchases ranging from pipes for restoration of sewer lines to computers and traffic signs. To promptly obtain goods and services, the department utilizes in excess of 800 commodity and service contracts. The branch is divided into three functional sections: procurement, contract management, and warehousing and distribution. This study was centered on the warehousing and distribution section.

The MMB has the responsibility for warehousing and distribution of general and automotive inventory items, from cradle-to-grave, for the department. Two central depots serve as staging locations for inventory that is distributed to a network of 10 general supply warehouses, 9 automotive warehouses, and many storerooms located throughout the city. The inventory consists of a variety of items, e.g., pipe, valves, fittings, office and janitorial supplies, etc.

33.4.4.2 Project Description

The MMB had been audited in previous years and the audits identified opportunities for improvement in the warehousing operations. The audits identified excess obsolete inventory, need to evaluate standard operating procedures, and labor productivity. Obsolete inventory is defined as inventory that has not had any requests for disbursement for over 1 year.

The current system contains 12 warehouses and 28 storerooms with an ongoing cost of $14.94 M. Upon inspection, it was estimated that the warehouses have a maximum of 30% space utilization. The current supply chain is shown in Figure 33.2. The city's public works owned $10.1 M of inventory within the MMB warehouses. The inventory that was deemed as obsolete was valued at $3.6 M or 35% of the total inventory.

The following modeling steps were followed to complete the analysis:

1. *Collect input data and establish baseline*: The current supply chain information was collected to form the input data for the network model. The inputs included costs for electricity, gas, data lines, and labor. Also, holding and transportation costs were estimated for each facility. The warehouses do not pay taxes or water costs since they are in a city building, so information as to lost water sales and lost taxes were also captured and used in the cost equations.
2. *Set optimization constraints*: The optimization constraints included the size limitations of each facility and the future demand at each facility. The facility size was collected from operations. The future demand on inventory was estimated to be the same as last year's value.

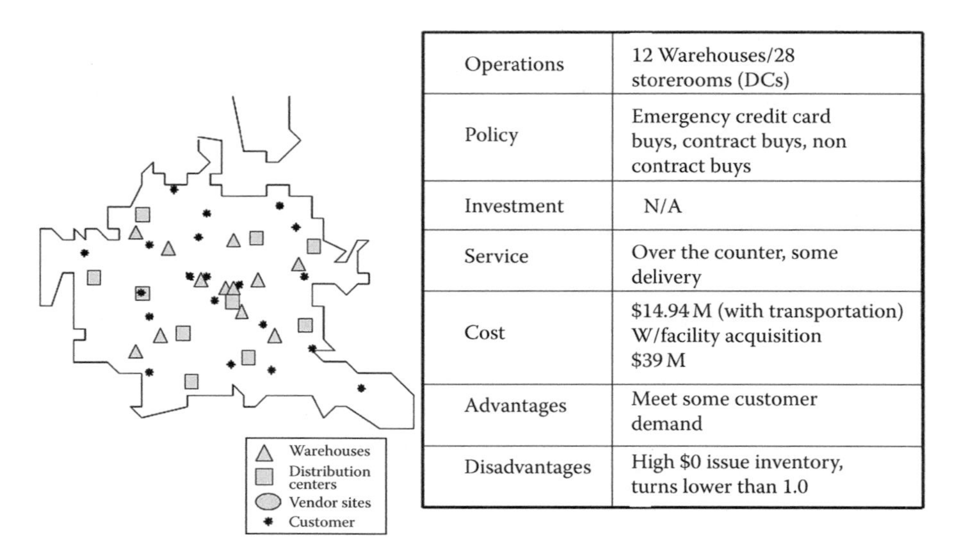

Operations	12 Warehouses/28 storerooms (DCs)
Policy	Emergency credit card buys, contract buys, non contract buys
Investment	N/A
Service	Over the counter, some delivery
Cost	$14.94 M (with transportation) W/facility acquisition $39 M
Advantages	Meet some customer demand
Disadvantages	High $0 issue inventory, turns lower than 1.0

FIGURE 33.2 Current supply chain.

3. *Run alternatives with the capacitated plant location model* (Chopra, 2004): The costs and data that were collected in steps 1 and 2 of the modeling were input into Chopra's model in Excel and the Solver Add-In was utilized to run alternatives of the least cost model.
4. *Show alternatives in revenue, savings, and customer service*: The different alternatives were then evaluated for revenue and savings with an ROI calculation assuming the project had a 5 year life. The customer service provided in each alternative was evaluated by a team from operations. Savings were associated with space reduction; taxes and depreciation are associated with closing facilities. The savings for closing the facilities was limited due to the fact the profit of selling the property was not included, only the savings of eliminating long-term leases. The city does not pay taxes on the property, so the savings were minimal compared to those in other industries. The team felt this provided a conservative and acceptable estimate for cost savings in governmental operations. In other industries, the taxes, depreciation, and probability of sales profit would be estimated in potential savings.
5. *Select an alternative*: The optimal solution contains two warehouses, which are centrally located as shown in Figure 33.3. This gives a reduction of 96,000 ft^2, which translates into $3.5 M over 5 years. This solution will increase the space utilization to 65% and reduce the obsolete inventory to 10% of the total value held within the warehouses.

As the city moves to the optimized model, the control phase of DMAIC will keep the improvements in place and running smoothly. The metrics that are given to continue the control are the key performance indicators given in Table 33.5. These should be measured and tracked utilizing SPC. These data could then be used to repeat an optimization in the future.

33.4.5 Lessons Learned

The lessons learned included model complexity changes and challenges and limitations, which could be better met. The model complexity was chosen to reflect a first look at the supply chain and a simple optimization. A more complex model may have been used if better original data had been available. The data that were available could not be validated because they came from the enterprise resource planning (ERP) system which was antiquated. The model was validated with site tours and sampling for volumetric data. We note that the model is only as good as the data it is provided.

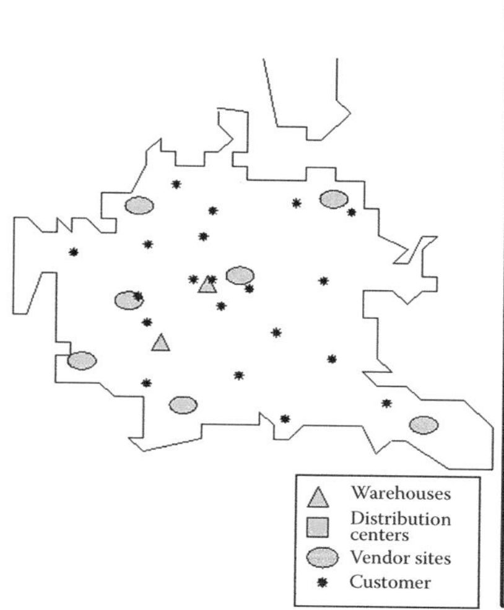

Operations	Two warehouse, no DCs
Policy	Vendor credit card, emergency credit card, contract, non-contract, internet
Investment	Warehouse management system (WMS), consolidation, delivery operation cost
Service	OTC, delivery service to site
Cost	Additional fleet maintenance, new racking, project manager
Advantages	Vendor managed inventory on noncritical fast moving items, critical items better managed, reduced labor, reduced facility cost

FIGURE 33.3 Optimized supply chain.

TABLE 33.5
Key Performance Indicators

Category	Metric	Definition
Service	Turns	Annual $ issued divided by average
Service	% obsolete inventory	# of commodity codes (CC) that have not been issued for over 1 year divided by the total # of CCs
Cost	Cost/pick	Total labor cost divided by total # of picks
Cost	Cost/order	Total labor cost divided by total # of orders
Asset management	Facility utilization	# of pallet positions utilized versus available
Future		
Transportation	Shipments/division	Track the number of deliveries to site for each division
Inventory	Velocity	Annual revenue/daily overhead
Labor	% productivity by area/task	Divide actual labor hours by the efficiency standard for the task and track by employee
Labor	Picks/hour	Number of pick issues and divide by pick labor hours
Purchasing	PCard spend by category	PCard spend by contract, noncontract, and emergency usage versus total PCard spend

A common term is "garbage in" to a model and you get "garbage out." If a more complex model was utilized, software other than Excel would be needed, as well as consultants.

It was difficult to reach agreement between divisions on what part of the cost data can be incorporated as reduced costs. For example, the portion of overall utility costs can be equated to warehouse space versus the other uses of the buildings today, and the percent of value can be used for holding cost because the city does not pay taxes and does not invest excess monies.

33.4.6 Implications for the Engineering Manager

An engineering manager can utilize this case study to better understand their own first steps in supply chain optimization projects. This information should encourage the manager to begin to look at their information internally, before hiring a consulting team. The first look may give a viable answer that can be implemented for increased efficiencies and savings.

An engineering manager can gather internal data and then use the methods in Excel Solver to create the appropriate supply chain model. The specific steps for using Excel can be found in different references; we recommend the steps outlined by Chopra. The engineering manager can justify many good solutions for the supply chain by further using current technical personnel. This may provide another tool for the engineering manager to justify his technical staff. If the internal team does not solve the problem, or the manager is ready for a second look at the issue, a consultant can be hired, with very little lost to the company.

This study provided an overall method for performing continuous improvement projects using the SS methodology. Further, this study shows how in the analyze stage an engineering manager can perform a supply chain analysis on current operations. The engineering manager can use this study as a guide for both.

33.4.7 Conclusions

This case study details a quick and relatively inexpensive way to perform an analysis for supply chain savings opportunities. The major benefit is that you utilize internal personnel who have been already budgeted for and may have a better understanding of operations than outside consultants. Also, the initial study may be modeled using existing spreadsheets before more costly software and consulting options are explored.

Also, the application of the SS methodology is not limited to city governments but can be duplicated in industry. This study introduced supply chain modeling as another technique that can be used in the SS methodology to help organizations improve operations. This additional technique may be considered in logistics, distribution, and other supply chain management projects which utilize transactional data to evaluate alternatives.

This simple analysis may not replace a complex supply chain analysis using some of the more advanced software that incorporates the CPLEX and ILOG modeling engines. These software companies allow for more constraints than the less powerful spreadsheet will allow. They have claimed to have 20%–50% more optimized results that may translate into more cost savings. This is contingent if consultants can better interpret data, future business strategies, and evaluate logistics networks including transportation traffic patterns.

From the case study application, we identified a workable set of challenges with lessons learned that can be valuable to organization when modeling the supply chain. The engineering manager and his team can be a valuable asset when doing both continuous improvement projects and providing valuable supply chain modeling expertise.

33.5 SIX SIGMA BEST PRACTICES

Companies often look for a "recipe for success" with the latest management initiatives such as Lean, total quality management, and business process reengineering. With these initiatives, companies hope to achieve success in their industry, as well as success when the general market is down. There are several ways that companies measure success such as profit, price of stock, or other various financial metrics. The impact on the bottom-line is a critical evaluation of the accomplishment of a management initiative. SS is a popular management strategy that affects that quality of a company's product. An explanation is needed for why some companies are successful in their SS implementation while others are not. A similar study of the evaluation of a company's successfulness was conducted by Jim Collins in his current management book *Good to Great*.

33.5.1 Good to Great

The popular management book *Good to Great* researches several companies to identify the characteristics that help a company transition from short-lived success to long-term results. Jim Collins identifies seven characteristics demonstrated by the companies that achieve long-term results: Level 5 Leadership, First Who...Then What, Confront the Brutal Facts, The Hedgehog Concept, A Culture of Discipline, Technology Accelerators, and the Flywheel and the Doom Loop (12–16). These characteristics were common attributes that Collins found in the companies that shifted from good companies to great companies. Each quality was determined through extensive research and interviews from key executives of the companies.

The methodology for evaluating successful companies and their characteristics was very systematic. Companies and comparison companies were chosen based on their cumulative returns and were observed over the same period of time to determine which companies transitioned from good to great (Collins, 219–229). Collins heavily researched the companies to find similar characteristics within the successful companies and similar characteristics within the unsuccessful companies. Using a similar methodology of employing financial metrics to evaluate companies that transition from short-term success to long-term results, one can evaluate the success of companies that implement SS.

33.5.2 Six Sigma's Effect on Profit

SS has become a popular management strategy that has provided success for some companies and failure for others. The initiative is a customer-focused strategy that focuses on improving quality while lowering costs. The methodology followed varies depending on the undertaken project, but the standard operational SS methodology is known as DMAIC. This methodology represents the five phases of SS: define, measure, analyze, improve, and control. SS takes a process that is operating at a three sigma level and increases the process to function at a much higher level, which in turn reduces the cost of poor quality. The most important benefit of this strategy is the impact SS has on bottom-line results. A typical SS project saves about $100,000; these projects last approximately 120 days, and several of these projects can be undertaken over a year. By improving cost of quality to an SS level, revenue is no longer wasted, which drives profits to increase (George, 31).

Companies that incorporate SS in their management strategy typically see benefits within the first year of implementation. The financial success of SS does not begin to materialize until after 18 months in a continual improvement implementation while an aggressive deployment sees benefits in 6–12 months; most companies realize benefits somewhere between 6 and 18 months (Snicker, 27). Figure 33.4 shows the typical time it takes a company to begin to see measurable benefits for two types of implementations: aggressive and continuous.

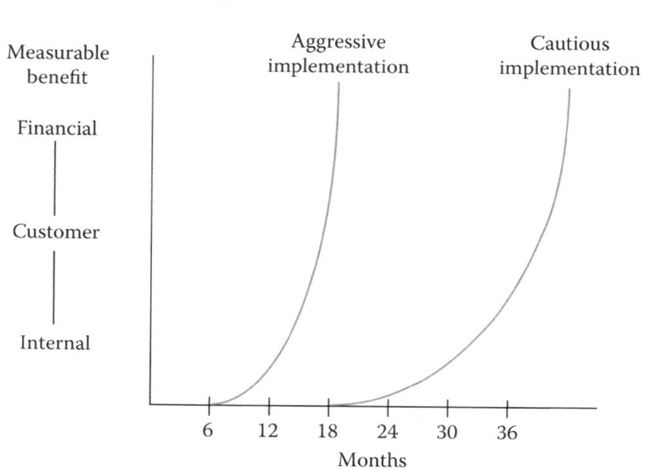

FIGURE 33.4 Effect of different implementations on benefits.

33.5.3 Six Sigma and Quality Awards and ISO Standards

SS's focus is very similar to the criteria of quality awards and ISO standards. One such quality award is known as the Malcolm Baldrige National Quality Award (MBNQA). The MBNQA is "given by the President of the United States to businesses [...] that apply and are judged to be outstanding in seven areas: leadership; strategic planning; customer and market focus; measurement, analysis, and knowledge management; workforce focus; process management; and results" (*Frequently*). The award focuses on the different perspectives of quality that make a business successful. To qualify for this award is almost as prestigious as winning the award. Using the Baldrige model coupled with SS "helps firms achieve higher levels of performance and customer satisfaction in each Baldrige category" (Mellat-Parast et al., 2007, p. 56).

Two important quality standards are ISO 9000:2008 (formerly ISO 9000:2000) and ISO 14000. These standards are not required by governments, but adhering to these standards and registering for certification demonstrate a company's need for excellent quality, and certification of these standards are recognized internationally. The ISO 9000 series consists of five specifications "designed to ensure all aspects of a business affecting product or service quality are addressed in the organization's quality program, and they are formally structured and documented in auditable form". In 2008, the ISO 9000 series was revised, but it is noted that it is the same standard (as the ISO 9000:2000) with several clarifications or modifications. Some modifications to ISO 9000:2008 from ISO 9000:2000 affect the management representative and process monitoring and measuring. The management that oversees the quality management systems (QMS) must be member of the organization, so outsiders are no longer allowed (*Praxium Research Group*, 2009). If a company wants to bring in a consultant to implement SS, this desire is no longer allowable; therefore, ISO 9000:2008 encourages the development of SS black belts in the business' management. Also, the impact of the QMS a company chooses to use must be considered when choosing a QMS (*Praxium Research Group*, 2009). No matter what type of management systems a company uses, there must be evidence that the impact the system has on quality is evaluated. According to the ISO Web site, the ISO 9000 series is "now firmly established as the globally accepted standard for providing assurance about the **quality** of goods and services in **supplier-customer relations**" (*In the global economy*, 2009).

As the world becomes more and more focused on the environment, global perspective of quality begins to extend to environmental aspects of quality. Increasing numbers of companies are making an effort to become environmental friendly as "going green" becomes increasingly popular. The second regulation, known as ISO 14000, focuses more on environmental management and responds to the changing atmosphere of quality by demanding more companies to be more environmentally conscious. The ISO 14000 standard focuses on regulating as well as improving the environmental impact of the processes of a company, and setting environmental goals that can be achieved by approaching the objectives systematically (*ISO 14000 Essentials*, 2009). Similarly to quality management techniques, it is important to strategically plan how processes can be environmentally friendly. Benefits of ISO 14000 include "reduced cost of waste management, savings in consumption of energy and materials, lower distribution costs, improved corporate image among regulators, customers and the public, and framework for continual improvement of environmental performance" (*Business benefits of ISO 14000*). Not only will registering for the standard improve quality management, but it will show other companies as well as customers throughout the world that the quality of a product is affected by the environmental factors as well.

33.5.4 Six Sigma Best Practices

Preliminary research in the RFSCL lab at the University of Nebraska-Lincoln has shown that not all SS deployments are created equal. A group of 50 companies were evaluated to determine whether there are differences in SS deployment. Results of this research have shown that unlike many management fads, SS does seem to have a long-term impact on financial results.

Within the companies that implemented SS, there remains a financial difference between companies that are successful in their deployment and companies that are unsuccessful. SS is known for its effects on profits due to its ability to lower costs and boost revenues. Companies that were successful in their implementation of SS saw profit increases after implementing SS.

Results also found that companies that increased their profits after SS implementation were more likely to be involved in quality awards, focus their mission statements around the core values of SS (customer-focused, lowering costs, improving quality), and be involved in LSS programs. The results of this study provide evidence that SS has a financial impact on companies, but SS implementation alone will not ensure the financial success of a company. Companies that were successful in their implementation were significantly more likely to incorporate other aspects into their deployment such as following MBNQA criteria, ISO 9000 and 14000 certification, company mission/values focused on the core values of SS, and Lean concepts; therefore, several other criteria must be met to ensure a successful implementation of SS strategies within a company.

33.6 SIX SIGMA IN MILITARY LOGISTICS

SS is management technique that is popular with several companies, but the initiative is not limited to the commercial industry. Since SS focuses on reducing costs, it is a valuable methodology for any organization to utilize, especially during a time such as a recession where there is a great need for cost savings. Although not yet common with the government sector, the military has utilized SS within their organization so they are able to improve their processes with the added benefit of saving costs. The importance of SS cost savings lies in the fact the government has a limited budget; therefore, resources are limited for the military. SS has the ability to provide efficient ways to make the best of those limited resources.

The U.S. Department of Defense (DOD) has adopted an LSS program that enables them to better understand the customers and improve their decisions. The Army Material Command (AMC) has utilized a Lean methodology since 2002 and by 2004, the Lean initiative evolved into an LSS program. This LSS program has allowed the AMC to realize over $110 M in cost savings. The military is already a disciplined organization, so streamlining processes with the use of LSS fits in well with their culture. The DOD has utilized the implementation of SS to achieve the highest level of cost-wise readiness.

As seen throughout the book, logistics is a very important aspect of the military. SS can be employed within the supply chain in a similar process as shown by the case study in this chapter. Utilization of SS allows for large cost savings and improved processes, which is important throughout the supply chain. Since this management technique fits well within the culture of the DOD, SS has evolved into a process not only utilized by the business sector but the government sector as well. The disciplined nature of the DOD combined with the use of IEs and SS best practices will provide a more successful implementation of SS and greater cost savings realized.

33.7 SIX SIGMA IN THE MILITARY: A CASE STUDY

The Joint Deployment and Distribution Enterprise (JDDE) must be capable of providing all command levels including regional component commanders and/or joint force commanders with the ability to rapidly and effectively move and sustain joint forces in support of major combat operations. In short, U.S. Transportation Command (USTRANSCOM) as the distribution process owner must develop and maintain the capability to insert, maintain, and sustain troops worldwide under any set of environmental conditions; provide for asset visibility in transit and in theater; and effectively manage the entire container fleet.

In support of the JDDE mission, the joint intermodal platform management system (JIPMS) capabilities-based assessment team, co-led by USTRANSCOM J5/4-AI and J6-IS recommended a need for the consolidation of existing automated information systems (AIS) into single AIS for use in managing intermodal platforms. Additionally, the business case analysis recommended that a joint J5/4-AI and J6-IS team research and evaluate the military utility of commercial off-the-shelf (COTS) advanced automated identification technology (AIT) devices (AADs) to enhance in-transit visibility (ITV) of the asset. The team defined an AAD as a non-nodal device requiring little to no fixed infrastructure.

Currently, DOD employs a variety of AIT devices to provide location, movement, and status data for cargo to support ITV; however, this collection of AIT is land based, infrastructure dependent, and limited to nodal visibility. To achieve true end-to-end visibility of intermodal assets requires augmentation to the RFID land-based status quo infrastructure. AADs could prove to be invaluable in austere areas of responsibility.

In order to minimize the risk in fielding COTS products that have not been tested to withstand the rigors imposed by the demands of the defense transportation system, a project to conduct climatologic, electromagnetic interference (EMI), and mechanical stress tests on AADs to determine their military utility is envisioned. Additionally, Phase III will identify their ability to interface with AISs and geographic information systems (GISs) to allow container managers and field commanders to track containers in austere infrastructure-deficient locales. To mitigate technology procurement and deployment risks, the team recommended testing AADs, to include satellite tags, to determine their ability to plug in-transit visibility (ITV)/total asset visibility (TAV) gaps, and support container management activities. The following outlines the proposed three-pronged testing process:

- *Phase I. Qualify tags'/military utility*: Accomplish controlled climatologic testing at a DOD facility to assess the physical integrity and transmission properties of devices when subjected to harsh weather conditions.
- *Phase II. EMRH survey and HERO certification*: The electromagnetic radiation hazard (EMRH) survey baseline ports' electromagnetic emanations and assess the AAD's ability to function within the port's electromagnetic footprint without interfering with port electromagnetic devices or being impacted by those devices. In addition, USTRANSCOM will sponsor vendors whose products successfully completed Phase I testing and the EMRH survey, for Hazards of Electromagnetic Radiation to Ordnance (HERO) certification to ensure each device can safely function when in proximity to ordnance.
- *Phase III. Field testing*: Includes operational field testing, employing containerized ammunition distribution system (CADS) containers; and integrating AADs with the JIPMS and intelligent Road/Rail Information System (IRRIS) via Radio Frequency In-Transit visibility (RF-ITV) server. Operational testing will include permanently affixing devices/tags to arms, ammunition, and explosive (AA&E) shipments to ensure location of CADS containers during sustainment operations. At this time, the team will permanently affix devices/tags to the government-owned fleet of containers. Subsequently, the Combatant Commands' (COCOM) will retrograde containers back to Continental United States depots.

33.7.1 Test Plan

To ensure scientific rigor, validity of results, procedural consistency, and compliance with MILSTD-810F, *DOD Test Method Standard for Environmental Engineering Considerations and Laboratory Tests*, engineers will develop test plans, conduct testing, interpret results, and document outcomes subsequent to each phase of testing. Should any anomalies be identified, all controlled and documented test conditions will be precisely replicated and the device(s) retested. Testing will be accomplished at a DOD facility. Test plans developed by the engineers will be based on parameters and procedures detailed in MIL-STD-810F to ensure procedural consistency and infuse the process with increased levels of scientific rigor.

In order to meet this scientific rigor, the test plan used will be the one used at the University of Nebraska-Lincoln Radio Frequency Identification Supply Chain Lab (RfSCL). The RfSCL utilize

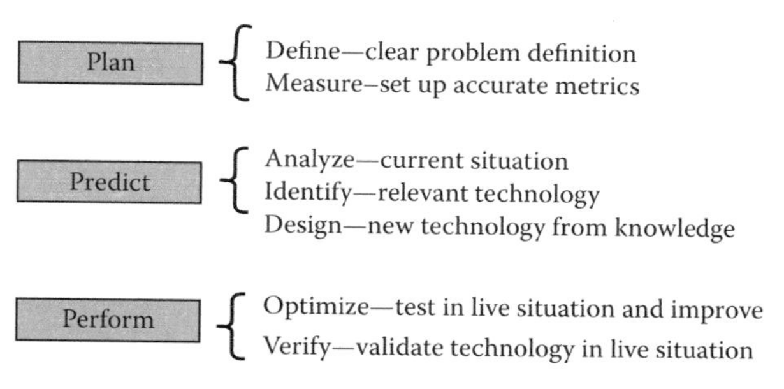

FIGURE 33.5 3P's process steps.

a plan derived within the lab called Design for Six Sigma Research (DFSS-R). It is based on a SS DMAIC (define, measure, analyze, improve, and control) approach. DFSS-R is organized into three themes: plan, predict, and perform (3P's) model is shown in Figure 33.5.

33.7.2 Plan

33.7.2.1 Define

In the define step, major test objectives are described, critical characteristics are detailed by measuring the identified metrics, and team resources are described.

33.7.2.1.1 Test Objectives

The major test objectives in this proposal are to

- Evaluate subject tags' ability to operate properly during and after exposure to varied climatic conditions in a controlled environment
- Examine the technology's ability to maintain structural integrity and accurately transmit data when submitted to mechanical stresses associated with typical container movements and storage
- Assess devices' ability to function in typical air- and seaport electromagnetic environments with no adverse impacts to port operations or no EMI from port devices with tags' functions

The metrics that will be tested are

- *Failure rate*: Rate at which devices from a given population are likely (or were found) to fail as a function of time (e.g., % per 1000h of operation). Failure rate = No. of failures/total no. of operating hours.
- *Mean time between failures (MTBF)*: Average time between failures for a particular device based on statistical or anticipated experience. Average time a device will function, expressed in operating hours, before failing. The inverse of the failure rate measured in hours: 1/failure rate.
- *Reliability*: The probability that an item will perform a required function without failure under stated conditions for a stated period of time.

33.7.2.1.2 Governmental Team

Prior to testing, the government will work with vendors to capture technical, support, maintenance, and functional data for their products; and solicit their support to provide AADs and supporting equipment for testing at no expense to the government.

33.7.2.2 Measure

On this step, the current process performance is assessed in order to have a baseline for future comparisons. Next, we use DOE approach in order to test the relevant factors that have an influence on the outcomes defined on the previous step.

33.7.2.2.1 Anti-Echoed Chamber and Other Lab Testing

Prior to baselining the products' physical, technical, and transmission parameters, devices will be inserted into a test chamber or other testing apparatus and subjected to the conditions and stresses identified in Table 33.6. For prescribed test conditions, threshold values will be incrementally adjusted to assess for impacts to structural integrity and identify any attenuation of baseline values for individual transmission characteristics, e.g., temperature increased in 10° increments to assess for structural and signal impacts. When the devices reach the prescribed time and test parameter limits, they will again be examined to authenticate deterioration in structural integrity, e.g., cracked seals, corroded electrodes, etc. Also, subsequent to each test event, engineers will employ a spectrum analyzer or other device to determine transmission property values and compare them to baseline parameters to assess levels of signal attenuation.

Devices unable to transmit or attain requisite transmission power levels will be physically inspected to catalog each failure and identify its cause(s). Test results will be documented in a phase report and integrated into a final report by a systems engineer. In addition, devices must maintain satellite coverage adequate to accurately transmit longitudinal and latitudinal coordinates IAW global positioning systems (GPS) lock criteria.

This approach provides for a high level of confidence in the military utility of the devices, validates product technical baselines, and defines product functional capabilities under adverse conditions.

33.7.2.2.2 Baseline Metrics

Prior to initiating testing activities, engineers will inspect the devices for structural defects and establish baseline values for transmission properties to include

- *Signal strength*: measure of the received radio frequency power
- *Levels of signal attenuation*: loss in power of electromagnetic signals between transmission and reception points
- *Signal-to-noise ratio*: comparison of the signal level relative to the noise level
- *Noise temperature*: measure of the amount of thermal noise present in a system or device; the lower the noise temperature, the better the performance

TABLE 33.6
Climatologic and Mechanical Stress Test Conditions

Climatologic Conditions	Mechanical Stress
High and low temperatures	Drop test
Thermal shock	Vibration
Blowing sand and dust	Shock
Rain	
Altitude	
High and low humidity	
Salt fog	
Snow	

- *Antenna gain*: measure of the change in an antenna's signal power amplification based on orientation or signal direction
- *Frequency shifts*: sudden change in the frequency of a signal

33.7.2.2.3 Design of Experiments

DOE is an approach that allows for setting up scientific testing situations effectively and it is commonly used in SS approaches. The approach allows for multiple applications that are dependent upon the number of factors (variables) that need to be tested and the level of confidence desired. The most common approaches are Latin square, full factorial design, 2^k factorial design, and 3^k factorial design. We envision a sequential DOE, which uses information learned from the first or previous experiments, to eliminate unnecessary or undesirable experimentation within a series of experiments. This method provides a powerful means to achieve breakthrough improvements in product quality and process efficiency. The DOE's testing of the equipment will be performed according to the requirements previously commented for the three phases:

33.7.2.2.3.1 Experimental Design 1 Controlled climatic environmental testing for Phase I (Qualify tags for military utility): evaluate subject tags' ability to operate properly during and after exposure to varied climatic conditions in a controlled environment.

This is the first step in a three-pronged approach to qualify satellite technology as capable of supporting TAV/ITV and container management objectives. Activities conducted during this phase will qualify devices as capable of operating under harsh environmental and stress conditions typically experienced during transportation and routine storage activities:

- Subject devices to adverse weather conditions to ensure the product maintains structural and physical integrity
- Simulate mechanical stresses generated during actual container movements and routine storage activities—vibration, dropping from a height, and shock
- Subsequent to submitting AADs to adverse weather and mechanical stresses, assess a device's capability to transmit at sufficient signal strength to trigger satellite systems

All tags submitted by vendors will conform to the following requirements:

- Each device will have a means for visually determining if it is functioning
- The device must be able to be non-obtrusively attached to the container
- Each device will have a detailed set of written programming and operating instructions
- The device must be field ready to operate with a container at time of testing

33.7.2.2.3.2 Experimental Design 2 EMRH and HERO testing for Phase II (EMRH survey and HERO certification): assess devices' ability to function in typical air- and seaport electromagnetic environments with no adverse impacts to port operations or no EMI from port devices with tags' functions.

Electromagnetic radiation is emitted from electrical and electronic devices as a by-product of normal operations causing unwanted signal interference or noise that interrupts, obstructs, or otherwise degrades effective performance. Among devices that may interfere with AADs are cell phones, wireless computers, other AIT equipment, etc. EMRH engineers will survey an aerial and water port to baseline typical port electromagnetic radiation footprints and threshold levels prior to introducing devices to the environment. Once introduced into the port's electromagnetic footprint, engineers will test to determine if AAD transmissions adversely impact port electronic devices or are unfavorably impacted by those devices. Electromagnetic radiations can overpower receivers, disrupting acquisition and tracking of satellite signals. Sources of interference can be cataloged and defined as

- *Incidental interference*: A device is unable to distinguish a desired signal due to the reception of a stronger signal. Devices could include wireless devices, mobile phones, and radar systems.
- *Adjacent channel interference*: Receiving a strong signal at a relatively close frequency to the selected signal.
- *Band congestion*: Overcrowding of frequency band can cause significant interference between devices operating within shared or closely spaced frequency bands.
- *Environmental interference*: Natural sources of electromagnetic radiation, e.g., lightning, can interfere with the operational effectiveness of electronic systems and communication data links.
- *Man-made interference*: May disrupt GPS signals, e.g., malfunctioning TV antenna preamplifier.

After successfully completing the EMRH survey, USTRANSCOM will sponsor individual devices for HERO certification in a DOD facility at vendor expense. HERO certification assesses the ability of devices to safely operate in proximity to ordnance with electrically initiated devices (EIDs) that could be inadvertently ignited or experience degraded performance characteristics after exposure to external electromagnetic emissions. Per MIL-STD-464, *Electromagnetic Environmental Effects Requirements for Systems,* critical safety functions will be verified to ensure devices are electromagnetically compatible within the system and the external operational environment prior to employment in that environment. Certification is accomplished using production representative systems.

33.7.2.2.3.3 Experimental Design 3 Field test CADS following MILSTD-810F for Phase III (Field Test): examine the technology's ability to maintain structural integrity and accurately transmit data when submitted to mechanical stresses associated with typical container movements and storage.

Field testing, employing CADS containers, will be conducted to exercise operational conditions and pass AAD data to the RF-ITV server to be input into JIPMS and IRRIS to support processing, reporting, and visualization of data via GIS capabilities. This phase will employ multi-COCOM operational testing efforts by

- Placing teams at depots in support of AA&E to sustain operations
- Permanently affixing tags/devices qualified from testing onto specific testing platforms
- Tracking each container throughout the movement cycle
- Integrating data into the JIPMS and IRRIS via the RF-ITV server

33.7.3 Predict

33.7.3.1 Analyze

Using the results of the previous step, an analysis will be performed in order to evaluate COTS vendors and attain insights about the factors that have influence on the outcomes, important interactions between variables, reliability analysis, failure rate, repeatability, reproducibility.

The analysis techniques that can be utilized are the analysis of variance, analysis of covariance, and linear regression. The choice of each of those mentioned techniques will depend upon the type of analysis expected from the data set. Also, other important qualitative factors will be integrated into the evaluation. These factors include

- Risk mitigation for fielding untested technology
- Qualified AADs with defined military utility, ruggedness thresholds, and vulnerabilities
- AADs able to support TAV/ITV and container management activities for CADS managers
- Implementation recommendations

33.7.3.2 Identify

This step is intended for identifying points that need to be improved based on the findings of the previous steps.

33.7.3.3 Design

Based on the analysis performed and on the points of improvement detailed, discussion with the tested equipment manufacturer in order to improve the performance can be done.

33.7.4 Perform

33.7.4.1 Optimize

The designed equipment is then tested in order to achieve the best results possible.

33.7.4.2 Verify

In this step, the validation of the equipment and the final real-life applications should be set up and verified for the full functionality.

33.7.5 Tools Utilized in Each Theme

Define	Measure	Analyze	Identify	Optimize	Verify
Interview process	Design of experiments	Reliability test	Interview process	Design of experiments	Control plans
Language processing	Hypothesis testing	*X–Y* mapping	Prioritization matrix	Regression	Visual systems
Prioritization matrix	Quality function deployment	FMEA	Stakeholder analysis	Process capability analysis	5-S
System map	Flow down	Multi-vari chart		Analysis of variance	TPM
Stakeholder analysis	Measurement system analysis	Chi-square		Data mining	Mistake-proofing
Thought process map	Graphical methods	Regression		Blocking	SPC/APC
Value stream mapping	Process behavior charts	Buffered tolerance limits		Response surface methodology	
		Process capability analysis		Multiple response optimization	
		Analysis of variance		Theory of constraints	

34 Case Study: Multichannel RFID

There are many examples on how radio frequency identification (RFID) technologies have been used in military applications. This chapter highlights two innovative ideas from two military contractors.

The first is using RFID technologies that have the possibility of switching from passive RFID technologies to active technologies. The other highlights the concepts of "RFID in the Box." The two companies that have provided the information are VerdaSee Solutions and Savi technologies, a division of Lockheed Martin, respectively.

34.1 BACKGROUND

As we discussed in our text, there are typically two types of RFID systems: passive and active RFID technologies. Traditionally, manufacturers seek to focus on one of those types of technologies in order to minimize manufacturing costs. Yet, in many military operations, the benefits of utilizing passive RFID technologies for inexpensive inventory and the use of active technologies for important, expensive, and hard-to-find assets are a common challenge for logistic planners. Oftentimes, the logistics planner would have to purchase multiple systems in order to satisfy the inventor objectives. This is also an operational challenge in that the warfighter may have to keep multiple units and excessive equipment in order to track asset and inventory. The idea for RFID units that can be utilized for multiple technologies has been proposed to the military, including a combined satellite and active tag readers.

There are other readers in the market that can be set up to read the different ultrahigh frequency (UHF) frequencies (only the UHF frequencies) such as 915 MHz for North America, 868 MHz for Europe, and 950 MHz for Japan. Though the readers have the ability to read the different frequencies, the manufacturer must "lock in," or choose one of these frequencies. The choice of frequency is driven by the restrictions dictated by the governmental organizations that control frequencies used in each country. Generally, the end use or commercial customers for the readers have no ability to change the frequency of the reader.

The Department of Defense (DOD) has the ability to get approval from the governmental organizations that control frequencies (through the FCC) to change the frequencies on readers. This ability allows DOD to use global positioning systems (GPS) to automatically determine the location of the reader and set (reset) the frequency of the reader remotely based on its location. This is a feature very important to the DOD for readers that are placed on ships and readers that can be deployed at various temporary ground locations around the world. This military ability has not been realized due to fact that there have not been readers available that enable the switching of RFID frequencies remotely by end user. A technology that has recently been developed to allow the warfighter as the end user to change these frequencies is the frequency-agile reader from VerdaSee Technologies.

34.2 VERDASEE TECHNOLOGIES FREQUENCY-AGILE RFID SYSTEM

The VerdaSee frequency-agile reader system is designed to read EPC standard RFID tags on different frequencies that can be set by the end user or warfighter in the field. The frequency-agile reader is being further developed to read more than four different RFID technologies and technology to be set by remote communications for military purposes only.

Within RFID, there are different technologies: active RFID, passive high frequency (HF) RFID, passive UHF RFID, passive low frequency (LF) RFID, and semi-passive RFID. Each technology has advantages for tracking different types of products and assets with some overlap for some items.

For instance, when tracking pallets and containers, the technology with the best read distance for the application and the lowest cost is the UHF technology. This is the technology that the DOD and various commercial retailers (i.e., Walmart) have specified that their suppliers use for tracking pallets and cartons. This technology and the protocol have been adapted almost worldwide except that the specific frequency within the UHF band is slightly different between North America, Europe, and Asia.

When tracking individual packages of product, one of the best technologies to use is HF RFID. Generally, the read distance required is short, the tag needs to be small, and in many cases the tag has to be able to be read when close to liquids. These are all features of HF although UHF technology has advanced to the point that it may be able to come close to the HF features. This technology is currently being used to track individual bottles of pharmaceuticals such as Viagra (Pfizer). The technology on the bottles is also being used to prevent counterfeiting.

Active RFID is a good technology to use on permanent assets or high-value products. The read distance on active tags is much greater than HF or UHF but the tag is considerably more expensive. This type of tag requires a battery but the battery generally lasts for up to 5 years; therefore, can be cost justified for placing on expensive product, or on assets.

Items that have a lot of liquid or some signal absorbing material can be tracked using the LF RFID technology. This technology has not been widely used but is used by some groups to track vehicle tires.

Semi-passive technology is used for asset tracking similar to active technology. It is particularly useful when the item being tracked should not be exposed to a constant signal transmission by the tag as one would see with an active tag. A semi-passive tag will operate the same as a passive tag but have additional read distance due to the battery that comes with the tag.

Since each RFID technology has advantages when used with different items being tracked, suppliers use different tags on items they ship to customers. Since each technology requires a different reader, customers that use RFID such as the DOD and receive many different types of items at one location are required to maintain a "bank" of different technology readers. For the DOD, the requirement on space conscience ships and ground staging areas is a problem.

Consolidating duplicate items in the different reader technologies into one box can create a reader that has the ability of reading different technologies without having the space taken by multiple readers. The cost of the frequency-agile reader is expected to be less than the cost of the combined individual readers when produced in volume. The technology has been proven through a prototype reader that VerdaSee has developed and has been tested by the Navy.

Within each of the technologies, different protocols for the technologies were tested to be sure they were read. The results are in Tables 34.1 and 34.2.

The importance of this type of system is that enabling efficient RFID data acquisition will speed up and improve the accuracy of the receipt, redistribution, and storing of materials for the military. The military will, as with commercial industry, have the benefit of having one reader for items

TABLE 34.1
ISO Standards Used Internationally

RFID Technology	Protocol/Specification	Read and/or Range
Low frequency (LF)	ISO 18000 part 2 (124–134 KHz)	Read
High frequency (HF)	ISO-18000 part 3 (13.56 MHz)	Read
Ultrahigh frequency (UHF)	ISO 18000-part 6 (EPC)[a]	Read—242 in.
Ultrahigh frequency (active)	ISO 18000-part 7[b]	

[a] ISO 18000 part 6 is mandated by DOD Passive RFID policy.
[b] ISO 18000 part 7 is mandated by DOD Active RFID policy.

TABLE 34.2
Previous RFID DOD Protocols Tested by VerdaSee

RFID Technology	Protocol/Specification	Read and/or Range
Low frequency (LF)	Atmel 5550	Read
	EM 4102	Read
High frequency (HF)	ISO-15693	Read
	Tag-It	Read
	I-Code	Read
Ultrahigh frequency (UHF)	ISO 18000-6B	Read—242 in.
	EPC—1 64 bit	Read—69 in.
	EPC—1 96 bit	Read—24 in.
	Philips U1.19	Read—51 in.
	Intellitag	Read—3 and 36 in. (on tire)
	EM 4222	Read—109 in.
MW semi-passive	Alien	Read—40 in.

that will be tagged with different RFID frequencies. This may be particularly true on ships where there is limited space, in the field where there are limited resources and other locations where quick simple setups are required.

The significance for the commercial world is the same as for the DOD—multiple frequencies of RFID tags will be used which, without the frequency-agile reader, will require multiple readers at any one location. This will be particularly true for commercial organizations that manufacture, assemble, and ship multiple types of products. An example would be a company that ships many different types of medical equipment (e.g., drugs, disposables, stretchers) to their customers. Each of those mentioned products could use a different RFID frequency but could be placed in one shipment to a particular customer. This frequency-agile technology will be strong support to future military operations.

34.3 MILITARY IN THE BOX

Savi Technologies, a division of Lockheed Martin and a leader in providing RFID technologies to the military, has used a concept for the warfighter that has developed with their experience in working with the military logistics. Concepts such as nested RFID design, last mile visibility, and military in the box have been utilized by the military using some of the Savi technologies.

The concept of nested RFID design has been described in this book but has been utilized as a strategy in prior military missions. The concept is now becoming more prevalent with multiple RFID technologies becoming more available to military operations. The idea is that multiple automatic identification technologies (AIT)/automatic data capture (ADC) technologies can be used for tracking capabilities. The most appropriate technology for the correct inventory or asset. So we can envision a bar code for labeling the item level and that information becoming linked to an active tag as the multiple RFID items are moved by container. The container would contain the active tag. This also folds into the strategy to use the appropriate technology based upon the asset value. For example, inventory that costs less than $25 should not be tagged with a $25 active tag. This concept is demonstrated in Figure 34.1.

34.4 LAST MILE VISIBILITY

The idea of last mile visibility has been utilized by the military with respect to AIT/ADC technologies. Utilizing this strategy along with the correct technology based on the importance of the asset mentioned in the aforementioned section is important. Figure 34.2 and Table 34.3 demonstrate this concept.

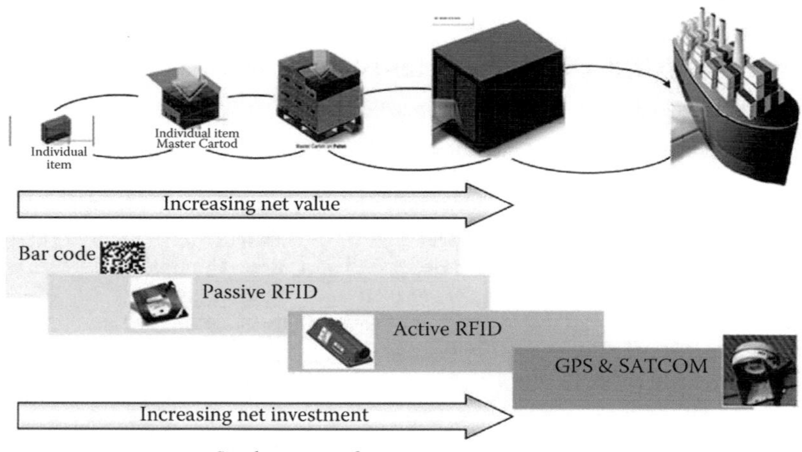

FIGURE 34.1 Increasing ADC value. (Courtesy of Savi Technologies, Lawrenceville, GA.)

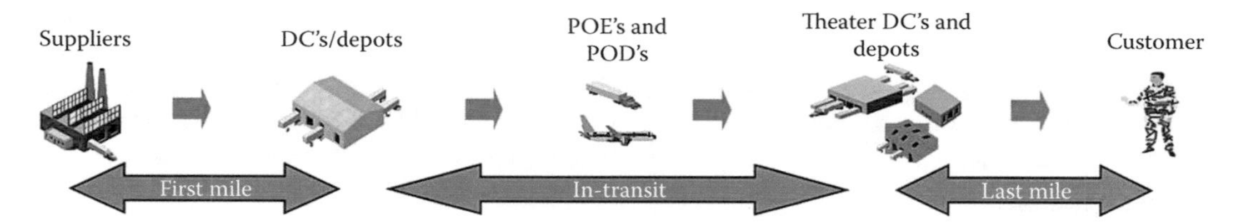

FIGURE 34.2 Last mile visibility soldier. (Courtesy of Savi Technologies, Lawrenceville, GA.)

TABLE 34.3
Last Mile Visibility Table

	Key Requirement	Technology Evolution
First mile	Automated inbound receipting	Passive RFID tagging by vendors
	Rapid shipment creation and validation	Automatic aggregation via RFID
	Locate and expedite built shipments	Shipment manifest on RFID tags
In-transit	Shared logistics with allied nations	Interoperable RFID technology
	Simple shipment de-/consolidation	Handheld updates of manifest tags
	Real-time inventory location	Integration of GPS and RFID
Last mile	Rapidly locate supplies	Deployable mobile RFID network kits
	Less reliance on central systems	Full searchable RFID manifests
	Simplified material issue	Passive RFID tags on issued material

Source: Courtesy of Savi Technologies, Lawrenceville, GA.

The concept of last mile visibility describes how the warfighter attains the right level of asset or item visibility based on the mission. The use of the technology continues to evolve with the focus being put on three key areas:

1. The first mile—Inbound to supply depots
   In surge-type operations, accuracy and throughput velocity to the operational area are critical as demand accelerates and emergencies unfold. This is often where bottlenecks happen or processes breakdown to get supplies moving in a hurry.
   Two areas of continued improvement and development are in
   a. *Inbound supplies from vendors*—We are beginning to see the evolution of passive RFID use to tag the source material. This development will speed up the receipting process through automatic receipting of goods as they pass through portals, at depots, and distribution centers. This will provide more granular inventory visibility to logistic commanders, as well as provide benefit to suppliers of speedier payment.
   b. *Rapid creation and verification of outbound shipments*—With the introduction of passive RFID tagged source supplies, the data aggregation and build of container shipment manifests with improved throughput can be further automated.
2. In-transit—From supply depots to deployed theater depots
   Two areas of continued improvement and development are in:
   a. *Shared logistics operations*—One of the developments in modern military operations is joint allied operations where there is a need to combine logistics operations with other militaries. In the past, these have tended to be duplicated; but, in NATO and humanitarian operations, typically different nations will manage different parts of the supply line. Afghanistan is a good example where NATO, Germany, and Denmark all manage different nodes in the logistic chain into Afghanistan. This is leading to the need to be able to share total asset visibility (TAV) type infrastructure so that visibility can be maintained regardless of the logistics or transportation provider. This "interoperability" is now being built into the TAV technology.
   b. *Real-time inventory location*—GPS location technology is already in use in the TAV system. These GPS devices are primarily used to track the movement vehicle (e.g., truck tractor unit), and not the containers within the vehicle and therefore cannot report what is inside the containers. A developing enhancement to this technology is the combining of GPS devices with mobile RFID readers so that the GPS unit not only knows "where" it is; it also can read the active RFID tag on the attached container or trailer and knows "what" it is carrying and can forward the information via SatCom or GSM/GPRS networks to the TAV application.
3. Last mile—Receipt and issue at forward depots

The last mile is often where logistics processes are often overlooked in the rush to get material and supplies deployed, resulting in the loss of accurate demand and consumption information.

The *deployable mobile RFID kits* and the ability to have *searchable manifest tags* from a handheld have gone a long way to aid this. This is demonstrated in Figure 34.3.

The next key development for the U.S. DOD is to be able to rely on the passive RFID tags affixed at the "first mile" to allow rapid recording of material issue without time/labor intensive process that will be overlooked at critical time. This will be the next key area of development once passive RFID tagging of material at source becomes more pervasive over the coming years.

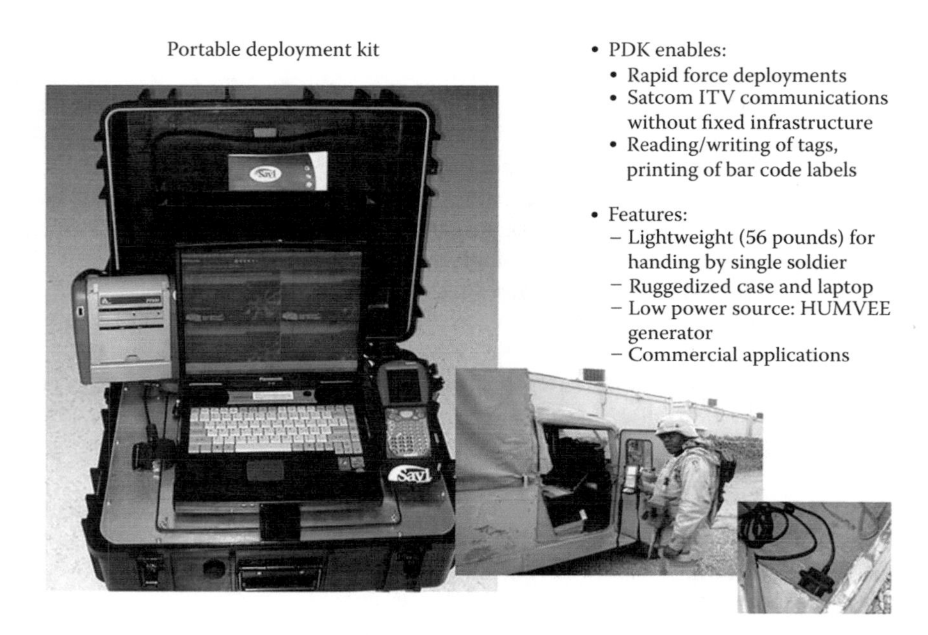

FIGURE 34.3 Deployable RFID kit. (Courtesy of Savi Technologies, Lawrenceville, GA.)

34.5 CONCLUSION

The military is using technology and strategies today that utilize the AIT/ADC such as RFID technologies in their deployment strategies. These technologies can be expanded and further developed to include technologies that become more useful in military operations. We introduced technologies such as the frequency-agile reader and the deployable RFID kit to demonstrate just two of the technologies that are being utilized today. We also demonstrated how they can be used with military planning and strategy for reasonable use in operations.

35 Radio Frequency Theory

35.1 INTRODUCTION

This section explores some of the basics of radio frequency (RF) communication, including wave propagation and electromagnetic field (EMF) theory. It is useful when beginning any radio frequency identification (RFID) project to understand the underlying physical properties that allow the system to operate. We have included four Sections. Section 35.2: Common Terms; Section 35.4: Radio Wave Theory; Section 35.5: Antenna Theory; and Section 35.6: Modulation.

35.2 COMMON TERMS

Cycle: A cycle is a complete crest-to-crest or peak-to-peak movement of a wave.

Period: A period is the time taken to complete a cycle. It is given the physics code T and measured in seconds.

Frequency: Frequency is the number cycles in a second. It is given the physics code f and measured in Hertz (Hz). $f = 1/T$.

Bandwidth: Bandwidth describes a range of frequencies. It equals the difference between the highest frequency and the lowest frequency of the device or application. There is a direct correlation between bandwidth and data carrying capacity. All RF components are classified as being either wideband or narrowband. Bandwidth is often measured by looking at standing wave ratio (SWR). We introduce an SWR reader in the design section of the text.

Resistance: Resistance is the ease with which electrons flow through a conductor. Current flow is proportional to the applied voltage and inversely proportional to the value of the resistance. This is Ohms law and this type of resistance often is referred to as ohmic. It applies whether the voltage is direct current from a battery or alternating current. It is given the physics code r and measured in ohms.

Reactance: Reactance is the form of resistance sensitive to the frequency. Two kinds exist: inductive and capacitive.

Impedance: Impedance is a measure of the total opposition to current flow in an alternating current circuit, made up of two components: resistance and reactance. It is given the physics code Z and usually is represented in complex notation as $Z = R + iX$, where R is the resistance and X is the reactance.

Decibels: Loss and gain are two fundamental concepts that affect all devices. If the signal coming out is bigger than the signal going in, the device exhibits gain. It is also known as an amplifier. If a signal coming into the device is 10 times bigger than the signal going out, the gain is said to be 10 dB. The signals leaving a transmitter can be 1 billion times bigger than those going in, so the multiplication and division of these numbers can be difficult. Engineers have come up with a mathematical means to express these big numbers with a convenient notation known as Decibels. Based on logarithms to the base 10, there are only three things you need to know:

Logs are always a ratio of two values and governed by the formula $10 \times \log$ (power out/power in)

Multiplication is the addition of logs while division is the subtraction of logs. To work with this concept, there are really only two dB conversions you must memorize:

(a) +3 dB means two times bigger
(b) +10 dB means 10 times bigger

There are two corollaries you must also know (if the numbers get smaller, the dBs are negative):

(c) −3 dB means ½ the size
(d) −10 dB means 1/10 the size

Example:

(e) If a signal experiences a gain of 8000 times, what is the gain in dB?
(f) Break up the gains into simpler figures $10 \times 10 \times 10 \times 2 \times 2 \times 2$
(g) Replace the multiplication by the dB factors by the addition of dB:
(h) 8000 = 10 dB + 10 dB + 10 dB + 3 dB + 3 dB+ 3 dB = 39 dB

If the number does not break down into factors of 2 and 10, then interpolate and you will be close enough.

dBm: notation represents a power level in decibels relative to 1 mW. dBW notation represents the power level in decibels relative to 1 W.

dB: notation is useful for representing gain or attenuation, where the output will always be related to the level of the input signal to a device. The input signal is either amplified or diminished by a certain factor, which is represented in decibels. Using dB notation simplifies power calculations in communications systems. For example, if the measured power at the input of an amplifier is 5 dBm and the gain of the amplifier is 20 dB, then the measured output power after the amplifier should be 5 dBm + 20 dB = 25 dBm (Figure 35.1).

Signals: Signals fall into two categories: analog or digital. In either case, the electrical energy contained in the signal is important. An analog signal varies over time. A sine wave is an example of an analog signal. Whether it is current flowing in the air or over a wire, a sine wave signal varies in a controlled manner. The intensity of a signal is characterized by a measure of power in watts. Typically, the antenna's power is referred to as ERP, or effective radiated power in watts.

Digital signals are commonly used in logic-processing machines such as computers. Unlike the analog signal, which varies over time, the digital signal varies in its transition from 0 to 1. For practical purposes, the digital signal has two states, which it uses to create a pattern to represent information. It is important to understand that digital signals do not carry information over the air. Only analog signals carry information over the air. When an analog signal is used for this purpose, it is referred to as a carrier. The process of combining digital information onto the carrier is called

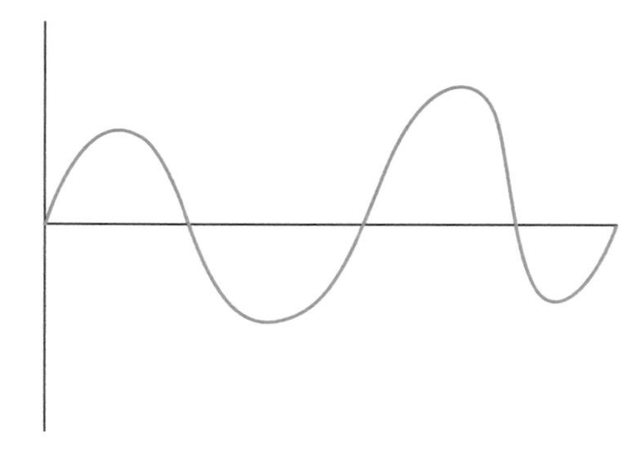

FIGURE 35.1 Signal.

TABLE 35.1
Typical Prefixes

Prefix	Definition	Example	Notation (Hz)
Milli	1/1,000th	100 MHz	0.1
Kilo	1,000	10 kHz	10,000
Mega	1,000,000	915 MHz	915,000,000
Giga	1,000,000,000	2.45 GHz	2,450,000,000

modulation. When a transmitter is always on, it is referred to as a continuous wave RF signal. When the transmitter is rapidly turned on and off, the signal is known as a pulsed RF signal.

Signal phase: For alternating voltages or currents, the relative timing of the signal is important. If the two waveforms A and B below represent an airborne signal arriving at an antenna in a different phase at the same time, you can dramatically affect the resultant signal. In one extreme, the waves may be added constructively to not only double the positive signal but also double the negative signal for a total increase of +6 dB or a quadrupling of the original signal. In the other extreme, the waves may cancel each other out completely. Impedance matching, voltage standing wave ratio (VSWR), and return-loss (*RL*) are factors that deal with phase shifts. Directional properties of antennas take advantage of combining signals from several directions being in phase in the desired direction and out of phase in the undesired directions.

35.3 UNITS AND PREFIXES

Common prefixes associated with RF and RFID technology are listed in Table 35.1.

35.4 WAVE THEORY

This section describes theoretical concepts that are related to RFID technology. We describe the basics of radio wave (RW) creation.

35.4.1 Electromagnetic Waves

RFID is a way to store and retrieve data through electromagnetic (EM) transmission to an RF compatible electronic circuit. Radio transmissions use electromagnetic waves (EMW) that are created when alternating currents flow through an antenna.

Electric fields are created by differences in voltage. The relationship can be described as the higher the voltage, the stronger the field. Magnetic fields are created when current flows. Also, it is known that the greater the current, the stronger the magnetic field. The difference is that electric fields exist even when current is not flowing, whereas magnetic fields exist only when current is flowing. When the two exist together, they are commonly referred to as EMF.

35.4.2 Radio Waves

RW are created by changing an electric current in a wire creates to a magnetic field. This can be demonstrated placing a compass very close to a wire with current flowing in it. The magnetic field will exert a force on the compass to move it from magnetic north. An electrical voltage between two points generates a field of electric force in the space between the two points. The electric field can be detected by the appropriate equipment almost as easily as the magnetic field.

Magnetic and electric fields exist because energy is temporarily transferred from the electrical circuit to the surrounding space and termed electromagnetic radiation (EMR). These fields of stored energy can be shown through calculation to reach an infinite distance in all directions where the concentration of energy per unit volume is always decreasing as you increase distance from the source.

Mathematically, EMR is a combination cross product of changing electric and magnetic fields perpendicular to each other, moving through space as a wave, effectively transporting energy and momentum. EMR is quantized as particles called photons. Any electric charge that accelerates, or any changing magnetic field, produces EMR. EM information about the charge travels at the speed of light. When any wire (or other conducting object such as an antenna) conducts alternating current, EMR is propagated at the same frequency as the electric current. Depending on the circumstances, it may behave as waves or as particles. As a wave, it is defined by three factors:

- Velocity (the speed of light)
- Wavelength and frequency
- Amplitude

When considered particles, they are known as photons, each with an energy related to the frequency of the wave given by Planck's relation $E = h\nu$, where E is the energy of the photon, $h = 6.626 \times 10^{-34}$ J is Planck's constant, and ν is the frequency of the wave. EMR is classified by wavelength into electrical energy, radio, microwave, infrared, light, ultraviolet, x-rays, and gamma rays, all of which are part of the EM spectrum. The behavior of EMR depends on its wavelength. When EMR interacts with single atoms and molecules, its behavior depends on the amount of energy per quantum it carries. One rule is always obeyed, regardless of the circumstances. EMR in a vacuum always travels at the speed of light, relative to the observer, regardless of the observer's velocity. (This observation led to Albert Einstein's development of the theory of special relativity). RW used in the air interface for RFID, carry information by varying amplitude and by varying frequency within a frequency band, and, in the case of electronic product code (EPC) standards for North America, must change or hop frequencies at a predetermined rate. When EMR impinges upon a conductor, it couples to the conductor, travels along it, and induces an electric current on the surface of that conductor. This effect is known as the skin effect and is used in antennas.

35.4.3 Frequency Spectrum

The power and variance of the EMF is vital to radio's successful operation. An important concept that helps define EMF is frequency. Imagine an ocean with a series of regular waves. The frequency simply describes the number of waves per second that crest at a static point of measurement. Engineers describe this as the oscillations or cycles per second. The term wavelength describes the distance between the crest of one wave and the next. Hence, wavelength and frequency are related: the higher the frequency, the shorter the wavelength.

A simple analogy that explains the concept of frequency would be the following demonstration: tie a long rope to a door handle and hold the free end. Moving it up and then down slowly will generate a single big wave; more rapid motion will generate a series of smaller waves. The length of the rope is constant; as you create more waves, you increase the frequency while making them shorter in wavelength (Figure 35.2).

Frequency is commonly known as Hertz in honor of radio pioneer, Heinrich Hertz. One cycle per second is 1 Hz. The frequency of oscillations ranges from 1 Hz to infinity and this entire range is known as the frequency spectrum.

Frequency spectrum is viewed as an important resource and is coordinated with legal and political governing bodies generating a plethora of complicated rules and regulations. RFID has specific frequencies for its use. Currently, they are

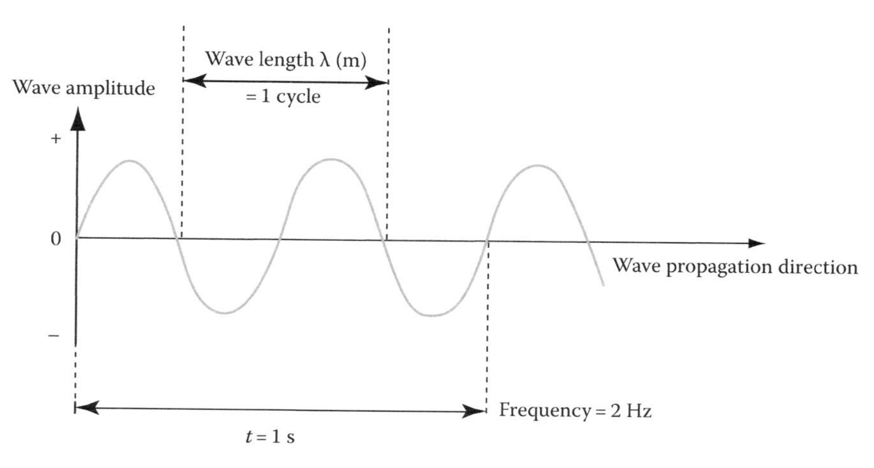

FIGURE 35.2 Frequency.

- LF (low frequency): 125, 134 kHz
- HF (high frequency): 13.56 MHz
- UHF (ultrahigh frequency): 868 MHz—Europe; 902–928 MHz—United States
- Microwave: 2.45 GHz

35.4.4 Measuring Power Loss

To measure the amount of power loss in the system, RF engineers use SWR, a measurement of match that represents how efficiently an RF signal is transferred from one point to another. The better the match, the less RF energy is wasted on leakage. A table of some key SWR values is shown in Table 35.2.

SWR may be understood by considering the voltage at various points along a cable driving a poorly matched antenna. A mismatched antenna reflects some of the incident power back toward the transmitter. Because this reflected wave is traveling in the opposite direction of the incident wave, there will be some points along the cable where the two waves are in phase and other points where the waves are out of phase (assuming a sufficiently long cable). If one could attach an RF voltmeter at these two points, the two voltages could be measured; their ratio would be the SWR. Identical results would be obtained by measuring the electrical current with an ammeter. By convention, this ratio is calculated with the higher voltage or current in the numerator, so the SWR is one or greater.

To better understand the calculation, look at the following examples. Consider a 5 V source driving a 50 Ω cable with a short on the end such that all the power is reflected back to the source. Because the reflected wave is as big as the incident wave, there will be points at which the two voltages cancel completely and other places at which the voltage will be 10 V. The ratio of 10/0 is infinity, the worst SWR possible. If, instead, the load was equal to 50 Ω, the characteristic impedance of the feed line, no power would be reflected and only a constant incident wave would appear at all points along the cable. The ratio of any two voltages would therefore be 1, the best SWR possible. The

TABLE 35.2
SWR Values

SWR	Meaning
1.0:1	Perfect match, no waste, cannot be realized in practice
1.4:1	Excellent match, very little leaking, often a design goal
2.0:1	Good match, acceptable amount of waste
10:1	Poor match, unacceptable performance

SWR for terminations between these two extremes may be calculated by considering the interaction of the reflected wave with the incident wave to determine the minimum and maximum voltages. The SWR is simply the ratio of the resistance of the termination and the characteristic impedance of the line. For example, a 75 Ω load will give an SWR of 1.5 when used to terminate a 50 Ω cable since 75/50 = 1.5. A 25 Ω resistor will give an SWR of 2 since 50/25 = 2.

The goal is that the impedance for the entire chain from the radio to the antenna is the same, and the standard for RFID radio equipment is an impedance of 50 Ω. If any piece of the chain fails to show 50 Ω impedance because of bad connections, incorrect antenna length, etc., the maximum power will not be transferred along the path. Instead, part or the entire wave is reflected back down the line. The amount of the wave reflected back depends on the level of the mismatch.

35.4.5 Smith Chart

One common way to visualize the VSWR is a polar plot called the Smith chart. From this plot, the VSWR value, the *RL*, and the impedance for the different frequencies can be derived. Therefore, it is an important tool for understanding RF transmission paths and antennas.

The Smith chart appeared in 1939 as a graphical method of simplifying the complex math of impedance (recall, calculations involving variables of the form [$Z = R + jX$]) needed to describe the characteristics of resistance and reactance. Although the Smith chart can look imposing, it is nothing more than a special type of 2D graph, much as polar and semilog and log–log scales constitute special types of 2D graphs.

Typically, the RF engineer wants to know what reflection coefficient would result from connecting a particular load impedance to a system having a given characteristic impedance. A Smith chart visually ties together the *RL*, the reflection coefficient, and VSWR for specified impedance. These are vital steps in creating an efficient air interface with an optimum RF uplink and downlink to support communications between the tag and the reader in an RFID system. End users need not worry about these issues so long as they comply with the manufacturers' recommendations for coaxial cable, connectors, antenna, and installation practices.

35.4.6 Return-Loss

Match and power transfer are vital to the performance of RF circuits. To augment the concept of VSWR is *RL*, which is measured in dB and has widespread use in the specification of signal cabling. Because it is a measure of the reflected energy of the signal, there is a direct correlation between VSWR and *RL*. Table 35.3 shows a few key conversion points.

Consider a simple cable assembly; there will be a mismatch when the connector is mated with the cable. There may also be an impedance mismatch caused by nicks or cuts in a cable. For RFID, the frequencies used by EPC tags are sufficiently high to be subject to additional problems, including material properties and the dimensions of the cable or connector, which plays an important role in determining the impedance match or mismatch. A high value of *RL* denotes better quality of the system under test (or device under test). For example, a cable with an *RL* of 21 dB is better than a similar cable with an *RL* of 14 dB, and so on. If 50% of the signal is absorbed by the antenna and 50% is reflected back, we say that the *RL* is −3 dB. A very good antenna might have a value of −10 dB (90% absorbed and 10% reflected). *RL* can be calculated using the following equation, where Z_i is internal impedance:

$$RL = 20\log\left|\frac{(Z_i - 100\,\Omega)}{(Z_i + 100\,\Omega)}\right|$$

TABLE 35.3
VSWR

VSWR	Return-Loss (dB)
1.0:1	Infinite
1.01:1	46.1
1.4:1	15.6
2.0:1	9.5
10:1	1.7
Infinite	0

35.4.7 Coupling

Inductive and capacitive coupling are important terms for RFID due to the fact that different tags use one of the two methods for transferring electromagnetic energy (EME). Inductive coupling is the transfer of the EME from one circuit to another as a result of mutual inductance between the circuits. Inductive coupling is created by matching the impedance of a transmitter or a receiver to an antenna to guarantee maximum power transfer. The tendency of a change in the current of one coil to affect the current and voltage in a second coil is called mutual inductance.

Capacitive coupling is the transfer of EME from one circuit to another through capacitance, which is the ability of a surface to store electrical energy. Capacitance is the measure of the electrical storage capacity between circuits. Capacitive coupling favors the transfer of higher frequency signal components, whereas inductive coupling favors lower frequency elements.

35.4.8 Polarization

Polarization is a process or state in which EMW exhibit different properties in different directions. EMW are composed of two plane waves. In most cases, the amplitude and phase of the plane waves define the character of the polarization of the product wave.

A plane EMW is said to be linearly polarized. The transverse electric field wave is accompanied by a magnetic field wave.

If the RW is composed of two plane waves of equal amplitude but differing in phase by 90°, then the light is said to be circularly polarized. If you could see the tip of the electric field vector, it would appear to be moving in a circle as it approached you. If, while looking at the source, the electric vector of the radiation coming toward you appears to be rotating clockwise, the light is said to be right-circularly polarized. The electric field vector makes one complete revolution as the radiation advances one wavelength toward you. Elliptically polarized waves consist of two perpendicular waves of unequal amplitude that differ in phase by 90°.

35.5 ANTENNAS

EMW are sent airborne or received through an antenna. The antenna is a critical component designed to radiate energy out into free space and/or collect radio energy from space. It is important to recognize that in doing this job the antenna is the most important part of the radio system—without it the system dies. In most systems, the antenna is common to both the transmitter and the receiver; therefore; any change in the antenna affects both transmission and reception.

The antenna changes radio energy from the transmission line into radiated energy and vice versa. The antenna's operation can be broken down into two fundamental modes of wireless communications:

1. Near-field communications, aka close proximity EM, aka inductive coupling
2. Far-field communications, aka propagating EMW

Far-field radiation is distinguished by the fact that the intensity is inversely proportional to the square of the distance. In reality, obstructions, absorption, and interference make the loss more severe. On the other hand, near-field radiation intensity is inversely proportional to the cube of the distance in the region that is less than 1/6 wavelength from a simple loop antenna. (For additional reference see: Lee, *Principles of Antenna Theory*, p. 231; the *ARRL Antenna Book*, pp. 2–8; and TI Literature Number 11-08-26-003).

In either mode, antennas have optimal sizes that relate to the frequency of the signal. Basically, the higher frequencies require smaller antennas because of shorter wavelengths. This becomes an even greater factor with RFID, as tags are typically designed so their geometric lengths are a fraction of the operational frequency's wavelength, e.g., quarter wavelengths. Deviating from the optimum geometry "detunes" the antenna. It likely will still function, but at a nonoptimized range. As an analogy, consider a portable radio. With its antenna mast fully extended, it likely will have optimum reception. With the mast retracted, the radio still will have reception, but not at the level previously achieved unless it is close enough to the broadcast tower.

Because the sizes of wavelengths vary, radio signals propagate differently through free space. Some are well suited to short ranges while others are good for transmitting over long distances. Typically, the higher the frequency, the shorter the distance the signal will travel. The strength of the radio signal diminishes rapidly as it moves away from the transmitter antenna.

35.5.1 Standard Antenna Impedance

What is the meaning of 50 Ω? It is the standard RF transmission line impedance. To understand this concept, we will think of the conductor and the devices as part of a garden hose wherein the RF signal is represented by the water inside. If an RF signal is to move through a conductor, then enter a device efficiently, there will be minimal leakage of signal. The size of the garden hose must be specified to allow the optimum amount of water to flow through the system. If interconnection is to work and for water to flow in the garden hose system, the diameter and connection type must be standardized. Like the garden hose, some of the water leaks out depending on the quality of the connection and how well the male and female couplings are matched. To standardize for RF, engineers have specified the size of the hose to carry the RF energy. When talking about an RF resistance, we use impedance. Remember, this is the measurement of resistance for the signal to move through a conductor (aka garden hose size). The standard for RF is 50 Ω of resistance. Impedance is dictated by the size of the conductor, the material, and the temperature.

35.5.2 Impedance Matching

Impedance matching is the lifeblood of RF circuit design. From the end user's point of view, all the work for impedance matching is necessary to create an interoperable RFID system. It can be shown mathematically that any source of power, for instance an RFID reader, will deliver its maximum possible power output when the impedance of the load is equal to the internal impedance of that source. The ideal condition provides only 50% efficiency because half the power is consumed in the source. This can be explained through the following example (Figure 35.3).

Choose a value for E_{source}, = 100 V and R_{source}, = 50 Ω. These will be held constant.

Use the formulas

$$I = \frac{E_S}{(R_S + R_L)}$$

$$W_S = I^2 \times R_S$$

$$W_L = I^2 \times R_L$$

Recalculate I, W_S, and W_L for each occurrence of R_L to generate the graph below. The intersection of the source and load lines proves the power to the load given by W_L is maximum when $R_L = R_S$.

From the graph we can infer that

1. Maximum power to load occurs when the source impedance = load impedance. At this condition, the source expels the same amount of power in heat as it delivers to the load.
2. As load impedance increases, the power converted to heat from the source decreases.

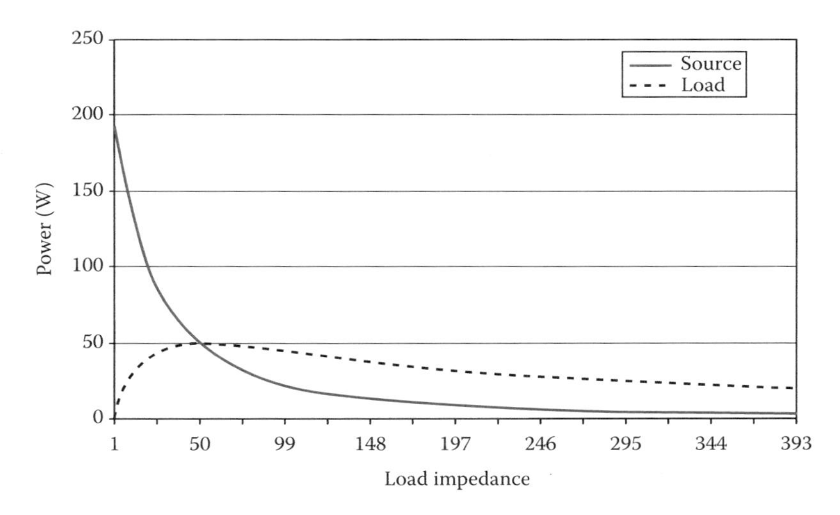

FIGURE 35.3 Effects of impedance matching.

35.5.3 Tuning

For each particular radio transmission, there will be a certain number of cycles completed each second. As defined earlier, the frequency of the transmission is the most basic property of RW. Tuning of a transmitter or receiver is achieved by changing frequency. Frequency is the property that controls the pitch of a sound wave, the color of a light wave, and the band and channel of a radio signal.

It is because of differences in frequencies that many radio signals can coexist in the airborne atmosphere and be sorted out by the radio's receiver circuitry. Differences of propagation are also largely related to the frequency and governed by the equation

$$\lambda = \frac{c}{f}$$

where
- λ is the wavelength
- c is the speed of light
- f is the frequency (cycles per second)

The wavelength for RFID in the United States is typically 13½ in. or 0.33 m.

In a vacuum, c is equal to the speed of light (299,793,077 m/s), but RW are slower when passing through other materials, hence the wavelengths are shorter. This is of great importance when designing antennas.

Isolation of frequencies is based on the fact that there are electrical components and circuits that respond differently to signals of different frequencies. In some cases, the selective response to a small range of frequencies can increase the signal rejected outside the pass band by a factor of up to 100 times compared to the desired signal. A series of these circuits may be arranged to work on the signal one after the other as the signal flows through the transmitter or receiver and can be used to provide the desired rejection of the unwanted signal to effectively make the desired signal clearly recognized.

Tuning may be performed by using a variable capacitor or inductor to change the resonant frequency of the circuit. This has the effect of electronically changing the center frequency of bandwidth for desired signals.

35.5.4 Range and Path Loss

Another key consideration is the issue of read range. It is important to realize that read range is a nonlinear relationship and is governed by the concept of path loss. As RW propagate in free space, power falls off as the square of range for the RW (E field) and the fourth power for the magnetic field (which is used in the 13.56 MHz RFID applications). For a doubling of range, power reaching a receiver antenna is reduced by a factor of four. Path loss reduces signal strength. In free space, this effect is caused by the spreading of the RW as they propagate. It can be calculated as follows:

$$L = 20\log_{10}\left(\frac{4\pi D}{\lambda}\right) \quad (35.1)$$

where
- D is the distance between receiver and transmitter
- c is the speed of light (3×108 m/s)
- f is the frequency (Hz)
- L is the path loss

The equation above describes line-of-sight (LOS) or free space propagation path loss. This is the best case for a RW as free space reduces the signal the least amount as it travels. On the other hand, earth bound RW travel faces such obstructions as buildings, trees, and mountains, among other things. Propagation losses indoors can be significantly higher because of building walls, glass, steel, and concrete. RFID may be even higher due to the combination of attenuation by walls and ceilings, conveyors, pallets, and blockage caused by forklift trucks, equipment, furniture, and even people. For example, a "2 × 4" wood stud wall with sheetrock on both sides results in about 6 dB loss per wall. Experience has shown that LOS propagation holds only for about the first 20 ft. Beyond 20 ft, indoor propagation losses increase at up to 40 dB per 100 ft in dense warehouses, office environments, or factories. This is a "rule-of-thumb" and will have to be measured for each use case. Actual propagation losses will vary significantly depending on building construction and layout.

35.6 MODULATION

35.6.1 Signal Propagation

Once an RF signal is airborne, it is attenuated by something called free space loss (FSL). The further away a receiver is from the transmitter, the smaller the signal is because of FSL. Power density is a measure of an airborne signal's strength and always is used in the RF world. Imagine a square that is 1 m on each side. The amount of RF energy passing through the square is watts per square meter or power density. This tells us how powerful the RF energy is at this location.

Just about everything an RF signal encounters while airborne changes it in some way. These changes tend to do one of three things to the signal: they make it smaller, change its direction, or create heat. Many of the "attenuators" the RF signal encounters are such common items as the air we breathe, rain, glass, steel, brick, wood, and even foliage. We may model these things as passive devices with some insertion loss. Insertion loss exhibited by nature is called absorption, because it absorbs the RF signal. Absorption explains how your microwave oven works. An RF signal is radiated inside the oven at a frequency that water likes to absorb. As the RF signal encounters water, it gets smaller, translating its energy into heat. This is the basis for microwave cooking.

Not everything encountered by the RF wave absorbs its energy. Some things have the ability to change the signal's direction, called reflection. The amount of reflection depends on the frequency of the signal and the material of the object. As a rule, RF waves tend to reflect off the objects at the same angle at which they encountered them.

35.6.2 Modulation

The purpose of radio signaling is to carry some intelligent information. If a steady RF carrier is transmitted such that the frequency and amplitude do not change, then no information is transmitted beyond the fact that a RW is present in the environment. To encode intelligence onto the RW, some property needs to change. A method of varying some property to encode intelligence is called modulation. It is the lifeblood of carrying information within RW.

RW make this easy to do as there are only two primary properties that may be changed:

- Adjusting the power of the signal over time. Known as amplitude modulation (AM), you will witness by listening to an AM station.
- Adjusting the frequency of the signal over time. Known as frequency modulation (FM), you witness this by tuning to an FM station.

Of course, there also exist a number of permutations and combinations of these two basic properties to increase the amount of information carried within the RW. A modulated signal does not exist as a separate entity from the RW. The signal must be processed through a detector or demodulator circuit to reverse the process and extract the information that was encoded. Some basic forms of modulation techniques are

- Amplitude modulation
- Frequency modulation
- Phase modulation
- Pulse modulation

Selection of modulation method determines system bandwidth, power efficiency, sensitivity, and complexity. For the purposes of link budget analysis, the most important aspect of a given modulation technique is the signal-to-noise ratio (SNR) necessary for a receiver to achieve a specified level of reliability in terms of bit error rate (BER).

A graph of E_b/N_o versus BER is shown in Figure 35.4. E_b/N_o is a measure of the required energy per bit relative to the noise power. Note that E_b/N_o is independent of the system data rate. To convert from E_b/N_o to SNR, the data rate and system bandwidth must be taken into account as shown in the following equation:

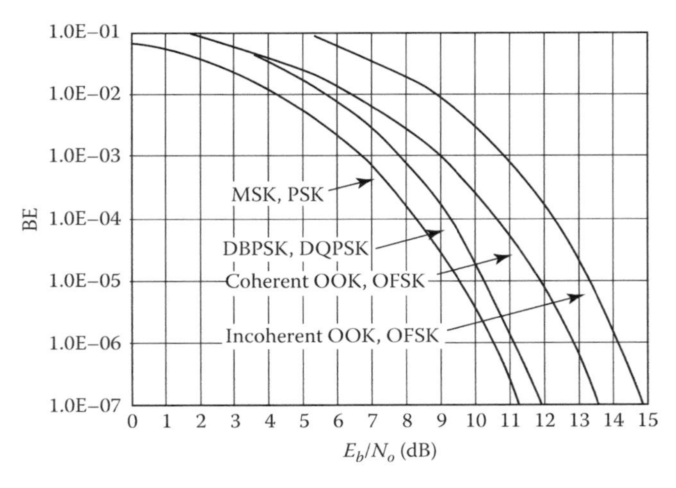

FIGURE 35.4 E_b/N_o versus BER.

$$SNR = \left(\frac{E_b}{N_o}\right) + \left(\frac{R}{B_T}\right)$$

where

E_b is the energy required per bit of information
N_o is the thermal noise in 1 Hz of bandwidth
R is the system data rate
B_T is the system bandwidth

35.6.3 Frequency Modulation

In FM, the frequency of the carrier wave is time varied based on the original signal to be modulated. Specifically, the change in frequency at any instant is proportional to the modulating signal that varies with time. Its principal benefit over AM is increased noise immunity and decreased distortion; however, this is achieved at the expense of requiring more bandwidth. The FM band has become the choice of music listeners' desire for quality and faithful reproduction of a musician's talents.

In analog applications, the carrier frequency is varied in direct proportion to changes in the amplitude of an input signal. In digital applications, the carrier frequency is varied in accordance with a set of discrete values—to encode a 0 or 1. This technique is known as frequency shift keying (FSK).

The main advantages of FM over AM are

- Improved SNR (about 25 dB) w.r.t. to man–made interference
- Smaller geographical interference between neighboring stations
- Less radiated power
- Well-defined service areas for given transmitter power

The disadvantages of FM are

- Much more bandwidth (as much as 20 times)
- More complicated receiver and transmitter

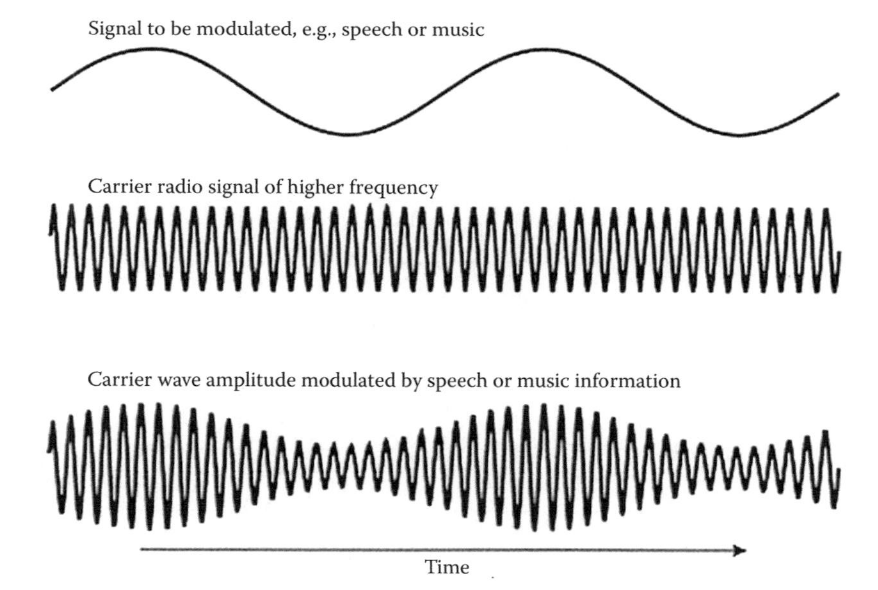

FIGURE 35.5 Amplitude modulation.

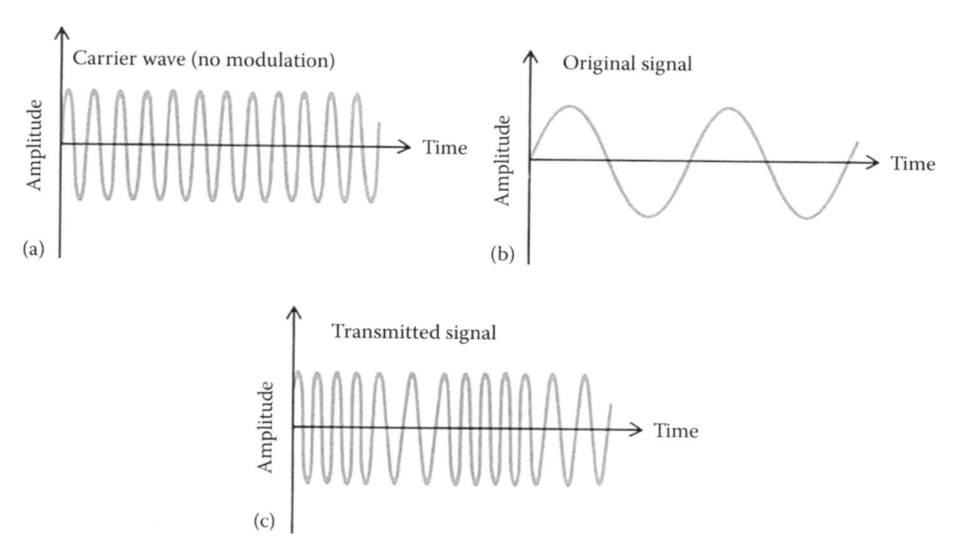

FIGURE 35.6 Frequency modulation.

35.6.4 Amplitude Modulation

AM is the modulation used in the AM radio broadcast band. In this system, the intensity, or amplitude, of the carrier wave varies over time based on the signal to be modulated, e.g., speech or music. This varying signal embeds the information in accordance with the modulating signal onto a carrier radio signal of higher frequency. When the RF carrier is AM modulated, the amplitude of the carrier RW varies in symmetry to the input signal. A fraction of the power is converted to sidebands extending above and below the carrier frequency by an amount equal to the highest modulating frequency. If the modulated carrier is rectified and the carrier frequency filtered out, the modulating signal can be recovered (Figures 35.5 and 35.6).

AM modulation is not an efficient way to send information, mainly because the power required is relatively large because the carrier, which contains no information, is sent along with the information.

In a variant of AM, called single sideband modulation (SSB), the modulated signal contains only one sideband and no carrier. The information can be demodulated only if the carrier is used as a reference. This is normally accomplished by generating a wave in the receiver at the carrier frequency. SSB modulation is used for long-distance telephony (as in amateur radio bands) and telegraphy over land and submarine cables.

35.6.5 Phase Modulation

Phase modulation, like FM, is a form of angle modulation (so called because the phase is shifted to the RW carrier is changed by the modulating wave).

The detector discerns the intelligence in the carrier wave by measuring the phase shift from the original carrier. This information may then encode digital data to represent a specific combination of 1s and 0s.

The two methods are similar in the sense that any attempt to shift the frequency or phase is accomplished by a change in the other. An RFID tag and reader use pulse modulation; the data are contained in changes in the phase of the carrier wave sent out by the reader.

35.6.6 Pulse Modulation

Pulse modulation involves modulating a carrier that is a train of regularly recurrent pulses. The modulation may vary the amplitude (pulse amplitude modulation (PAM)), the width (pulse width

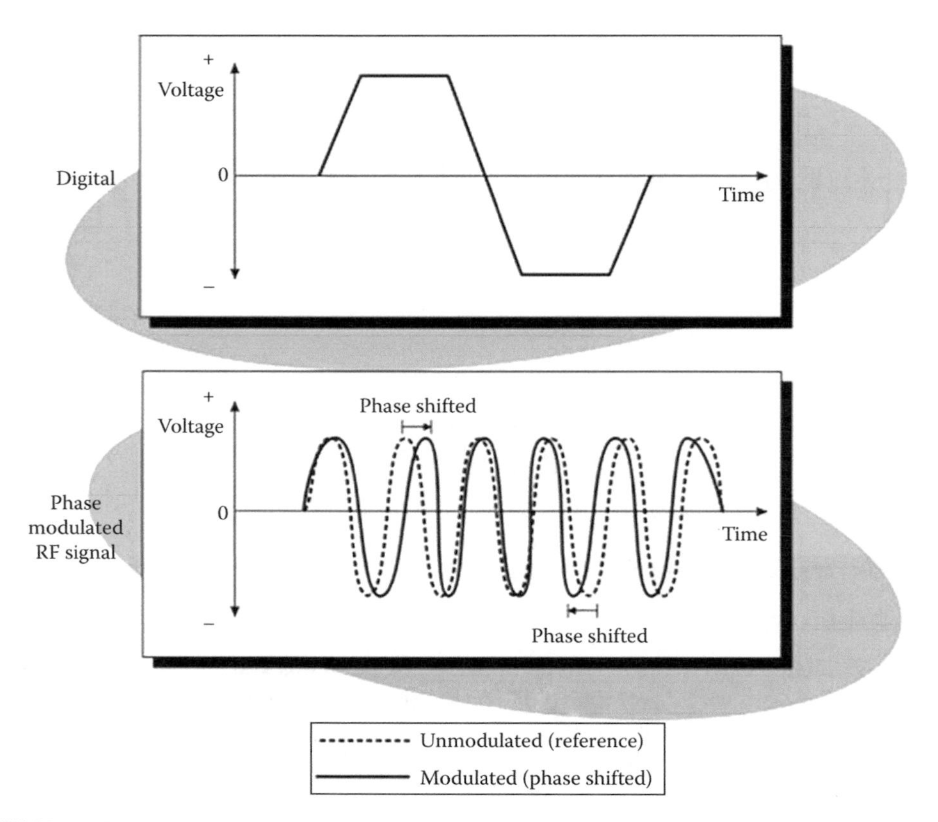

FIGURE 35.7 Pulse modulation.

modulation (PWM)), or the presence of the pulses (pulse code modulation (PCM)). PCM is the most important form of pulse modulation because it can be used to transmit information over long distances with little interference or distortion. Although PCM transmits digital instead of analog signals, the modulating wave is continuous (Figures 35.7 and 35.8).

35.6.7 Amplitude Shift Keying

Amplitude shift keying (ASK), is useful in transmitting RFID tag data because it is simple and effective for digital communications.

One of the disadvantages of ASK is that it has a varying envelope, making power amplification more difficult. However, this makes it easy to demodulate with an envelope detector. ASK, in the context of digital communications, is a modulation process that provides two or more discrete amplitude levels to a sinusoidal signal. For a binary message sequence, there are two levels, typically 1 and 0. The modulated waveform looks like bursts of a sinusoid. There are sharp discontinuities shown at the transition points. Using a submultiple of the carrier frequency is typical for the data rate.

35.6.8 Handshaking

Once the radio engineering link is operational, you may consider how it transports information from one location to another. To explore a high-level description of the typical transmitting sequence for a generic tag-antenna-reader system, you start with the greeting. When you meet someone, you usually shake hands. An analogous situation occurs in electronics with a system handshake. The typical handshake for a passive tag is as follows.

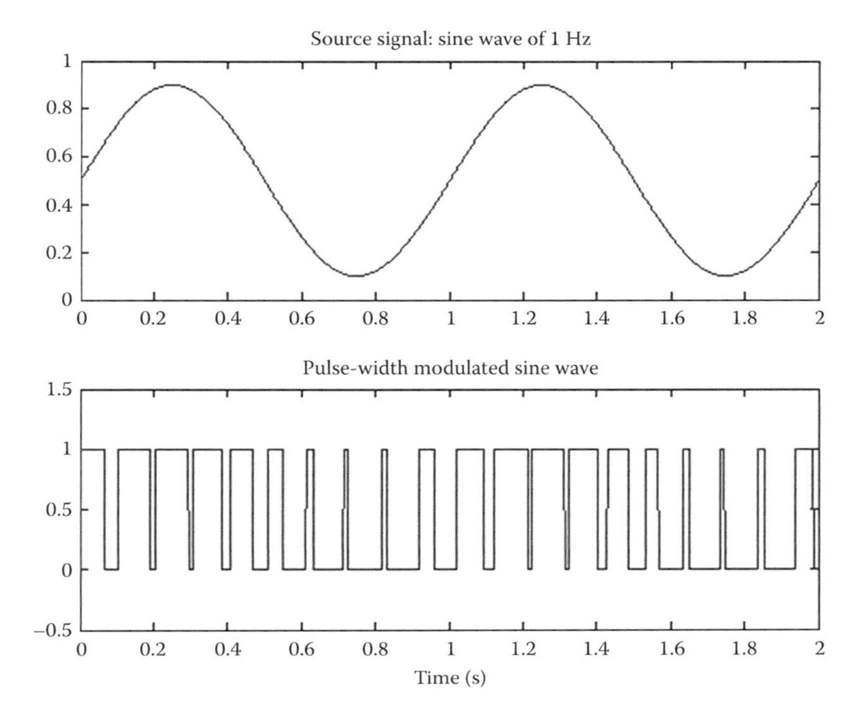

FIGURE 35.8 Pulse width modulation.

First, a reader looks for modulation of its RF sine wave to indicate the presence of a tag. When a tag's antenna captures the EMF generated by the reader's antenna, it initiates a process to respond with a data stream, which is encoded in the carrier frequency.

With a successful handshake, the system begins transmitting information as follows: the tag typically starts clocking its data messages against an output transistor, which is connected across coil inputs. In this case, the RF link behaves like a transformer, wherein the tag is the primary coil and the reader the secondary coil.

As the tag's output transistor shunts the coil, it effectively modulates the carrier to experience a momentary voltage drop. The pattern of voltage drops corresponds to the information to be uploaded from the tag to the reader.

The reader must detect these small voltage drops, which represent the modulation. This requires a sensitivity that can pick up 1/1000 of a change from the original carrier wave's amplitude as sent by the reader. By constantly monitoring these voltage drops, the reader detects and decodes a bit stream according to the modulation being used. To further complicate things, the modulation scheme also incorporates algorithms to affect error recovery, bandwidth, synchronization, and other system needs.

In addition to these basic tasks, the system needs to handle collision avoidance during the simultaneous reading of several tags in the same RF field. In this case, the tag and reader must be intelligent enough to detect that more than one tag is present. Failure to recognize this condition leads to all the tags modulating the carrier frequency at the same time; these multiple waveforms arrive at the reader only to create a garbled signal. This is referred to as a collision. No data would be transferred to the reader when this happens.

Consider the problem of having a telephone conversation with three people. If everyone talks at once, it is impossible to understand the conversation. Some order will allow everyone to be understood. The RFID radio interface requires arbitration so only one tag transmits data at one instance in time.

While it is possible to transmit all the data from the tag to the reader through AM, the practical electronic circuit's modulation of data bits is enhanced by using other methods:

- Frequency shift keying (FSK): two different frequencies are used for data transfer. A 0 is transmitted as amplitude modulated clock signal with a different frequency, while a 1 is sent on a different amplitude modulated frequency.
- Phase shift keying (PSK): similar to FSK, except only one frequency is used and the shift between 1s and 0s is accomplished by shifting the peak and trough of the wave forms.

Data are also sent using non-return-to-zero (NRZ), differential biphase, and Manchester coding schemes. They are used to improve noise immunity, interference, and efficiency.

These factors make the air interface the most complicated component of system design. In other words, the radio channel is the weakest link in the system and requires sophisticated engineering for peak performance; however, once that task is complete, the radio link operates with the greatest efficiency and highest reliability.

For RFID, the interrogator sends information to one or more tags by modulating an RF carrier using double-sideband amplitude shift keying, single-sideband amplitude shift keying, or phase-reversal amplitude shift keying, using a pulse-interval encoding format.

Tags receive their operating energy from this same modulated RF carrier. An interrogator receives information from a tag by transmitting an unmodulated RF carrier and listening for a backscattered reply. Tags communicate information by backscatter modulating the amplitude and/or phase of the RF carrier. The encoding format, selected in response to interrogator commands, is either FM0 or Miller-modulated sub-carrier. The communications link between interrogators and tags is half-duplex, meaning that tags will not be required to demodulate interrogator commands while backscattering. A tag will not respond using full-duplex communications to a mandatory or optional command.

35.6.9 Frequency Hopping

The term "spread spectrum" simply means that the energy radiated by the transmitter is spread out over a wider amount of the RF spectrum than would otherwise be used. By spreading out the energy, it is far less likely that two readers sharing the same spectrum will interfere with each other. This is an important consideration in an unlicensed band and explains why the FCC-imposed spread spectrum requirements on part-15 radios that transmit over −1 dBm (about 0.75 mW). It also explains why dense reader environment for RFID is defined as 50 readers operating within 1000 ft of each other.

In the United States, these bands are collectively designated as industry, science, and medicine (ISM) bands. Operation in these bands with approved devices does not require an FCC license. By waiving licensing requirements, these bands have been made generally accessible to virtually anyone. This is why the ISM bands are so important for commercial and consumer applications. It is critical for end users to realize that RFID does operate in a dedicated band and may catch interference from other devices operating within the band. As mentioned above, radios employing spread spectrum methods are allowed to radiate up to 1.0 W (30 dBm) of RF energy, as compared to less than 1 mW for non-spread radios.

There are two common types of spread spectrum systems. The easiest to understand is frequency-hopped spread spectrum (FHSS), used in RFID. In this method, the carrier frequency hops from channel to channel in some prearranged sequence. The receiver is programmed to hop in sequence with the transmitter. If one channel is jammed, the data are retransmitted when the transmitter hops to a clear channel. The major drawback to FHSS is limited data rate. In the 2.4 GHz band, FCC regulations require that the maximum occupied bandwidth for any single channel is 1 MHz. This effectively limits the data rate through this type of system to about 1 Mbps.

35.7 VALUE PROPOSITION: WAVE OF THE FUTURE

Since MIT commercialized RFID with the formation of EPCglobal, there has been a great deal of interest in the application of RFID technology in the supply chain, pharmaceuticals, and health care, to name prominent vertical markets. Several vendors, including Alien, Symbol, TI, Intermec, Samsys, and Impinj, offer products that comply with the U.S. FCC regulations for unlicensed operation in the 902–928 MHz spectrum. These regulations permit radiated RF power of up to 4 W when spread spectrum modulation techniques are used.

QUESTIONS

35.1 What units measure radio frequency?
- (a) Hertz
- (b) Amps
- (c) Volts
- (d) None of the above

35.2 How do you create radio waves?
- (a) When ohms flow through an antenna
- (b) When alternating currents flow through an antenna
- (c) When direct current flows through an antenna
- (d) None of the above

35.3 What word(s) refer to the signal direction for an antenna?
- (a) Horizontal pattern
- (b) Vertical pattern
- (c) Read range
- (d) None of the above

35.4 What units measure the antenna signal strength?
- (a) Voltage
- (b) Current
- (c) Watts
- (d) None of the above

35.5 What determines an RFID system's write range?
- (a) Antenna pattern
- (b) Transmitter output power
- (c) Link budget
- (d) All of the above

35.6 What does polarization mean?
- (a) The direction of the antenna relative to the ground
- (b) The direction of the radio energy relative to the ground
- (c) Both a and b

35.7 What frequencies does passive RFID use?
- (a) 125 kHz
- (b) 13.56 MHz
- (c) 902–928 MHz
- (d) 2.45 GHz
- (e) Both b and c

35.8 What does frequency hopping mean?
- (a) Changing frequencies from one location to another to avoid interference
- (b) Changing frequencies from one channel to another
- (c) Both a and b

35.9 Which of the following is true about multipath signals?
- (a) Multiple versions of the signal reach the receiver at different times due to the varying lengths of their paths.
- (b) When multiple versions of the signal arrive at different times, they create more work for the receiver because it has to decode the information more then once.
- (c) Multiple versions of the same signal may arrive out of time synchronization and recombine at the antenna to cancel each other out and significantly reduce the signal.
- (d) Multiple versions of the same signal may arrive out of time synchronization and recombine at the antenna to amplify each other and significantly increase the signal.
- (e) None of the above.

35.10 Why is the write operation to a tag more resource intensive than a read operation in consuming resources?
- (a) To write a tag requires the following operations "Read/Write/Verify." As a result of the additional steps, the write operation will take longer.
- (b) Write operation requires a different wavelength for the radio channel and the tuning to that channel takes additional time.
- (c) To latch the memory requires more energy than a read and therefore, the signal strength requirements are higher to give the tag access to more power to perform the necessary write operations to the onboard memory.
- (d) A read operation does not need to perform the binary exclusive or computation and as a result, it is faster.
- (e) None of the above.

35.11 What does 3 dB gain do to a signal?
- (a) It changes the phase of the signal and this increases the power significantly.
- (b) It doubles the original power.
- (c) It triples the original power.
- (d) It cuts the original power in half.
- (e) None of the above.

35.12 What is the tag attenuation factor?
- (a) A method of noting how much gain a tag has relative to its peers
- (b) A method of noting how much attenuation a tag has relative to its peers
- (c) A method of measuring when a tag ceases to respond when optimally presented
- (d) The amount of path loss before a tag fails to respond
- (e) None of the above

35.13 What important information does SWR tell you?
- (a) Measures antenna output power
- (b) Measures the voltage to the antenna
- (c) Measures the gain of the antenna
- (d) Confirms the characteristic impedance is 50 Ω and the system is providing maximum output power to point of measurement
- (e) None of the above

35.14 What is modulation?
- (a) The "to-and-from" pattern of a wave cycle
- (b) The length of a wave
- (c) The additional information added to a wave
- (d) The height of a wave
- (e) The distance from one wave to another
- (f) None of the above

36 Future RFID Applications

RFID is an emerging technology that has the potential to be used in various applications. In this book, we have described methods of using RFID in military applications, but one must be aware of the fact that there exists a much broader scope for this technology. As seen throughout the book, logistics is a wide-ranging subject that allows for RFID integration to make the supply chain more efficient. With military applications, the future of RFID could lead to investigation and development of manufacturing processes that support the embedding of RFID/AIT systems into military systems. To the question, what lies in the future for RFID technology? the answer is that it has endless applications that only need to be explored.

36.1 RFID AND AIT CENTERS

Several universities are exploring the future applications of RFID. Universities conduct research in several fields where RFID technology can be integrated. A particular type of RFID research center is known as "RAC," which is RFID and AIT centers that bring together faculty from industrial engineering and related fields across multiple universities with common interests in enabling automated technologies to support research that improves life expectancy with improved (1) quality of life, (2) security, and (3) well-being. A particular emphasis of the center is researching, designing, and innovation of RFID and AIT systems to support the aforementioned goals. In order to support this vision, the partnered universities have aligned certain areas with the themes of quality of life, security, and well-being such medical drug tracking, retail, transportation, etc. These areas can be seen in Figure 36.1.

36.2 RFID IN RESEARCH

The research focus areas in RFID have the ability to construct better-engineered systems in healthcare, transportation, and infrastructure assets. The research plan followed by those conducting research in RFID is aligned with three goals RFID and AIT systems, enabling technologies, and fundamental knowledge. To better align themselves with these goals, individuals should seek systems to solve their life and well-being problems. These issues need to be translated into basic knowledge through research, which then investigates the enabling technologies, which leads to development of RFID and AIT systems. A typical strategic research plan focuses on these three areas that have subareas that are available for further research as seen in Figure 36.2.

By utilizing the research in RFID and AIT, conducted across different universities, RFID is transformed by the diverse research methods and algorithms and enhances the fields that utilize RFID, GPS, GPRMS, bar codes, and other ADC technologies. The fundamental knowledge that enables the technologies will lead to additional inventory control models, sensor application models, and more enhanced computing algorithms. These discoveries can be utilized to innovate areas such as healthcare, information security, and transportation. Furthering partnerships with leading organizations, innovation centers, and practitioners will lead to knowing entrepreneurial thinking. This will support the creation of new product manufacturing and new companies that facilitate innovation.

Utilization of the strategic plan is important for the new creations to support the goals such as wellness, security, and quality. One must also identify the barriers within each level to allow for more successful creations. Within the first level of fundamental knowledge, the barriers that one

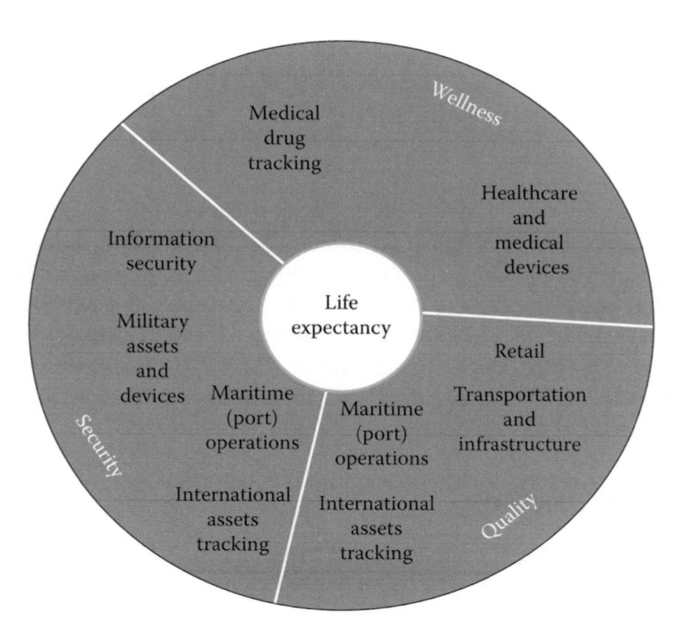

FIGURE 36.1 Goals of RTC.

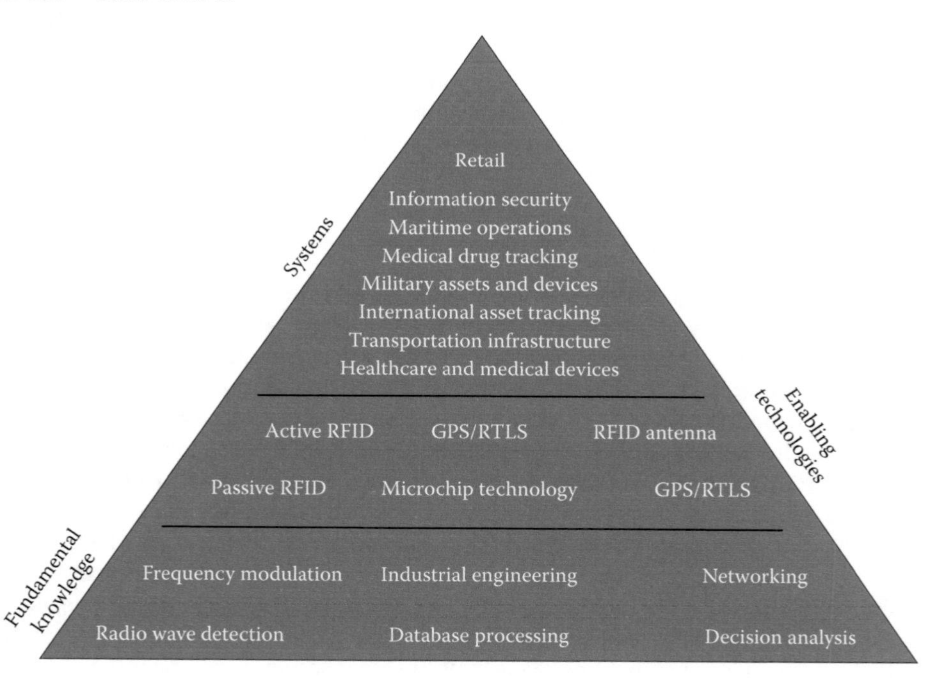

FIGURE 36.2 Design of strategic plan.

must overcome include material interference, information acquisition, processing power, and communication. The barriers identified for the second level, the enabling technologies level, include cost, life expectancy, standards, and performance. Finally, the barriers for the top level of the strategic framework include regulation, universal acceptance, backward compatibility, and cost. The barriers at each level of the strategic framework can be found in Figure 36.3.

Research also exposes students to discovery in the increasingly important research areas of logistics; transportation and RFID/ADC, which by their definition have vital international components.

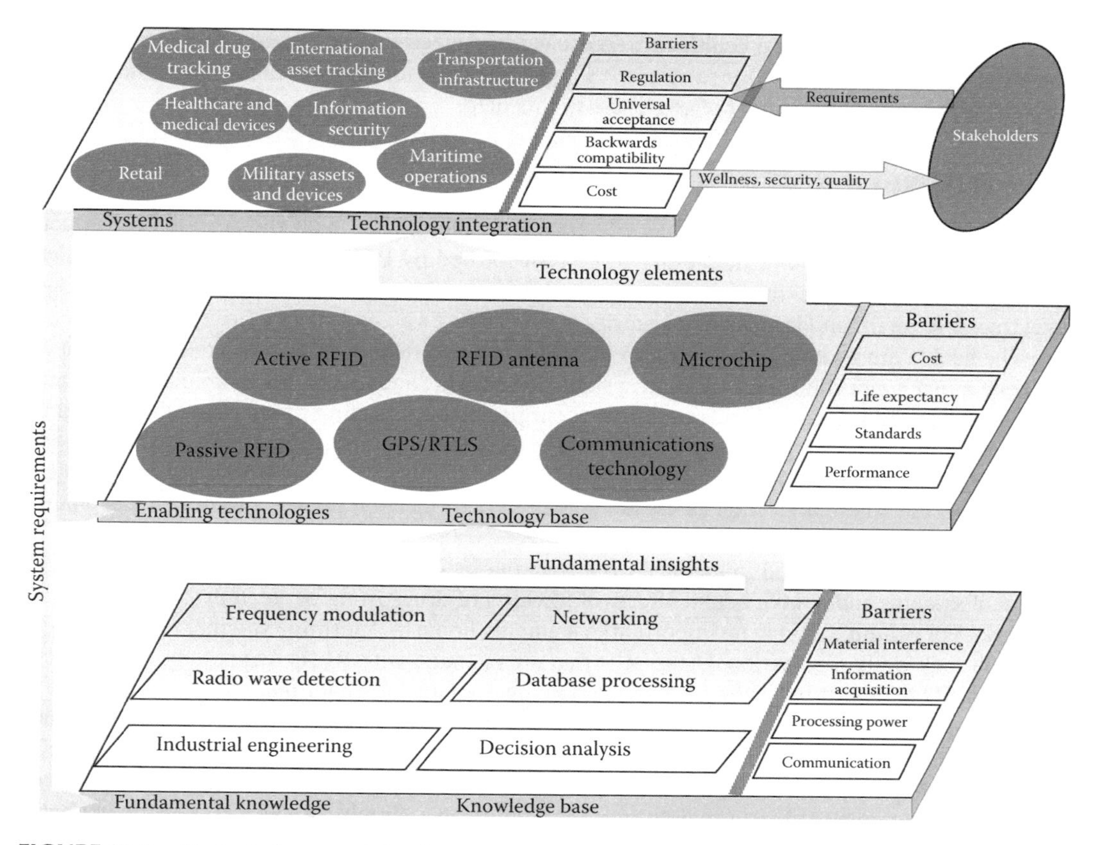

FIGURE 36.3 Strategic framework barriers.

Educators are preparing the next generation of engineers by exposing students to different cultures prior to entering the workforce. This education is compatible with the National Science Foundation's mandate to better prepare "future generations of U.S. scientists and engineers to gain professional experience beyond the U.S. borders early in their careers."

36.3 BROADER IMPACT

If these studies conducted in universities are proven successful, they will create new tools that companies can use successfully to integrate RFID technologies into several areas. Subsequently, these models will enable smaller companies to explore the benefits of RFID in distribution operations. The strength of having RAC, universities focus on researching these new technologies and providing real-world examples on integration into live operational applications is an effective way of supporting the NSF ERC mission. The mission is accomplished with universities providing support to industry in defining research that can be effectively used to assist companies in improving operations.

Compilation of the results of future research will provide new tools with which companies and researchers can have more success in integrating RFID and AIT technologies to support the increasing life expectancy through better quality of life, general health and well-being, and safety and security. These breakthroughs and models will enable organizations like the Department of Defense, Homeland Security, Health and Human Services, NASA, and FEMA, which provide crucial services, to explore benefits of this emerging technology for their asset, personnel, and client tracking needs. The need for expanding this technology has been defined by initiatives from FAA to FDA in industries including healthcare, aerospace, and electronic commerce. The discovery and education components of this future research support the need for better understanding, integration into operations, and development of engineers in this field.

36.4 FUTURE OF RFID IN SPECIFIC AREAS

RFID has a future in several fields and will transform these fields with the research being conducted. Some fields that have the ability to be transformed by RFID are healthcare and medical devices, medical drug tracking, transportation and infrastructure, maritime operation, international asset tracking, retail, and information security.

In the healthcare and medical devices field, RFID can investigate and develop real-time location algorithms that support hospital location systems for patients and assets. For example, RFID technology can be used to locate objects as simple as surgical sponges within a patient's body. A surgical sponge with an embedded RFID tag could be found by a doctor by simply scanning a patient's body to determine whether or not a sponge has been left inside. Other uses include asset management within a hospital where important assets are tagged so their location can be determined when these assets are needed by the hospital staff.

RFID can also be utilized in medical drug tracking by investigating the development and integration of systems and in vivo health effects of tracking of drugs from lot creation to ingestion confirmation. For example, it may be important in a nursing home to determine whether or not a patient has taken their daily medication. Using pills that are encoded with RFID technology, a technician has the ability to determine if the medication was ingested and at what time.

In transportation and infrastructure, RFID technology will aid in the investigation of manufacturing processes related to integrating these technologies in transportation such as commercial vehicles and infrastructures such as bridges and roads. Objects such as signs and guardrails are very important in monitoring the traffic flow and this monitoring not only provides efficient travel but it also provides a safe travel environment. By using RFID technology, the future of RFID allows for these important assets to be tracked; therefore, the proper authorities are notified when these assets have been damaged. Other technologies could allow for RFID tags to be embedded in license plates in order to track where a particular vehicle is located or what kind of cargo is within a particular type of vehicle.

RFID technology may also be important in maritime (port) operations because it can be used to investigate and develop automatic technologies that can work at ports given the challenges of the marine environment to radio signal technologies. RFID research will aid in overcoming the challenges imposed by marine environments and future RFID technology may become available for these areas.

International asset tracking is another field that future RFID technology has the ability to transform. Future research could investigate the technologies that support the tracking of international assets including medication and workers. New research could allow for an international company or countries to track workers throughout the world. These new technologies would allow for RFID to not only have an impact locally but also globally.

RFID has already made an impact on the retail industry through companies such as Walmart that utilize RFID in the inventory management strategies. Future research with RFID may further transform the retail industry by eliminating the need for bar codes. Retail stores in the future would have tagged merchandise that when put in a cart will be identified by the RFID reader attached to the cart. Instead of manually reading the bar code of each item, the cashier would only need to use the reader to determine the total cost of items being purchased.

Finally, RFID technology has the ability to transform the information security field by investigating the techniques and algorithms that secure RFID and AIT information. RFID is an emerging technology, but it is important for the information encoded in these tags to be secure. New research would allow for a more secure encoding to protect sensitive tag information.

RFID future research is not limited to the fields mentioned above; instead, the future research mentioned above merely gives the reader a sample of the future path of RFID technology. The various applications of RFID can be seen and these applications are the reasons that RFID is an up-and-coming technology that must be closely watched and heavily researched. RFID has the ability to transform the logistics field in so many ways and only time will tell the future of RFID technology.

Bibliography

CHAPTER 1

Bacon, B., *Sinews of War: How Technology, Industry, and Transportation Won the Civil War*, Presidio Press, Novato, CA, 1997.

Dolan, A., Lovingly restored Lord Admiral Nelson's letter reveals how raisins helped him win the Battle of Trafalgar, http://www.dailymail.co.uk/news/article-1089269/ Lovingly-restored-Lord-Admiral-Nelsons-letter-reveals-raisins-helped-win-Battle-rafalgar.html#ixzz0X3MYdKTa

Grau, L. and Jalali, A., The campaign for the caves: The battles for Zhawar in the Soviet-Afghan War, *Journal of Slavic Military Studies,* 14(3), 2001.

Holiday, L. P. and Gurfield, R. M., *Viet Cong Logistics*, Rand Corporation, Santa Monica, CA, 1968.

Hudson, C., *Falklands Mission Impossible*, http://tailspinstales.blogspot.com/2009/04/falkland-mission-impossible.html?widgetType=BlogArchive&widgetId=BlogArchive1&action=toggle&dir=open&toggle=MONTHLY-1241161200000&toggleopen=MONTHLY-1246431600000

Jomini, L., Précis de l'Art de la Guerre: Des Principales Combinaisons de la Stratégie, de la Grande Tactique et de la Politique Militaire, Brussels: Meline, Cans et Copagnie, 1838.

Maley, W. and Amin, S., *The Soviet Withdrawal from Afghanistan*, Cambridge University Press, Cambridge, U.K., 1989, p. 16.

Merle, R., Census counts 100,000 contractors in Iraq, The Washington Post, December 5, 2006, http://www.washingtonpost.com/wp-dyn/content/article/2006/12/04/AR2006120401311_pf.html

Paparone, C. R., *How Does the Gulf War Measure Up*? http://www.almc.army.mil/alog/issues/SepOct98/MS309.htm

Paulus, R. D., *From Santiago to Manila: Spanish–American War Logistics*, http://www.almc.army.mil/alog/issues/JulAug98/MS305.htm

Schechter, D. and Sander, G., *Delivering the Goods: The Art of Managing Your Supply Chain*, John Wiley & Sons, New York.

Smith-Dorrien, H., *Memories of Forty-Eight Years' Service*, John Murray, 1925—Sir Horace's autobiography. (Republished as Smith-Dorrien: Isandlwhana to the Great War Leonaur, 2009, ISBN-10: 1846776791 ISBN-13: 978-1846776793.)

Tokar, J., *Logistics and the British Defeat in the Revolutionary War*, http://www.almc.army.mil/alog/issues/SepOct99/MS409.htm

Webb, K., The continued importance of geographic distance and boulding's loss of strength gradient, *Comparative Strategy*, 26(4), 295–310, 2007, doi:10.1080/01495930701598607.

CHAPTER 2

Federal Supply Groups and Classes, SB708-21, Defense Logistics Information Service, Battle Creek Michigan 49017-3084, http://www.wingovcon.com/downloads/fscH2_2002.pdf

U.S. Army Field Manual 4-0, Combat Service Support, Headquarters, Department of the Army, Washington, DC, August 29, 2003.

CHAPTER 3

United States Department of Defense Suppliers' Passive RFID Information Guide, version 13.0, http://www.acq.osd.mil/log/rfid/guide/DoD_suppliers_passive_RFID_guide_v13.pdf

CHAPTER 4

Bacheldor, B., Dow and chemtrec's RFID-based rail safety project, *RFID Journal*, www.rfidjournal.com, April 6, 2007.

Bacheldor, B., Orbit one launches satellite-based RFID service, *RFID Journal*, 2008, www.rfidjournal.com, February 7, 2008.

Bacheldor, B., Unipart launches insight service for real-time shipment tracking, *RFID Journal*, www.rfidjournal.com, February 5, 2007.

Chopra, S. and Meindl, P., *Supply Chain Management: Strategy, Planning, and Operation*, Pearson, Upper Saddle River, NJ, 2007.

Dommety, G. and Jain, R., *Potential Networking Applications of Global Positioning Systems (GPS)*, Department of Computer and Information Science, The Ohio State University, 1996, January 2007 <http://arxiv.org/abs/cs.NI/9809079>.

Feng, S. and Law, C. L., Assisted GPS and its impact on navigation and intelligent transportation systems, *IEEE 5th International Conference on Intelligent Transportation Systems*, pp. 926–931.

Johnson, J. C., Wood, D. F., Wardlow, D. L., and Murphy, P. R., *Contemporary Logistics*, Prentice Hall, Upper Saddle River, NJ, 1999.

Novack, A. R., Langley Jr., J., and Rinehart, L. M., *Creating Logistics Value: Themes for the Future*, Council of Logistics Management, Oak Brook, IL, 1995.

O'Connor, M. C., Hammer combines RFID, GPS, mapping and sensor technologies, *RFID Journal*, www.rfidjournal.com, April 20, 2007.

O'Connor, M. C., Wherenet, Identec Solutions offer GPS tracking, *RFID Journal*, 2007, www.rfidjournal.com.

Reade, W. and Lindsay, J. D., Cascading RFID tags, *IP.Com*, 2003, January 2007, <http://www.jefflindsay.com/rfid3.shtml>.

Shapiro, J. F., *Modeling the Supply Chain*, Thomson, Belmont, CA, 2007.

Stalling, W., *Data and Computer Communications,* 8th edn., Pearson/Prentice Hall, Upper Saddle River, NJ, p. 123.

Stockman, H., Communication by means of reflected power. In *Proceeding of IRE*, pp. 1196–1204, October 1948.

Want, R., Enabling ubiquitous sensing with RFID, *Computer*, 37: 84–86, 2004, IEEE Xplore, January 2007.

Wessel, R., Paz energy uses RFID for fuel security and inventory efficiencies, *RFID Journal*, 2007, www.rfidjournal.com

CHAPTER 9

FCC OET, Bulletin 56 *Hazards of Radio Frequency and Magnetic Fields* and Bulletin 65 *Human Exposures to Radio Frequency and Electromagnetic Fields.*

http://www.rfidgazette.org/2005/12/rfid_regulation.html

United States Department of Defense Suppliers' Passive RFID Information Guide Version 1.0 12 3.6.1.

CHAPTER 12

Barber, L. and Weinstein, M., *Work Like Your Dog: Fifty Ways to Work Less, Play More, and Earn More*, Villard, New York, 1999.

Bassman, E. S., *Abuse in the Workplace, Management Remedies and Bottom Line Impact*, Quorum Books, Westport, CT, 1992.

Bauch, G. T. and Chung, C. A., A statistical project control tool for engineering managers, *Project Management Journal*, 32(2), 37–44.

Bohlen, G. A., Lee, D. R., and Sweeney, P. A., Why and how project managers attempt to influence their team members, *Engineering Management Journal*, 10(4): 21–28, 1998.

Bureau of Labor, *Labor Letter*. U.S. Department of Labor, Washington, DC, 1966.

Cherniss, C., *Staff Burnout: Job Stress in Human Services*, Sage Publications, Beverly Hills, CA, 1980.

Chung, C. A., *Simulation Modeling Handbook: A Practical Approach*, CRC Press, New York, 2003.

Cook, J. D., Hepworth, S. J., Wall, T. D., and Warr, P. B., *The Experience of Work*, Press Limited, San Diego, CA, 1989.

Cronbach, L. J., Coefficient alpha and the internal structure of tests, *psychometrika*, 16: 297–334, 1951.

Evans, J. R. and Lindsay, W. M., *The Management and Control of Quality*, West Publishing Company, St. Paul, MN, 1993.

Fisher, K. and Fisher, M. D., *The Distributed Mind: Achieving High Performance through the Collective Intelligence of Knowledge Work Teams*, Wiley, New York, 1987.
Foster, T., *Managing Quality: An Integrative Approach*, Pearson, NJ, 2003.
Gaudine, A. P. and Saks, A. M., Effects of an absenteeism feedback intervention on employee absence behaviors, *Journal of Organizational Behavior*, 22: 15–29, 2001.
Golembiewski, R. T. and Munzenrider, R. F., *Phases of Burnout, Development in Concepts and Applications*, Praeger Publishers, New York, 1982.
Leach, F. J. and Westbrook, J. D., Motivation and job satisfaction in one government and development environment, *Engineering Management Journal*, 12(4): 3–10, 2000.
Locke, E. A., The nature and causes of job satisfaction. In *Handbook of Industrial and Organizational Psychology*, Dunnette, M. D., ed., Rand McNally, Chicago, IL, 1976.
Lofquist, L. H. and Dawis, R. V., *Adjustment to Work: A Psychological View of Man's Problems in a Work-Oriented Society*, Appleton-Century-Cofts, New York, 1967.
Maslach, C., Burn-out, human behavior. In *Job Stress and Burnout: Research, Theory, and Intervention Perspectives*, Vol. 5, Sage, Beverly Hills, CA, 1976, pp. 16–22.
Mobley, W. H., *Employee Turnover: Causes, Consequences, and Control*, Addison-Wesley, Reading, MA, 1982.
Mowday, R. T., Porter, L. W., and Steers, R. M., *Employee-Organization Linkages*, Academic Press, New York, 1982.
Nunnally, J. C., *Psychometric Theory*, 2nd edn., McGraw-Hill, New York, 1978.
Spector, P. E., *Summated Rating Scale Construction: An Introduction*, Sage, New York.
Turner, W. C., Mize, J. H., and Case, K., *Introduction to Industrial and Systems Engineering*, Prentice-Hall, NJ, 1987.

CHAPTER 13

Atmel, *Antenna Matching for UHF–RFID Transponer ICs*, Atmel Corporation, Heilbronn, Germany, 2005, pp. 1–11.
Clampitt, H. G., *RFID*, 1st edn., PWD Group Inc., Houston, TX, 2006, pp. 1–280.
Lee, Y., *Antenna Circuit Design for RFID Applications*, Microchip Technology Inc., 2003, pp. 1–50, September 10, 2006, http://ww1.microchip.com/downloads/en/AppNotes/00710c.pdf#search=%22antenna%20design%20RFID%22
Nikitin, P. V. and Seshagiri, R., Power reflection coefficient analysis for complex impedances in RFID tag design, *IEEE Transactions on Microwave Theory and Techniques*, 53: 2721–2724, 2005, September 15, 2006.
Olsson, T., Research, Electronics Design Division, Sweden University, 2004, September 18, 2006, http://www.itm.mh.se/forskning/elektronik/research/groups/system/rfid.htm
Rao, K., Nikitin, P. V., and Lam, S. F., *Impedance Matching Concepts in RFID Transponder Design*, Intermec Technologies Corporation, Intermec, pp. 1–4.
Seshagiri, R. and Lam, S. F., Antenna design for UHF RFID tags: A review and a practical application, *IEEE Transactions on Antennas and Propagation*, 53: 3870–3876, 2005. September 8, 2006.
Sanford, J. R., *Antenna Design Considerations for RFID Applications*, Cushcraft Corporation, pp. 1–5. September 13, 2006, http://www.cushcraft.com/comm/support/pdf/Antenna%20Design%20for%20RFID%20app.pdf#search=%22antenna%20design%20RFID%22
Sure, P., The Silver ink printed antenna. In *The World of RFID*, 2005, pp. 70–72.
Vuong, T. P. and Beroulle, V., *Antennas for RFID Tags*, Grenoble, France, 2005, pp. 19–22.

CHAPTER 14

Banerjee, P., Zhou, Y., and Montreuel, B., Genetically assisted optimization of cell layout and material flow path skeleton, *IIE Transactions*, 29(4): 277–291, 1997.
Breyfogle, F. W., *Implementing Six Sigma: Smarter Solutions using Statistical Method*, 2nd ed., Wiley, New York, 2003.
Carbon, T. A., Measuring efficiency of semiconductor manufacturing operations using data envelopment analysis (DEA). In *IEEE SEMI Advanced Semiconductor Manufacturing Conference*, 2000.
Gary, M., Gaukler, G. M., Özer, Ö., and Hausman, W. H., *RFID and Order Progress Information: Improved Dynamic Emergency Ordering Policies*, July 10, 2006.
Gleixner, S., Young, G., Vanasupa, L., Dessouky, Y., Allen, E., and Parent, D., Teaching design of experiments and statistical analysis of data through laboratory experiments. In *32nd Annual Frontiers in Education Conference*, Boston, MA, Vol. 1, November 6–9, 2002.

Gotsman, C. and Koren, Y., Distributed graph layout for sensor networks, *Lecture Notes in Computer Science (LNCS)*, Vol. 3383, 2005.

Jones, E. C., Volakis, J., and Verma, V., How RFID reliability effects inventory control accuracy. *Antennas and Propagation Society International Symposium,* IEEE 2007, 9–15 June, Honolulu, HI, 2007.

Kleijnen, J. P. C., Sensitivity analysis and optimization in simulation: Design of experiments and case studies, In *Simulation Conference Proceedings*, Winter, 1995, pp. 133–140.

Lee, Y. M., Cheng, F., and Leung, Y. T., Exploring the impact of RFID on supply chain dynamics. In *Proceedings of the 2004 Winter Simulation Conference*.

Pan, J., Tonkay, G., and Quintero, A., Screen printing process design of experiments for fine line printing of thick film ceramic substrates, *Journal of Electronics Manufacturing*, 9(3): 203–213, 1999.

Penttila, K., Sydeimo, L., and Kivikoski, M., Performance development of a high-speed automatic object identification using passive WID technology. In *Proceedings of the International Conference on Robotics and Automation*, New Orleans, LA, 2004, pp. 4864–4868.

Rao, K. V. S., Nikitin, P. V., and Lam, S. F., Antenna design for UHF RFID tags: A review and a practical application, *IEEE Transactions on Antennas and Propagation*, 53(12): 3870–3876, December 2005.

Tompkins, J. A., White, J. A., Bozer, Y. A., Frazelle, E. H., Tanchoco, J. M. A., and Trevino, J., *Facility Planning,* 2nd edn., John Wiley & Sons, Inc., New York, 1996.

Wehking, K. H., Seeger, F., and Kummer, S., RFID transponders: Link between information and material flows. How reliable are identification procedures? *Logistics Journal*, reviewed Publications—ISSN 1860–7977, 2006.

Zhang, Y., Liu, J., and Zhao, F., Information-directed routing in sensor networks using real-time reinforcement learning. In *Combinatorial Optimization in Communication Networks*, 2006, pp. 259–288.

CHAPTER 15

Caglar, D., Li, C. L., and Simchi-Levi, C., Two-echelon spare parts inventory system subject to a service constraint, *IIE Transactions*, 36: 655–666, 2003.

Graves, S. C., A multi-echelon inventory model for a repairable item with one-for-one replenishment, *Management Science*, 31: 1247–1256, 1985.

Johnson, J. C., Wood, D. F., Wardlow, D. L., and Murphy, Jr. P. R., *Contemporary Logistics*, 7th edn., Prentice-Hall, NJ, pp. 586, 1999.

Lee, C. B., Multi-echelon inventory Optimization. *Evant White Paper Series*, 2003.

Muckstadt, J. A., A model for a multi-item, multi-indenture inventory system, *Management Science*, 20: 472–481, 1973.

Muckstadt, J. A. and Thomas, L. J., Are multi-echelon inventory methods worth implementing with low demand items? *Management Science*, 26: 483–494, 1980.

Schnetzler, M. J., Sennheiser, A., and Schonsleben, P., A decomposition-based approach for the development of a supply chain strategy, *International Journal of Production Economics*, 105: 21–42, 2007 (in process).

Sherbrooke, C. C., METRIC: A multi-echelon technique for recoverable item control, *Operations Research*, 16: 122–141, 1968.

Simon, R. M., Stationary properties of a two-echelon inventory model for low demands, *Operations Research*, 19: 761–777, 1971.

Wang, Y., Cohen, M. A., and Zheng, Y. S., A two-echelon repairable system with restocking-center-dependant depot replenishment lead times, *Management Science*, 46: 1441–1453, 2000.

CHAPTER 16

Blanchard, B., *Logistics of Engineering and Management*, 2nd edn., Practice Hall, Englewood Cliffs, NJ, 1992.

Clemen, R. T., *Making Hard Decisions*, 2nd edn., Duxbury Press, Pacific Grove, CA, 1996.

Collins, J., New two-frequency RFID system, *RFID Journal*, September 2004, pp. 1105.

Eschenbach, T. G., *Cases in Engineering Economy*, Wiley, New York, 1989.

Eschenbach, T. G., *Engineering Economy: Applying Theory to Practice*, 2nd edn., Oxford University Press, Oxford, U.K., 2003.

Eschenbach, T. G., Technical note: Constructing tornado diagrams with spreadsheets, *The Engineering Economist*, 51(2): 195–204, 2006.

Evans, J. L., Zhang, D., and Nathan, V., Investment analysis for automotive electronics manufacturing: A case study, *The Engineering Economist*, 49: 159–183, 2004.

Hazen, G. B., A new perspective on multiple internal rates of return, *The Engineering Economist*, 48: 31–51, 2003.
Kaliski, B., RFID blocker tags. In *Dr. Dobb's Journal: Software Tools for the Professional Programmer*, 29, 2004.
Nobel, C., *Sun, Sybase, IBM tackle RFID*, *eWeek*, 21, 2004, pp. 14–16.
RFID Wizards, Inc., *RFID Equipment Test Results Manufacturing*, RFID Wizards, Inc., 2003.
ID TechEx, *The Need for Total Asset Visibility*, ID TechEx, Cambridge: U.K., 2004.
United States Department of Defense, *United States Department of Defense Suppliers' Passive RFID Information Guide*, Version 1, Updated August 31, 2004.

CHAPTER 17

Nahmias, S., *Production and Operations Analysis*, 4th edn., McGraw-Hill, New York, 2001. Print.

CHAPTER 18

Bowman, E. H., Consistency and optimality in managerial decision making, *Management Science*, 9: 310–321, 1963.
Bowman, E. H., Production scheduling by the transportation method of linear programming, *Operations Research*, 4: 100–103, 1956.
Chung, C. and Krajewski, L. J., Planning horizons for master production scheduling, *Journal of Operations Management*, 389–406, 1984.
Cohen, M. A., Fisher, M. L., and Jaikurmar, J., International manufacturing and distribution networks. In *Managing International Manufacturing*, Ferdows, K., ed., North Holland, Amsterdam, the Netherlands, 1989, pp. 67–93.
Erenguc, S. and Tufekci, S., A transportation type aggregate production model with bounds on inventory and backordering, *European Journal of Operations Research*, 35: 414–425, 1988.
Hax, A. C. and Candea, D., *Production and Inventory Management*, Prentice Hall, Englewood Cliffs, NJ, 1984.
Hax, A. C. and Meal, H. C., Hierarchical integration of production planning and scheduling, In *TIMS Studies in Management Science*, Vol. 1, *Logistics*, Geisler, M., ed., Elsevier, New York, 1975.
Hillier, F. S. and Lieberman, G. J., *Introduction to Operations Research,* 5th edn., Holden Day, San Francisco, CA, 1990.
Holt, C. C., Modigliani, F., Muth, J. F., and Simon, H. A., *Planning Production, Inventories, and Workforce*, Prentice Hall, Englewood Cliffs, NJ, 1960.
Kogut, B. and Kulatilaka, N., Operating flexibility, global manufacturing, and the option value of a multinational network, *Management Science*, 40: 123–139, 1994.
McGrath, M. E. and Bequillard, R. B., International manufacturing strategies and infrastructural considerations in the electronics industry. In *Managing International Manufacturing*, Ferdows, K., ed., North Holland, Amsterdam, the Netherlands, 1989, pp. 23–40.
Nahmias, S., *Production and Operations Analysis*, 4th edn., McGraw-Hill, New York, 2001. Print.
Schrage, L., *Linear, Integer, and Quadratic Programming with LINDO*, Scientific Press, Palo Alto, CA, 1984.
Schwarz, L. B. and Johnson, R. E., An appraisal of the empirical performance of the linear decision rule for aggregate planning, *Management Science*, 24: 844–849, 1978.
Silver, E. A. and Peterson, R., *Decision Systems for Inventory Management and Production Planning*, 2nd edn., John Wiley & Sons, New York, 1985.

CHAPTER 19

Morris, W. T., *Engineering Economic Analysis*, Reston Publishing, Reston, VA, 1976.
Ranson, G. M., *Group Technology*, Pergamon, New York, 1970.
Subramaniam, S., Design evaluation and cost estimation expert system, Master Thesis in Industrial Engineering, Department of Industrial Engineering, North Carolina State University, Raleigh, NC, 1991.
White, J. A., Case, K. E., Pratt, D. B., and Agee, M. H., *Principles of Engineering Economic Analysis*, 4th edn., John Wiley & Sons, New York, 1998.
Whitney, D. E., DeFazio, T. L., Gustavson, T. E., Graves, S. C., Abell, T., Coopride, K., and Pappu, S., Tools for strategic product design, First Report, SCDL-R-2115, MIT and The Charles Stark Draper Laboratory, Cambridge, MA, November 1988.

CHAPTER 22

Chopra, S. and Meindl, P., *Supply Chain Management: Strategy, Planning, and Operation*, Pearson, Upper Saddle River, NJ.

Dommety, G. and Jain, R., *Potential Networking Applications of Global Positioning Systems (GPS)*. Department of Computer and Information Science, The Ohio State University. 1996. January 2007 <http://arxiv.org/abs/cs.NI/9809079>.

Feng, S. and Law, C. L., Assisted GPS and its impact on navigation and intelligent transportation systems. In *IEEE 5th International Conference on Intelligent Transportation Systems*, pp. 926–931.

Johnson, J. C., Wood, D. F., Wardlow, D. L., and Murphy, P. R., *Contemporary Logistics*, 7th edn., Prentice Hall, Upper Saddle, NJ, 1999.

Novack, A. R., Langley, J. Jr., and Rinehart, L. M., *Creating Logistics Value: Themes for the Future. Oak Brook*, Council of Logistics Management, 1995.

Reade, W. and Lindsay, J. D., Cascading RFID tags, IP.Com (2003), January 2007 <http://www.jefflindsay.com/rfid3.shtml>.

Shapiro, J. F., *Modeling the Supply Chain*. Thomson, Belmont, CA, 2007.

Want, R., Enabling ubiquitous sensing with RFID, *Computer*, 37: 84–86, 2004. IEEE Xplore. Jan. 2007.

CHAPTER 23

Banerjee, P., Zhou, Y., and Montreuel, B., Genetically assisted optimization of cell layout and material flow path skeleton, *IIE Transactions*, 29(4), 277–291, 1997.

Carbon, T. A., Measuring efficiency of semiconductor manufacturing operations using data envelopment analysis (DEA). In *IEEE SEMI Advanced Semiconductor Manufacturing Conference*, 2000.

Gary, M., Gaukler, G. M., Özer, Ö., and Hausman, W. H., *RFID and Order Progress Information: Improved Dynamic Emergency Ordering Policies*, Stanford University, July 10, 2006.

Gotsman, C. and Koren, Y., Distributed graph layout for sensor networks, *Lecture Notes in Computer Science (LNCS)*, p. 3383, 2005.

Tompkins, J. A., White, J. A., Bozer, Y. A., Frazelle, E. H., Tanchoco, J. M. A., and Trevino, J., *Facility Planning*, 2nd edn., John Wiley & Sons, Inc., New York, 1996.

Wehking, K.-H., Seeger, F., and Kummer, S., RFID transponders: Link between information and material flows. How reliable are identification procedures, *Logistics Journal*, 2006.

Zhang, Y., Liu, J., and Zhao, F., Information-directed routing in sensor networks using real-time reinforcement learning. In *Combinatorial Optimization in Communication Networks*, pp. 259–288, 2006.

CHAPTER 24

Stevenson, W. J., *Operations Management*, 10th edn., McGraw-Hill, Boston, MA, 2009, p. 581.

CHAPTER 25

Bacheldor, B., Fort hood to RFID-tag medical records, *RFID Journal*, http://www.rfidjournal.com/article/articleprint/2536/-1/1

Fee, J. and Schmack, A., Improving RFID technology, *Army Logistician*, March–April 2005, http://www.almc.army.mil/ALOG/issues/MarApr05/rfid.html

Ferguson, R., Army Taps 3M for RFID Tracking of Medical Records, http://www.eweek.com/c/a/Mobile-and-Wireless/Army-Taps-3M-for-RFID-Tracking-of-Medical-Records/, July 25, 2006.

Granata, J. P., Tracking materiel from warehouse to warfighter, *Army Logistician*, 37(5).

HealthNEWS Team, 3M deploys RFID system for medical records management at US army base, http://www.healthnewsdirect.com/?p=358, June 30, 2008.

http://www.idautomation.com/rfid_faq.html#DOD-96_UID

O'Conner, U.S. army uses UWB to track trainees, *RFID Journal*, November 15, 2005, www1.rfidjournal.com/article/view/1987/

Plinsky, J. and Rogers, J., Enhanced logistics tracking and monitoring through sensor technology, http://www.almc.army.mil/alog/issues/JulAug08/enhancelog_w_sensortech.html

Swedberg, C., U.S. army gun-monitoring RFID prototype gets upgrade, *RFID Journal*, 2007, http://www.rfidjournal.com/article/articleview/3643/1/1/

Swedberg, C., U.S. army developing RFID system to track weapons usage, *RFID Journal*, November 9, 2006, http://www.rfidjournal.com/article/articleview/2806/1/1/

CHAPTER 26

Granata, J. P., Tracking materiel from warehouse to warfighter, *Army Logistician*, 37(5).

Headquarters, United States Marine Corps, Deputy Commandant, Installation and logistics, USMC Radio Frequency Implementation Plan, July 27, 2006.

Marine corps base manages personnel records using 3M RFID tracking system, http://www.morerfid.com/details.php?subdetail=Report&action=details&report_id=5753&display=RFID

CHAPTER 27

Savi Technology, http://www.savi.com/products/SensorTag_676.pdf

Sullivan, L., U.S. ports tackle security with technology, http://www.informationweek.com/news/showArticle.jhtml;jsessionid=GC3RUMNDZEYDKQSNDBOCKH0CJUMEKJVN?articleID=177105452&pgno=1, January 30, 2006.

U.S. Customs and Border Protection, *Maritime Cargo Security in the Age of Global Terrorism*, http://www.customs.gov/xp/cgov/newsroom/full_text_articles/trade_prog_initiatives/cargo_security.xml

CHAPTER 28

Bacheldor, B., Yakka uses RFID to size N.Z. military, *RFID Journal*, January 18, 2007, http://www.rfidjournal.com/article/articleview/2982/1/1/

French army tracks emergency equipment with TI-RFid tags, http://www.ti.com/rfid/docs/news/eNews/enewsvol30.htm#story3

New Zealand: Yakka apparel uses RFID tags to reduce errors on uniform specs, December 7, 2006, http://www.fibre2fashion.com/news/textiles-technology-news/newsdetails.aspx?news_id=27172

O'Connor, M. C., Spanish military rolls out RFID, http://www.rfidjournal.com/ article/articleview/2142/1/1/

Sahu, B., Yakka Apparel embeds RFID tags in garments, December 11, 2006, http://www.rfidblog.org/page

Spanish Army Realizes the Importance of RFID, http://www.rfid-weblog.com/50226711/ spanish_army_realizes_the_importance_of_rfid.php

Yakka Apparel in NZ First, http://www.istart.co.nz/index/HM20/PC0/PVC197/EX236/CS27955

CHAPTER 29

AMB Identification and Timing, http://www.amb-it.com/

ChampionChip, www.championchipusa.com

Wyld, D. C., Sports 2.0: A Look at the Future of Sports in the Context of RFID, http://www.thesportjournal.org/2006Journal/Vol9-No4/Wyld.as

CHAPTER 31

Laczniak, S., A VeriChip on Society's Shoulder: Positive and Negative Implications of the VeriChip.

Sahu, B., VeriChip provides its VeriTrace System to FEMORS and Hawaii Health Department, http://www.rfidblog.org/page/16/, September 25, 2006.

Verichip Corporation, http://www.verichipcorp.com/index.html

CHAPTER 32

Coleman, F., Eck, R. W., and Russell, E. R., Railroad-highway grade crossings: A look forward, In *Transportation in the New Milliennium,* Transportation Research Board, Committee on Railroad-Highway Grade Crossings, Washington, D.C., 2009.

Clampitt, H. G. and Jones, E. C., *RFID Certification Textbook,* PWD Group Inc., Houston, TX, 2006.

Intelligent Transportation Primer, The Institute of Transportation Engineers, 2000, *Library of Congress*.

Roess, R. P., Prassas, E. S., and McShane, W. R., *Traffic Engineering,* 2nd edn., Pearson Prentice Hall, Upper Saddle River, NJ, 2004.

Siegemund, F. and Florkemeier, C., Interaction in pervasive computing settings using bluetooth-enabled active tags and passive RFID technology together with mobile phones. In *IEEE International Conference on Pervasive Computing and Communications 0-7695-1893-1/03*.

Singh, J. P., Bambos, N., Srinivasan, B., and Clawin, D., Wireless LAN performance under varied stress conditions in vehicular traffic scenarios, 0-7803-7467-3/02/$17.00 ©2002 IEEE.

Southwest Research Institute, *Railroad Delay Advance Warning System*, Texas Department of Transportation, TransGuide, March 25, 1998.

Xing, K., Ding, M., Cheng, X., and Rotenstreich, S., Safety warning based on highway sensor networks. In *IEEE Communications Society 0-7803-8966-2/05*.

CHAPTER 33

Cater, D. J. and Pasqualone, R.G., *IEEE Transactions on Industry Applications*, ISO 9000 – A Perspective on a Global Quality Standard, 31(1), January 1995.

Chang, Y. and Makatsoris, H., Supply chain modeling using simulation, *International Journal of Simulation*, 2(1): 24–30, 2004.

Chopra, S. and Meindl, P., *Supply Chain Management: Strategy, Planning, and Operation*, Pearson, 2004.

Collins, J., *Good to Great – Why Some Companies Make the Leap and Others*, HarperCollins Publishers Inc., New York, pp. 219–229, 2001.

George, M. L., *Lean Six Sigma*, McGraw-Hill, 2003.

In the global economy, International Organization for Standardization, September 1, 2009, <http://www.iso.org/iso/iso_catalogue/management_standards/iso_9000_iso_14000/in_the_global_economy.htm>

ISO 14000 essentials, International Organization for Standardization, September 1, 2009, <http://www.iso.org/iso/iso_catalogue/management_standards/iso_9000_iso_14000/iso_14000_essentials.htm>

ISO 9001 2008 vs ISO 9001 2000, *Praxiom Research Group Limited.* September 1, 2009, <http://www.praxiom.com/iso-new.htm>

Jones, E. C. and Hain, J. A., Using what you have, *Six Sigma Forum Journal*, 4(3): 23–28, 2005.

Jones, E. C., *Six Sigma Background*, White Paper of Radio Frequency and Supply Chain Lab, University of Nebraska-Lincoln, 2008.

Jones, E. C. and Hain, J. A., *A Case Study of a Supply Chain Management Network Model in Government Public Works Department*, White Paper of Radio Frequency and Supply Chain Lab, University of Nebraska-Lincoln, 2008.

Jones, E. C. and Riley, M. W., The value of industrial engineers in lean six sigma organizations. In Submitted to *Institute of Industrial Engineers Conference 2010, Cancun Mexico*, 2010.

Mellat-Parast, M., Jones, E. C., and Adams, S. G., *Six Sigma and Baldrige: A Quality Alliance*, *Quality Principles*, 2007. In Print.

Pyzdek, T., *The Six Sigma Handbook: The Complete Guide for Greenbelts, Blackbelts, and Managers at All Levels*, The McGraw-Hill, 2003.

Shapiro, J. F., *Modeling the Supply Chain*, Duxbury, Thomson Brook/Cole, 2001.

Snicker, R., *Implementing Six Sigma: A Planning Guide for Executive Teams*. Oriel Inc., 2004. In Print.

CHAPTER 34

Kenney, B., RFID as an enabler of networked logistics. In *Proceeding of RFID World*, Dallas, TX, March 1, 2006.

Vasquez, R., Frequency agile reader, Whitepaper, VerdaSee Solutions, 2007.

CHAPTER 35

Kai Fong Lee, *Principles of Antenna Theory*, John Wiley & Sons, Hoboken, NJ, 1984, p. 231.

Index

N

O

P

R

S

T

U

V

W

For Product Safety Concerns and Information please contact our
EU representative GPSR@taylorandfrancis.com Taylor & Francis
Verlag GmbH, Kaufingerstraße 24, 80331 München, Germany